Wicked Problems of Water Quality Governance

This book explores the many dimensions of water quality problems in different parts of the globe, with focus on problems of governance, from legal frameworks to social discourses and compensation measures.

Sustainable Development Goal (SDG) 6.3 on Water and Sanitation emphasizes the centrality of improving water quality to attain sustainable development. Yet the obstacles to achieving this goal are significant. This book explores the variety of difficult, possibly intractable "wicked" problems of water quality governance around the world. Cases include the challenge of managing water from source to sea, exploring why attempts to do so have come up short in limiting harm to the Great Barrier Reef; differing social discourses on market based instruments in Canada; efforts to bring to closure the human legacies of Minamata methyl mercury poisoning half a century ago in Japan; current problems of mercury use in Andean mining; misalignment of established Eastern European water laws with those of the EU; water quality markets in China; the impacts of service coverage and quality on low income households in countries from New Zealand to Bangladesh and Malawi; the importance of perceptions, ranging from the use of treated wastewater by farmers in the MENA region to consumers in Fukushima and to users of the artificial river in Beijing's Olympic Park; and finally the confluence of wicked problems in refugee camps facing COVID.

The chapters in this book were originally published in the journal, *Water International*.

James E. Nickum, Fellow, Global Reach Awardee, and former Vice-President of the International Water Resources Association (IWRA), is the Editor in Chief of *Water International*, non-resident Professorial Research Associate at the School of Oriental and African Studies (SOAS), University of London, and Adjunct Senior Research Fellow at the East-West Center, Honolulu, Hawaii.

Raya Marina Stephan is Fellow and Former Director of IWRA. She is an expert in water law, and an international consultant in water related projects with international organizations. She is the Deputy Editor in Chief of *Water International*

Henning Bjornlund is Fellow and Vice President of the International Water Resources Association and Research Professor in Water Management and Policy at the University of South Australia. He has been researching water management and policy issues in Australia since 1993, in Canada since 2005, and in southern Africa since 2013 working with partners in Mozambique, Tanzania, Zimbabwe, and South Africa.

Routledge Special Issues on Water Policy and Governance

Edited by: Cecilia Tortajada *(IJWRD) – Institute of Water Policy, Lee Kuan Yew School of Public Policy, NUS, Singapore*, and **James Nickum** *(WI) – International Water Resources Association, France*

Most of the world's water problems, and their solutions, are directly related to policies and governance, both specific to water and in general. Two of the world's leading journals in this area, the *International Journal of Water Resources Development* and *Water International* (the official journal of the International Water Resources Association), contribute to this special issues series, aimed at disseminating new knowledge on the policy and governance of water resources to a very broad and diverse readership all over the world. The series should be of direct interest to all policy makers, professionals and lay readers concerned with obtaining the latest perspectives on addressing the world's many water issues.

A River Flows Through It
A Comparative Study of Transboundary Water Disputes and Cooperation in Asia
Edited by Selina Ho

Putting Water Security to Work
Addressing Global Sustainable Development Challenges
Edited by Chad Staddon and Christopher A. Scott

Global Water Resources
Festschrift in Honour of Asit K. Biswas
Edited by Cecilia Tortajada and Eduardo Araral

Source-to-Sea Management
Edited by Josh Weinberg, Qinhua Fang, Sarantuyaa Zandaryaa, Greg Leslie and James Nickum

Strengthening Cooperation over Transboundary Groundwater Resources
Edited by Gabriel Eckstein and Alice Aureli

Wicked Problems of Water Quality Governance
Edited by James E. Nickum, Raya Marina Stephan and Henning Bjornlund

For more information about this series, please visit:
www.routledge.com/Routledge-Special-Issues-on-Water-Policy-and-Governance/book-series/WATER

Wicked Problems of Water Quality Governance

Edited by
James E. Nickum, Raya Marina Stephan and Henning Bjornlund

Routledge
Taylor & Francis Group
LONDON AND NEW YORK

IWRA

Designed cover image: © Shutterstock
First published 2023
by Routledge
4 Park Square, Milton Park, Abingdon, Oxon OX14 4RN

and by Routledge
605 Third Avenue, New York, NY 10158

Routledge is an imprint of the Taylor & Francis Group, an informa business

British Library Cataloguing in Publication Data
A catalogue record for this book is available from the British Library

ISBN13: 978-1-032-36334-9 (hbk)
ISBN13: 978-1-032-36335-6 (pbk)
ISBN13: 978-1-003-33143-8 (ebk)

DOI: 10.4324/9781003331438

Typeset in Minion Pro
by Newgen Publishing UK

Publisher's Note
The publisher accepts responsibility for any inconsistencies that may have arisen during the conversion of this book from journal articles to book chapters, namely the inclusion of journal terminology.

Contents

Citation Information

The following chapters were originally published in various volumes and issues of the journal *Water International*. When citing this material, please use the original page numbering for each article, as follows:

Chapter 1

IWRA's water quality project, including the report Developing a Global Compendium on Water Quality Guidelines
Heather Bond
Water International, volume 43, issue 3 (2018), pp. 327–335

Chapter 2

Wicked problems facing integrated water quality management: what IWRA experts tell us
James E. Nickum, Henning Bjornlund and Raya Marina Stephan
Water International, volume 43, issue 3 (2018), pp. 336–348

Chapter 3

Water quality management from source to sea: from global commitments to coordinated implementation
Birgitta Liss Lymer, Joshua Weinberg and Torkil Jønch Clausen
Water International, volume 43, issue 3 (2018), pp. 349–360

Chapter 4

Adaptive or aspirational? Governance of diffuse water pollution affecting Australia's Great Barrier Reef
Poh-Ling Tan and Fran Humphries
Water International, volume 43, issue 3 (2018), pp. 361–384

Chapter 5

The social discourses on market based instruments to manage non-point-source water pollution in the Oldman River basin, southern Alberta
Gillian Kerr and Henning Bjornlund
Water International, volume 43, issue 3 (2018), pp. 385–403

Chapter 6

Minamata: how a policy maker addressed a very wicked water quality policy problem
Hikaru Kobayashi
Water International, volume 43, issue 3 (2018), pp. 404–423

Chapter 7

Mercury pollution in Colombia: challenges to reduce the use of mercury in artisanal and small-scale gold mining in the light of the Minamata Convention
Farith A. Diaz, Lynn E. Katz and Desmond F. Lawler
Water International, volume 45, issue 7–8 (2020), pp. 730–745

Chapter 8

Water laws of Georgia, Moldova and Ukraine: current problems and integration with EU legislation
Yuliya Vystavna, Maryna Cherkashyna and Michael R. van der Valk
Water International, volume 43, issue 3 (2018), pp. 424–435

Chapter 9

The evolution of markets for water pollution allowances in China: a case study of Jiaxing
Hao Wang, Sander Meijerink, Erwin van der Krabben and Huaguo Yang
Water International, volume 47, issue 3 (2022), pp. 361–382

Chapter 10

Service levels for the four billion people with piped water on premises
Neil S. Grigg
Water International, volume 43, issue 4 (2018), pp. 531–547

Chapter 11

Water and sanitation in Dhaka slums: access, quality, and informality in service provision
Sabrina Sharmin Haque, Monica Yanez-Pagans, Yurani Arias-Granada and George Joseph
Water International, volume 45, issue 7–8 (2020), pp. 791–811

Chapter 12

Poor water service quality in developed countries may have a greater impact on lower-income households
Anna Robak and Henning Bjornlund
Water International, volume 43, issue 3 (2018), pp. 436–459

Chapter 13

Quality matters: incorporating water quality into water access monitoring in rural Malawi
Sarah L. Smiley
Water International, volume 42, issue 5 (2017), pp. 585–598

Chapter 14

Radiation knowledge and willingness to buy bottled water from regions near the Fukushima Daiichi Nuclear Power Plant
Kentaka Aruga
Water International, volume 45, issue 7–8 (2020), pp. 812–823

Chapter 15

Farmer perceptions regarding irrigation with treated wastewater in the West Bank, Tunisia, and Qatar
Anne Dare and Rabi H. Mohtar
Water International, volume 43, issue 3 (2018), pp. 460–471

Chapter 16

Issues and challenges of reclaimed water usage: a case study of the dragon-shaped river in the Beijing Olympic Park
Xiaoqin Zhou, Zifu Li, Chad Staddon, Xuejun Wu and Han Song
Water International, volume 42, issue 4 (2017), pp. 486–494

Chapter 17

The potential impact of water quality on the spread and control of COVID-19 in Syrian refugee camps in Lebanon
Issmat I. Kassem and Hadi Jaafar
Water International, volume 45, issue 5 (2020), pp. 423–429

Chapter 18

Exploring challenges in safe water availability and accessibility in preventing COVID-19 in refugee settlements
Nazifa Rafa, Sayed Mohammad Nazim Uddin and Chad Staddon
Water International, volume 45, issue 7–8 (2020), pp. 710–715

Notes on Contributors

Yurani Arias-Granada, Department of Agricultural Economics, Purdue University, West Lafayette, IN, USA.

Kentaka Aruga, Graduate School of Humanities and Social Sciences, Saitama University, Saitama, Japan.

Heather Bond, Swimways Coordinator, Wetlands International, Ede, Gelderland, Netherlands.

Henning Bjornlund, Water Management and Policy, University of South Australia, Adelaide, Australia.

Maryna Cherkashyna, Environmental Law Department, Yaroslav Mudryi National Law University, Kharkiv, Ukraine.

Anne Dare, Middle East Regional Cooperation (MERC), United States Agency for International Development (USAID), Washington, USA.

Farith A. Diaz, Department of Civil, Architectural and Environmental Engineering, Center for Water and the Environment (CWE), University of Texas at Austin, Austin, TX, USA.

Neil S. Grigg, Department of Civil and Environmental Engineering, Colorado State University, Fort Collins, USA.

Sabrina Sharmin Haque, Department of Environmental Health Sciences, Rollins School of Public Health, Emory University, Atlanta, GA, USA.

Fran Humphries, School of Law, Queensland University of Technology, Brisbane, Queensland, Australia.

Hadi Jaafar, Department of Agriculture, Faculty of Agricultural and Food Sciences, American University of Beirut, Beirut, Lebanon.

Torkil Jønch Clausen, Scientific Programme Committee of the World Water Week, Stockholm, Sweden; UNDP-SIWI Water Governance Facility.

George Joseph, Water Global Practice, World Bank, Washington, DC, USA.

Issmat I. Kassem, Center for Food Safety, Department of Food Science and Technology, University of Georgia, Griffin, Georgia, USA.

Lynn E. Katz, Department of Civil, Architectural and Environmental Engineering, Center for Water and the Environment (CWE), University of Texas at Austin, Austin, TX, USA.

Gillian Kerr, LogicalOutcomes, Toronto, Canada.

Hikaru Kobayashi, Graduate School of Media and Governance, Keio University, Shonan-Fujisawa, Japan.

Desmond F. Lawler, Department of Civil, Architectural and Environmental Engineering, Center for Water and the Environment (CWE), University of Texas at Austin, Austin, TX, USA.

Zifu Li, School of Energy and Environmental Engineering, Beijing Key Laboratory of Resource-oriented Treatment of Industrial Pollutants, University of Science and Technology Beijing, PR China.

Birgitta Liss Lymer, International Unit, Swedish Environmental Protection Agency; UNDP-SIWI Water Governance Facility.

Sander Meijerink, Institute for Management Research, Department of Geography, Planning and Environment, Radboud University, Nijmegen, Netherlands.

Rabi H. Mohtar, Department of Biological and Agricultural Engineering and Zachry Department of Civil Engineering, Texas A&M University, College Station, TX, USA.

James E. Nickum, School of Oriental and African Studies (SOAS), University of London, UK and the East-West Center, Honolulu, Hawaii, USA.

Nazifa Rafa, Environmental Sciences Program, Asian University for Women, Chattogram, Bangladesh.

Anna Robak, Department of Civil Engineering, University of New Brunswick, Canada; Research & Innovation, WSP, Canada.

Sarah L. Smiley, Department of Geography, Kent State University at Salem, Ohio, USA.

Han Song, Beijing I & Earth Environmental Protection Engineering Co., Ltd, Beijing, PR China.

Chad Staddon, Department of Geography & Environmental Management, University of the West of England, Bristol, UK.

Raya Marina Stephan, Water law expert, International consultant, Deputy Editor-in-Chief, Water International.

Poh-Ling Tan, Griffith Law School, Law Futures Centre and Australian Rivers Institute, Griffith University, Brisbane, Australia.

Sayed Mohammad Nazim Uddin, Environmental Sciences Program, Asian University for Women, Chattogram, Bangladesh.

Erwin van der Krabben, Institute for Management Research, Department of Geography, Planning and Environment, Radboud University, Nijmegen, Netherlands.

Michael R. van der Valk, Hydrology.nl – Institute for International Hydrology and Water Resources, Amsterdam, Netherlands.

Yuliya Vystavna, Biology Centre CAS, Institute of Hydrobiology, České Budějovice, Czech Republic.

Hao Wang, Institute for Management Research, Department of Geography, Planning and Environment, Radboud University, Nijmegen, Netherlands.

Joshua Weinberg, Policies, Innovations and Networks for China Europe Water Cooperation (PIANO) and Nature Insurance value: Assessment and Demonstration (NAIAD), UNDP-SIWI Water Governance Facility.

Xuejun Wu, Beijing I & Earth Environmental Protection Engineering Co., Ltd, Beijing, PR China.

Monica Yanez-Pagans, Poverty Global Practice, World Bank, Washington, DC, USA.

Huaguo Yang, Department of Law, School of Humanities and Law, Jiaxing University, Jiaxing, China.

Xiaoqin Zhou, School of Energy and Environmental Engineering, Beijing Key Laboratory of Resource-oriented Treatment of Industrial Pollutants, University of Science and Technology Beijing, PR China.

Introduction: The Wicked Problems of Water Quality Governance

James E. Nickum, Raya Marina Stephan and Henning Bjornlund

Water managers face a variety of complex and interconnected problems. Continued population and economic growth coupled with increased pollution of scarce freshwater sources and climate change create multiple wicked problems for water managers. Hence, there is an urgent need to implement drastic measures, if we are to meet future demands for human consumption, economic use and ecosystem health. Continuing business as usual, supplying new freshwater, treated for human consumption, to meet all domestic, municipal, agricultural and industrial needs is not an option. Increased demand for and pollution of freshwater have also made this option economically impossible. One obvious measure is to reuse water several times and use water of different characteristics to meet the needs for which each quality is suitable. This can increase available supply, reduce demand on new freshwater and reduce the cost of water for each use.

This book is a considerably expanded outgrowth of a special issue of *Water International* (43.3, April 2018) by the same title that in turn arose out of the work of the International Water Resources Association (IWRA). In the first chapter of this volume, Heather Bond provides a more detailed record of this previous IWRA work and summarizes the results and findings from the report *Developing a Global Compendium on Water Quality Guidelines: Which Quality for Which Use?*

To set the scene, in Chapter 2 we invited six world water experts who are IWRA members to identify one wicked problem each of their choice that is likely to beset water quality management over the next 20 years and provide a brief elaboration. The results represent the range of the challenges presented by water quality: (1) defining water quality parameters for each water use; (2) putting each water quality to its most beneficial use; (3) the soil and water nexus; (4) pharmaceuticals in waters; (5) how to monitor and enforce water quality standards; and (6) balancing economic development and water protection.

Lymer et al. (Chapter 3) then provide a useful exploration of water quality management in the context of Theme 3.3 of the Seventh World Water Forum, "Ensuring Water Quality from Ridge to Reef" for which IWRA was the thematic champion (www.iwra.org/water quality). Freshwater, marine waters and coastal zones are most often separate analytical and policy domains – yet they are interconnected in reality, as coastal dead zones and floating continents of detritus have clearly illustrated. The authors draw on experiences from the Danube River and Black Sea, East Asian seas, and the Baltic Sea to identify issues that need addressing when implementing the 2030 Agenda for Sustainable Development in order to achieve water quality objectives for fresh, coastal and marine waters.

Tan and Humphries (Chapter 4) also look at the importance of ensuring water quality from ridge to reef in the context of the Great Barrier Reef in Australia, a World Heritage site. This reef has great economic and cultural values for Queensland in particular and Australia, in general. The reef is under threat due to diffuse water pollution from a complex set of land uses in the catchment, especially agricultural activities resulting in runoff of nutrients, pesticides, and sediment. As a result, the World Heritage Committee has threatened that it may deem the reef a World Heritage in danger. The chapter examines the range of environmental, social, political and legal conditions that have enabled or hindered the success of diffuse water quality governance associated with managing the impacts of nonpoint source pollutants on the reef.

Market-based instruments have often been promoted as effective tools to manage water pollution in a catchment and thus to help manage water quality 'from source to sea,' as discussed by Lymer et al. and Tan and Humphries, but their use for the management of nonpoint source pollution has been scarce and their success even scarcer. Kerr and Bjornlund (Chapter 5) explore the reasons for this lack of success in the context of an intensive livestock-producing region of southern Alberta, Canada. They identify four distinct discourses representing groups of stakeholders: incentive orienteers, rural advocates, honest brokers and progressive producers; each providing a unique perspective and advice that the government of Alberta needs to consider when integrating market-based instruments into water management.

Point-source pollution into coastal waters, especially when toxic, can have long-term blowback effects. In Chapter 6, Hikaru Kobayashi, retired Administrative Vice-Minister, Ministry of Environment, Japan, provides a unique insider's view of the policy, legislative and judicial processes associated with managing the long-term impact of water quality issues in relation to industrial pollution, specifically the notorious methyl mercury poisoning originally identified in Minamata, Japan in the 1950s. The chapter explains the drawn out and complex process of settling compensation claims associated with Minamata disease, which caused often significant neurological damage to affected, marginalized people.

Internationally, the Minamata Convention on Mercury was adopted in 2013. Colombia was one of the countries ratifying this convention. In Chapter 7, Diaz et al. discuss mercury pollution in Columbia resulting from legal and illegal gold mining, which has affected more than 200 rivers. Columbia has a long tradition of illegal and informal gold mining with an increasing number of miners depending on this work, a large number of municipalities where illegal mining is practiced, illegal trade of mercury, the remote location of many mining districts, obstacles to legalizing miners, and the involvement of illegal armed groups. These characteristics present challenges that the Colombian government faces in its attempts to control the use of metallic mercury in artisanal and small-scale gold mining. The Columbian case shows that a significant decrease in the use of mercury seems unlikely in the short term, particularly in developing countries. It also indicates that different countries will require different strategies to eliminate the use of mercury in mining activities, even if it is possible in the long-run.

Policy implementers need consistent laws to manage water quality across jurisdictions. Vystavna et al. (Chapter 8) examine the complex issues of adapting the water laws of three countries, which previously were part of the Soviet Union, into the EU water legislation framework, with a focus on water quality, pollution and management. They identify three areas in which significant discrepancies exist between the current water laws of the three countries and the EU framework, which need to be addressed in future law revisions: (1) fragmentation resulting in poorly defined responsibilities; (2) different laws on property rights and natural resource users' responsibilities; and (3) outdated water quality standards. These discrepancies are not simply an issue of terminology but more conceptually about how to organize and protect water resources.

Wang et al. (Chapter 9) provide a case study of the performance of markets for water pollution allowances in the municipality of Jiaxing, China, over the period from 2002 to 2018. Lying on the Grand Canal not far from Shanghai, Jiaxing has a relatively long history of developing these markets, claiming good achievements, despite underperformance during the development phase. The authors evaluate the relationship between changes in institutional design and performance, finding they have resulted in positive adaptations. They identify some links between institutional design and performance and conclude that Jiaxing demonstrates that government can play a positive role in the development of such markets through providing an effective administrative and legal system and ensuring the consistency and improvement of the markets, while developing good coordination among the counties within the municipality. The authors remind us that market design is a continuous learning process.

While more than half of the people in the world have access to piped water on premises, no global assessment has been made of the availability, safety, reliability and affordability of service levels. The range of situations is challenging, but assessments based on performance can raise awareness and equip officials to promote higher and safer piped water service levels. The chapter by Grigg (Chapter 10) explores explanatory variables for service levels and uses descriptive information and expert opinion to triangulate towards conclusions about how service levels vary. Research needs are identified for development of protocols and data classes for a piped water services rating system.

Haque et al. (Chapter 11) deal with access to WASH in Dhaka, Bangladesh slums. They explore a number of dimensions such as types of infrastructure, reliability, accessibility, sharing status, perceived quality and fecal sludge management They also examine some of the mechanisms of service provision. The authors conclude that although access to improved water is nearly universal in the studied slums, that water is mostly supplied by non-regulated providers, so its safety is not assured. The population relies highly on shared toilets for sanitation. Significantly, not all slums are alike; interventions must be tailored to each context.

Robak and Bjornlund (Chapter 12) explore another wicked water quality issue facing many small suppliers of potable water. These suppliers face two contradictory pressures: increasingly stringent water quality guidelines, which amplify the cost of potable water supply; and more and more pressure to provide affordable supply of portable water. This raises two issues: (1) Would it be economically more efficient to lower the quality of service, with respect to water quality parameters not affecting human health, and leave it to households to take actions to customize their water quality to their desired level?; and (2) Is there a difference in the extent to which richer and poorer households are burdened by levels of service that are lower than desired? The findings suggest that there is a high level of willingness to customize supply, so that it might be viable to improve affordability by lowering quality standards. However, the actions taken by poor households to customize their water quality are cheap in the short term but much more expensive in the long term, as they cannot afford the up-front cost of options that are less costly in the long run. Hence, a government wanting to pursue this option might need to consider subsidizing such investments for poor households.

The Sustainable Development Goals offer an ambitious plan to achieve universal access to water that is safe and affordable. Smiley (Chapter 13) uses data from 139 household surveys from 13 villages in rural Malawi, together with Petrifilm quality testing of 27 drinking water sources to highlight areas where the goals' proposed monitoring framework might not fully capture water quality. Households make complex water decisions and sometimes choose to use unsafe sources. They generally do not use water treatment but universally practice water storage. This chapter argues for improved monitoring to more effectively and accurately measure sustainable water access.

Aruga (Chapter 14) surveys factors affecting the consumer willingness to buy bottled drinking water from regions near the Fukushima Daiichi Nuclear Power Plant, which released substantial amounts of radioactive materials after the 2011 tsunami. He examines how differences in the types of radiation knowledge affect consumer's willingness to buy. This study illustrates the value of education programs on radiation to reassure consumers of the safety of bottled water from the impacted areas.

Dare and Mohtar (Chapter 15) address the issue of acceptability in the use of treated wastewater for irrigation. Both irrigators and consumers often perceive such use to be undesirable and unsafe. This study confirms that users generally view treated wastewater as unsafe for agricultural uses and perceive existing monitoring and oversight to be insufficient. The authors suggest that to improve this view and develop the use of treated wastewater for irrigation, more extensive information is needed about the real risk and benefits of this practice. Improving the transparency of wastewater management will increase the success and safety of future use of treated wastewater.

Increased use of reclaimed water could be one of the solutions to Beijing's growing water shortage, particularly for non-potable (e.g. landscaping) purposes. The dragon-shaped river, a large artificial waterscape built on the site of the 2008 Beijing Olympic Games, offers a useful case study of the issues and challenges attendant on wastewater reclamation and reuse. Of particular interest is the use of phytoremediation techniques for bioremediation of nutrient loads. Chapter 16, by Zhou et al., presents the results of ongoing monitoring which indicate that phytoremediation is succeeding in preventing eutrophication, though some challenges remain, notably management of aquatic biomass.

To close this volume, we conclude with two short chapters on the unique problems of water quality in refugee settlements raised by the COVID-19 epidemic. Kassem and Jaafar (Chapter 17) focus on the specific case of Syrian refugees who live in crowded camps in Lebanon. A disproportionate number of them are in high-risk categories, such as children, elderly, and fragile individuals. Yet they have limited access to basic needs such as clean water and sanitation, robust medical support, or safe and nutritious foods – all necessary to stem the spread of the virus. Of additional concern, given that the coronavirus has been tracked in sewage elsewhere, is the authors' detection of fecal coliform in the rivers adjoining some of the camps.

Finally, Rafa et al. (Chapter 18) broaden the scope to consider the water-related challenges to the prevention of COVID-19 in refugee settlements in general, where proper access to health, water, or sanitation facilities is absent. The authors identify water availability, accessibility, and quality as the primary challenges to be addressed in prevention strategies for refugees. Water availability becomes particularly acute in areas suffering from water stress, and where even the host country population does not receive enough water. Even when water is found and sent on its way, it does not necessarily reach the refugees. When it does reach them, quality is often a problem, both to prevent direct infections by COVID-19 and to avoid the weakening of the immune system by other waterborne diseases that can accelerate the spread of a COVID-19 infection.

This volume covers a wide area, both geographically and thematically, illustrating the wide range of wicked water quality problems in the world. We hope that the material herein contributes to improved understanding of the complex issue of water quality management, and that it informs the debate over these critical issues.

IWRA's water quality project, including the report *Developing a Global Compendium on Water Quality Guidelines*

Heather Bond

IWRA's water quality project

Sustainable Development Goal (SDG) 6, which aims at ensuring safe and sustainable water for all, recognizes the need to improve global water quality in target 6.3. Water is used in a variety of ways throughout society and these different uses also have different water quality requirements. Water reuse, increasingly imperative in water-scarce regions, requires attention to the quality demands of the next users, including the environment. With more demanding pressures on available resources, water quality management is increasingly essential for a more balanced approach to the management of water resources.

In recent years, the International Water Resources Association (IWRA) has actively supported global initiatives related to water quality. IWRA acted as the coordinator and thematic champion for the Implementation Roadmap Thematic Process for Theme 3.3, Ensuring Water Quality from Ridge to Reef, at the 7th World Water Forum in Daegu, Republic of Korea, in 2015. This process included activities beyond the forum, such as collaborating on events and meetings with organizations in Korea, as well as Edinburgh, Stockholm and Cancun. IWRA continues its initiative in this work through acting as the topic lead for the 8th World Water Forum in Brasilia, Brazil, where Ensuring Water Quality from Ridge to Reef is a topic under the theme of Ecosystems – Water Quality, Ecosystem Livelihoods and Biodiversity.

Water quality topics have also played a significant role at the two most recent IWRA World Water Congresses. The 2015 XV World Water Congress, in Edinburgh, Scotland, discussed water quality in research and in practice under the themes Key Vulnerabilities and Security Risks, and Non-conventional Sources of Water. Two years later, the dedicated theme Water Quality, Wastewater and Reuse appeared at the XVI World Water Congress, in Cancun, Mexico. Six sessions focused on this theme, including two special sessions, as did 41 research papers. At the special session Towards SDG Implementation – Ensuring Water Quality from Source to Sea, IWRA worked with the Source to Sea Platform, an initiative hosted and supported by the Stockholm International Water Institute, to coordinate and provide content for the event, with the IWRA's executive director, Callum Clench, participating as a panellist. For more information on the papers and presentations from these and past World

Water Congresses, please visit the Congress Proceedings at https://www.iwra.org/congresses-archives/.

With this background and expertise, IWRA is well equipped to continue leading knowledge-sharing projects on water quality. Since late 2017, with support of the World Water Council, it is engaged in several streams of work directed towards advancing global knowledge and collaboration to improve water quality and efficient water management practices. These streams include:

- finalising and publishing the report *Developing a Global Compendium on Water Quality Guidelines: Which Quality for Which Use?* developed by IWRA and AFB (Agence française pour la biodiversité) over the past five years (see below);
- scoping the creation of an online interactive database to hold the contents of this report, so that it may be used as a larger-scale living resource;
- collaborating with other international organizations to create synergy in our separate work on water quality regulations and guidelines, to support the development of a joint online resource;
- developing a thematic working group of IWRA members on water quality to channel expertise for the above tasks, as well as to cultivate connections between professionals in the field;
- providing general coordination for the 8th World Water Forum, in Brasilia in March 2018; and
- creating synergies with other IWRA projects in water management, including the ongoing Smart Water Management and Science-Policy Interface activities.

The thematic working group was initiated at the end of 2017 with IWRA's call for panellists to join a task force dedicated to IWRA activities related to water quality, which elicited a great number of applications from IWRA members. From these, a gender-balanced team of 13 experts was selected, with geographical diversity representing Peru, Chile, the United States, Canada, Costa Rica, the United Kingdom, Germany, India, the Philippines and Ghana. Their backgrounds vary from academic to consulting and the private sector, and from central government to local councils. Altogether these IWRA members provide an impressive mixture of expertise on water quality guideline research, practice and knowledge.

Overview of *Developing a Global Compendium on Water Quality Guidelines: Which Quality for Which Use?*

While the IWRA executive office and the task force will work on a variety of projects over the coming year, the first goal is to complete the report *Developing a Global Compendium on Water Quality Guidelines*. This report will highlight the value of guidelines on targeted water quality for different uses, and of guidance for policymakers around the world in setting the most efficient water quality parameters. The document will set the framework for a more complete, user-friendly online compendium of policy guidelines and regulations that could expand on this and other papers to also include more detailed analysis of existing guidelines, while also promoting further research and development in the field by raising pertinent questions and identifying knowledge gaps.

What is this report?

Developing a Global Compendium on Water Quality Guidelines collects and examines examples of existing recommendations for influent water quality, as applied to various human and ecosystem uses. It provides a non-exhaustive global list of non-binding water quality guidance frameworks and reviews these case studies to determine their utility. Building on the case studies, the report explores new perspectives, and raises pertinent questions for future work on the topic. Its primary objective is to lay the groundwork for improving access to examples of water quality guidelines in order to facilitate a better understanding of how water quality demand and supply can contribute to appropriate and economical multi-sectoral water resource management.

Who will use this report?

The primary intended audience is national and international organizations working on water quality issues that wish to support water management decision-makers, policy-makers and regulators in both the public and private spheres – the ultimate users of the proposed Compendium. This includes those who manage water in primary and secondary industries, as well as those monitoring and regulating drinking water and environmental water. A wider audience in academia and water law may also find this report useful. The report could be used as a learning reference for a wider audience including for students, and professionals in academia and water law.

Background and justification

Human well-being and environmental sustainability depend on access not only to sufficient water but also to water of appropriate quality. Access to safe drinking water, the highest priority, remains a significant global challenge (WHO, UNICEF, 2017). Simultaneously, there is need for water of adequate quality in other domains, such as agriculture, ecosystems and industry of various types, including power generation. If we only look at water availability in quantitative terms, conflicts among these uses are inevitable. Water quality is an important determinant of water scarcity, defined as 'the point at which the aggregate impact of all users impinges on the supply or quality of water, to the extent that demand by all sectors, including the environment, cannot be satisfied fully' (UN Water, 2006). As not all uses require the same quality of water, a finer-grained approach to water management recognizes that variations in quality criteria for specific purposes can mitigate potential conflicts and reduce water scarcity.

Ensuring that the correct water quality is used has several important benefits. Primarily, it improves water security, 'the availability of an acceptable quantity and quality of water for health, livelihoods, ecosystems and production' (Grey & Sadoff, 2007). This security is provided through ensuring adequate drinking water and sustainable agricultural and industrial sectors, as well as sufficient water of an appropriate quality for ecosystems and biodiversity. More intelligent application of available resources is required to achieve water security, making full use of advances in available knowledge and technologies. To succeed in this, it is recommended that: (1) transparent information flows at the science–policy interface involve all affected sectors and stakeholders; (2) new technologies and management practices incorporate more effective Smart Water Management; and (3) the most appropriate water quality is used to make the most effective use of the water resources, especially taking advantage of the growing capacity of safely reusing wastewater.

In recent times, research, policy and governance have shifted from a singular focus on water quality and infrastructure to a more balanced and multidimensional approach that reflects the diverse and complex nature of water management. This approach acknowledges that water quality requirements are variable across a broad range of uses, including domestic, agriculture, industry and ecosystems. This variability offers the potential to put water of different qualities to their most beneficial uses (Enderlein, Enderlein, & Williams, 1997). Yet, tools and criteria for specifying water quality targets are not readily defined or globally accessible for the full range of water-use applications (Biswas, Tortajada, & Izquierdo, 2009). Until decision-makers have better benchmarks for the water quality requirements of each particular use, water resources will continue to be governed sub-optimally.

In preparation for the 6th World Water Forum, a stock-taking report by AFEID (French National Committee of the International Commission on Irrigation and Drainage), supported by several other French organizations (Académie de l'eau, SHF, ASTEE, ONEMA), outlined a starting point for global guidance on water quality for different uses, particularly for alternative water sources (AFEID, 2012). The IWA and UN-Water have also developed a compendium that compiles and analyzes global instruments and policies regulating water quality for different uses (UN-Water, 2015). The IWRA report supplements the regulatory instruments explored in these initiatives by focusing on guidelines – not mandatory regulatory limits, but non-binding recommended water quality targets. IWRA's focus is also on the input side for economic, human and environmental use. Influent quality guidelines are less well-documented than discharge guidelines for effluent exiting a system after use (WWAP, 2012).

Management of water quality has traditionally focused on requirements for human health, e.g., quality requirements for human consumption and domestic use, and standards for wastewater and effluent treatment, following the World Health Organization's Drinking-Water Quality Guidelines (Meybeck & Helmer, 1996; WHO, 2011). However, extensive degradation of waterbodies and aquatic ecosystems has increased the cost of treating water to standards appropriate for human consumption and made it difficult to meet the demand from all users. At the same time, recognition has been emerging of the ability of different user domains, such as agriculture, industry and ecosystems, to use water of different qualities (Medgal, 2012). This has highlighted the critical need for a more integrated approach to water quality management. *Developing a Global Compendium on Water Quality Guidelines* seeks to find the best way to promote such integration.

Objectives

General objective

In a context of global environmental, social and economic changes and limited water resources, this project ultimately intends to contribute to more efficient water management through linking water resources quality to the needs of different uses.

Overarching objectives

(1) Reduce water demand conflicts by encouraging consideration of variable water quality needs when allocating water resources.
(2) Support decision-making by water management authorities who are concerned with managing different uses and different water qualities.
(3) Contribute to improving awareness and access to information on existing water quality guidelines to increase efficient water use and support further development of water quality frameworks.
(4) Improve knowledge concerning certain water uses where international guidance on water quality does not yet exist.
(5) Enrich the debate on water quality, particularly in relation to multidisciplinary aspects (technical, economic, legal, institutional and social) and across different geographic scales (international, regional, national).

Structure of the report

The document will be composed of a general introduction, followed by five chapters which present and analyze selected case studies of water quality guidelines, and finishing with a cross-sectoral discussion and conclusion. The contents and outputs of the report are oriented around three axes: water use, geographical scale and multidisciplinary perspectives.

At present, there are few water quality guidelines that address the full range of uses. Therefore, this report divides selected guidelines into a number of categories, to test how the proposed compendium might be structured. The five main categories of water use in the report are domestic, agriculture, industry, energy and ecosystems. These five sectors were chosen based on categories in water quality and management reports from international organizations, including the United Nations and the European Environment Agency, to offer a comprehensive range of sectors that are considered essential when considering water quality applications (WWAP, 2012).

Domestic

For the purposes of this project, domestic water use encompasses drinking water and water for household uses, including food preparation, washing and personal hygiene. Guidelines for these applications typically have very specific criteria due to direct human consumption, water contact that spans a lifetime, and the sensitive life stages when people, including infants and the elderly, are more at risk (Enderlein et al., 1997; WHO, 2011).

Agriculture

The agricultural sector covers water used for food, fibre and feed production, including spray, flood and drip irrigation, as well as water for livestock and aquaculture (WWAP, 2012). This sector alone accounts for 70% of global water withdrawals and an even greater proportion of net consumption, and is thus a consequential sector when considering water consumption and its implications for water resource quality (WWAP, 2012). International guidelines must acknowledge the different needs of crops, livestock, and human health, which are particularly important given the widespread use of wastewater in agricultural processes.

Industry

Industry incorporates a wide range of sectors, including food and beverage, medical and pharmaceutical, machinery and equipment, technology and electronics, mining, and construction. Generally, water quality standards range from a lesser quality of water for industry types able to use recycled and reclaimed water, to the high-quality water required for food, technological and pharmaceutical processes (WWAP, 2012). While it has been estimated that industry accounts for 20% of freshwater withdrawals, standards and guidelines on water quality requirements for industrial processes are not yet readily defined (WWAP, 2012). Despite the varying water quality requirements within the industry sector, water of a higher quality than necessary for a specific use is often used in this sector due to cost or ease of supply (WWAP, 2012); this highlights the need and opportunity to define water quality standards across various industrial processes, and the potential for significant improvements in water use.

Energy

Water for energy is used in both primary (extracted, captured and cultivated) and secondary (conversion of primary energies) generation methods, particularly those that undergo a transformation process, e.g., petroleum products, biofuel production, thermal processes, hydropower, solar/photovoltaic, and wind (Øvergaard, 2008). The links between energy and water are complex and widely acknowledged (e.g., in the water–energy–food nexus); but the water quality requirements of this water-intensive sector are not readily defined, and guidelines are hard to identify. The delineation of energy use as a category separate from industry reflects this sector's specific role in society and its distinct consumption of water as compared to other industrial uses. It also recognizes the major growth this sector will undergo in upcoming decades, with global energy consumption predicted to increase by 28% between 2015 and 2040, a pertinent consideration for water consumption (EIA, 2017).

Ecosystems

Guidelines designed for aquatic ecosystems must consider the protection and maintenance of ecosystem structure (e.g., the distribution of species, biotic and abiotic factors), function (e.g., energy flow and cycles) and dependent species (e.g., primary producers to secondary consumers) (Carr & Neary, 2008). Water quality specific to ecosystem use is an emerging research and policy focus in the water field, with the importance of preventing and controlling water pollution in aquatic ecosystems highlighted under the UN Thematic Priority Area on Water Quality by UN-Water in September 2010 (Yillia, 2012). This recognition, along with the inclusion of ecosystems as a use category in this report, reflects better understanding and emerging focus on the importance of environmental considerations in water resource governance, as well as the essential role that water quality plays in ecosystem protection and restoration (WWAP, 2012).

Summary of key findings

The report compiled and analyzed water quality guidelines according to the above water uses. A representation of this list can be found in Table 1. Furthermore, key findings from the analysis and discussion are presented below.

- This is a sectoral analysis of water qualities; however, it is clear that a nexus exists between water and all the main sectors explored. Decision-makers must also collaborate with other sectors for an integrated and broader approach.
- While several water quality guidelines from international and government bodies exist for domestic and agricultural water purposes, few refer to industrial uses, which are mainly governed by private groups.
- There are few international or national guidelines on water uses for energy. Given the innovations and requirements for this sector in biofuel, tidal power generation, hydropower, nuclear, solar and wind, as well as the continued use of non-renewable sources, it is worth developing guidelines for water quality used for these purposes.
- Water inflow into ecosystems consists increasingly of water discharges from the four other water uses examined in this report. Water is cyclical, so a cross-sectoral perspective, including water discharges from domestic, agriculture and industry sectors, needs to be considered to safeguard environmental systems.
- As the global community works towards meeting the Sustainable Development Goals for 2030, further widespread use of the guidelines in this report will contribute to meeting the key performance indicators:
  - 6.3.2 Proportion of bodies of water with good ambient water quality
  - 6.4.1 Change in water-use efficiency over time
  - 6.4.2 Level of water stress: freshwater withdrawal as a proportion of available freshwater resources
  - 6.5.1 Degree of integrated water resources management implementation (0–100)
  - 6.6.1 Change in the extent of water-related ecosystems over time
    - The usefulness and success of guidelines developed for water quality recommendations could be reflected by the number of regulatory bodies which take up the content during the creation of binding policies.
    - With quickly developing technologies and innovations in water management, new knowledge should be incorporated into water quality guidelines and considered by regulators. This is especially true when it comes to combatting water quality issues from emerging micro-pollutants.

Conclusion

Water quality remains a key consideration for global water management, not only for addressing low-quality discharges that affect other uses and have environmental impacts, but also for examining the most efficient water quality for a specific purpose. This report will identify a number of existing guidelines directing water quality parameters across domestic, agriculture, industry, energy and ecosystem uses. International, regional and national guidelines outlining recommendations for water quality for these purposes

Table 1. Selected water quality guidelines mentioned in *Developing a Global Compendium on Water Quality Guidelines.*

Scale	Name	Date	Location
Domestic water use			
International	WHO Guidelines for Drinking-Water Quality, fourth edition incorporating the first addendum	2017	All
International	Global Drinking Water Quality Index	2007	All
Regional	(a) Taking policy action to improve small-scale water supply and sanitation systems. Tools and good practices from the pan-European Region (b) Guidelines on the Setting of Targets, Evaluation of Progress and Reporting *(UNECE/WHO Protocol on Water and Health, 1999)*	2016 2010	Europe
National	Guidelines for Canadian Drinking Water Quality	2009	Canada
National	Code of Practice on Piped Drinking Water Sampling and Safety Plans	2008	Singapore
National	Australian Drinking Water Guidelines (with amendments)	2017	Australia
Agricultural water use			
International	FAO Water Quality for Agriculture	1994	All
International	WHO-FAO Guidelines for the Safe Use of Wastewater, Excreta and Greywater, Volume 2, Wastewater Use in Agriculture	2006	All
International	Codex Alimentarius: Code of Hygiene Practice for Fresh Vegetables	2003	All
National	South African Water Quality Guidelines, Volumes 4–6	1996	South Africa
National	US EPA Guidelines for Water Reuse	2012	United States
Secondary industrial water use			
International	WHO-FAO Codex Alimentarius: Food Hygiene (Basic texts), Fourth edition	2009	All
International	WHO Good Manufacturing Practices: Water for Pharmaceutical Use	2012	All
National	South African Water Quality Guidelines, Volume 3	1996	South Africa
Ecosystem water use			
International	UNEP International Water Quality Guidelines for Ecosystems	2016	All
Regional	EU Water Framework Directive – The Environmental Quality Standards Directive (2008)	2000/2008	European Union
National	Australian and New Zealand Guidelines for Fresh and Marine Water Quality	2000	Australia, New Zealand
National	Canadian Water Quality Guidelines for the Protection of Aquatic Life	2011	Canada
National	South African Water Quality Guidelines for Aquatic Ecosystems	1996	South Africa

provide practical and targeted instruction. The aim of the report is to set the structure for a more comprehensive online compendium of water quality guidance according to water use. Promoting the integrated use of these guidelines, while acknowledging the water cycle as a whole and that all outputs are eventually released to the environment, will contribute to smarter water management. This will help relieve stress on scarce water resources and ensure adequate quality in water inputs to various applications, with the end goal of contributing to water security.

Disclosure statement

No potential conflict of interest was reported by the author.

Web links

The full document *Developing a Global Compendium on Water Quality Guidelines* and related materials will be published in 2018 on the IWRA website, and a link will be added to the project webpage. For more information on any of these water quality activities that IWRA is involved in, please visit the Water Quality Projects page on the IWRA website: https://www.iwra.org/waterquality/.

References

AFEID (L'Association Française pour l'Eau l'Irrigation et le Drainage) Office Association. (2012). *Which quality for which use? A regulators' and practitioners' perspective*. AFEID, Académie de l'Eau, SHF, ASTEE and ONEMA.

Biswas, A., Tortajada, C., & Izquierdo, R. (2009). *Water management in 2020 and beyond*. Berlin: Springer.

Carr, G. M., & Neary, J. P. (2008). Water quality for ecosystem and human health. In *United nations environment programme global environment monitoring system (GEMS/Water program)* (2nd). Ontario. Retrieved from http://hdl.handle.net/20.500.11822/12217

EIA (US Energy Information Administration). (2017). *International energy outlook 2017: Highlights*. Washinton, DC: Office of Energy. Retrieved from https://www.eia.gov/outlooks/ieo/

Enderlein, U. S., Enderlein, R. E., & Williams, W. P. (1997). Water quality requirements. In UNEP/WHO (Eds.), *Water pollution control – A guide to the use of water quality management principles*. London: E & FN Spon.

Grey, D., & Sadoff, C. W. (2007). Sink or swim? Water security for growth and development. *Water Policy*, *9*, 545–571. doi:10.2166/wp.2007.021

Megdal, S. B. (2012). The role of the public and private sectors in water provision in Arizona, USA. *Water International*, *37*(2), 156–168. doi:10.1080/02508060.2012.662732

Meybeck, M., & Helmer, R. (1996). An introduction to water quality. In D. Chapman (Ed.), *Water quality assessments – A guide to use of biota, sediments and water in environmental monitoring* (2nd ed., pp. 19–39). London: UNESCO/WHO/UNEP, E&FN Spon.

Øvergaard, S. (2008). Issue paper: Definition of primary and secondary energy. Prepared as input to chapter 3: Standard international energy classification (SIEC). In the *International Recommendation on Energy Statistics (IRES)*. Oslo: Oslo Group of Energy Statistics.

UN-Water. (2006). *Coping with water scarcity: A strategic issue and priority for system-wide action*. United Nations 'Retrieved from: http://www.un.org/waterforlifedecade/pdf/2006_unwater_coping_with_water_scarcity_eng.pdf.

UN-Water. (2015). *Compendium of water quality regulatory frameworks: Which water for which use?* Retrieved from: http://www.iwa-network.org/which-water-for-which-use.

WHO (World Health Organisation). (2011). *Guidelines for drinking-water quality* (4th ed.). Geneva: World Health Organisation Press.

WHO (World Health Organisation) and the United Nations Children's Fund (UNICEF). (2017). *Progress on drinking water, sanitation and hygiene: 2017 update and SDG baselines*. Geneva: WHO, UNICEF. Retrieved from http://apps.who.int/iris/bitstream/10665/258617/1/9789241512893-eng.pdf?ua=1

WWAP (World Water Assessment Programme). (2012). *The united nations world water development report 4: Managing water under uncertainty and risk*. Paris: UNESCO Headquarters.

Yillia, P. T. (2012). *Scoping study – International water quality guidelines for aquatic ecosystems (Project no. 2011-CPL-5060-3B41-1131-2201)*. Vienna: Vienna University of Technology, Institute for Water Quality, Resources and Waste Management.

Wicked problems facing integrated water quality management: what IWRA experts tell us

James E. Nickum, Henning Bjornlund and Raya Marina Stephan

Every year edge.org poses a 'big question' to the world's intellectual community.[1] Taking inspiration from them, at a much more modest scale, we asked a number of IWRA members who are international scholars and practitioners concerned with water quality to give us their reflections on the question, 'What is a wicked problem facing integrated water quality management over the next two decades?' In the end we received six responses spanning a gamut of concerns, from problems of allocation, management and monitoring, to ensuring sustainable development in global agriculture and in rapidly urbanizing China, to the troubling presence of pharmaceutical wastes. It is increasingly obvious that water supply and water quality are not separate domains, and cannot be managed in isolation if there is to be any hope of addressing these wicked problems.

Like the present special issue, this solicitation of members' responses to this question is part of the IWRA program on water quality. For further information, see https://www.iwra.org/waterquality/.

The complexity of water quality is an old problem for water managers, and will remain so

Blanca Jiménez Cisneros
Secretary, International Hydrological Programme, Water Science Division, UNESCO

Water quality, although seemingly a universally well-understood concept, is a tricky one, as it only takes on a practical meaning when it is associated with specific uses of water. For instance, the quality of water that is needed to flush toilets, irrigate crops or make bricks is not necessarily one that is fit for human consumption or that would allow trout to live in a river. To be able to express this difference, and to ensure the required quality of water for any given use, we need to associate various parameters and define critical values for each of them. This is a challenging task, because water, the universal solvent, can contain more than 70,000 identified artificial substances and compounds, any of which can render water deleterious for one use or another.

The process of judging the quality of water is complex. Take 'drinking water'. When we ask whether water is safe to drink, we may hope for a simple answer: yes or no. In the mind of a water utility staff or a policy maker, however, the answer is far from clear. Each country uses a number of parameters, typically 40 to 70 in number, to define the drinking water standard. A clear yes would mean that the water you wish to drink meets all the defined parameters, so by the given standard, is considered to be of appropriate quality so that a human body can safely consume it. But in many cases not all the parameters relevant to human health have been measured in the water that fills the glass in our hands.

Scientists know that water can contain tens of thousands of substances and compounds that are generally not controlled by a standard or norm. Yet to create a standard that can be implemented, local authorities and experts have to limit the parameters they monitor – first, to those that most frequently create a quality problem (e.g., water-related health issues) in the given country's setting (this is most often considered at national level); and second, to those that are technically and economically (and sometimes politically) measurable and controllable. Both restrictions often leave out substances and compounds that may be toxic at the local level or under specific conditions. To complicate matters further, the actual health effects of many if not most substances are not well understood. Advances in medical knowledge can lead to significant revisions in allowable thresholds.

At the same time, it is clear that water for consumption must always contain certain molecules beyond its basic H_2O. Our body needs some of the compounds commonly found in water to assimilate the water and to function well. Hence purified water is not good for human health.

Now let's look at the water of a country that is used for purposes other than drinking. Without defining a specific use of water, how could we provide a valid statement on a matter this complex? In this case, water practitioners usually consider that water of good quality is water that can sustain aquatic ecosystems (defined as a 'use' that has to be protected). This is a good approach when we talk about the quality of surface water, like lakes and rivers. In this case, however, as with drinking water, it is often not enough to define a list of parameters and values that should not be exceeded; we also have to identify the substances that have to be present and define the critical threshold concentrations, below which the quality of water would be improper for its purpose. For instance, if levels of dissolved oxygen are too low (normally below 4 mg/L), fish die.

The quality of groundwater is a bit different. In aquifers, water is in close contact with different types of soil. And one of the properties of water that we know well is that it is an excellent solvent. Thus, the water in aquifers tends to dissolve compounds of the soil. Even though this is a natural process, in some cases aquifers are naturally 'polluted' with substances, such as arsenic or fluorides, at levels well above the concentrations that are acceptable for drinkable water. According to the World Health Organization, arsenic is naturally present at high levels in the groundwater of a number of countries, including Argentina, Bangladesh, Chile, China, India, Mexico and the US. If this contaminated water is not treated, it can be a major health risk, whether it is used as drinking water, for irrigation or in food preparation. Long-term exposure to inorganic arsenic can lead to chronic arsenic poisoning, characterized most often by skin lesions and skin cancer. The number of people affected can be very high. In Bangladesh alone, it is estimated that more than 45 million people are at risk of being exposed to arsenic concentrations that are higher than the national standard.

Ensuring that the water we drink will not harm us and that lakes, rivers and oceans can sustain ecosystems have been wicked management problems to date. They will remain so for at least the next two decades.

Unlocking the potential for using different water qualities as a complementary resource

Tom Soo
Past Executive Director, IWRA

There are two main ways the world can cope with the looming shadow of water scarcity. One is the traditional method of developing new sources of freshwater of adequate quality. The other is by better managing and using water quality, including by matching different qualities to different uses. According to reports by the OECD and the United Nations, water withdrawals are going to be increasingly unable to keep up with demand. In response to growing pressures from changing demographics, climates, sectoral demand and other changes inherent in the earth's cycles, water managers must adapt the ways we manage water resources. If they do not, our current trajectory will lead to an unacceptable situation of stress on the planet and many of our social and economic systems. The Water Resources Institute estimates that 33 countries will face extremely high water stress by 2040 (Luo, Young, & Reig, 2015). Increasing water volumes on its own is not enough: an abundance of water is of little benefit if its quality means that we impose limits on our ability to use it. The Indo-Gangetic Basin, for example, accounts for a quarter of the world's groundwater withdrawals, and although its water levels are stable or rising, extensive groundwater contamination means that less and less is available for potable uses (MacDonald et al., 2016). Viewing treated or untreated water as a potential resource for some uses according to its varying levels of quality is thus an essential way to complement our efforts to combat water scarcity.

More efficient use and waste reduction can increase water productivity. However, better matching water quality to each type of use can unlock further potential. Leaders and water managers no longer have the luxury of overlooking the potential of systematically and intelligently using water of different qualities as part of supply strategies to satisfy demand from multiple uses. Yet, we know very little about how to go about this, and are therefore a long way from being able to fully benefit from this approach.

The potential of wastewater reuse in agriculture is arguably the most evident example. Agriculture currently uses 70% of global water resources, yet in a context of competing demands between the urban and rural sectors, and where a high volume of wastewater is generated by cities, wastewater is reported as accounting for only a small fraction (1.5–6.6%) of the quantity used in irrigation (Sato, Qadir, Yamamoto, Endo, & Zahoor, 2013). Adding further complexity to the scenario is the fact that while less than 10% of the world's collected wastewater receives any treatment, many developing countries use untreated drainage and wastewater for irrigation anyway, often because there is no alternative. Consequently, while these waters may provide farmers with perceived or real advantages, they may also carry dangers to individuals' health, as well as social, economic and environmental risks (Wichelns & Drechsel, 2011). Besides social and economic reasons, part of the blame lies with the lack of coherent frameworks to mobilize wastewater. Many countries, even relatively wealthy ones, do not have a sufficiently coherent institutional setting to address this problem. In southern Europe, for example, only 4 out of 15 countries have wastewater reuse regulations (Lavrnić, Zapater-Pereyra, & Mancini, 2017).

It is difficult to claim that there are sufficient global water resources when, as a result of low quality, many millions still lack access to drinking water of an appropriate level. Action to improve the world's use of water will not be easy due to the many financial, political and institutional challenges. However, matching water qualities from other sources such as greywater or non-conventional water to different uses can help address scarcity. Rapidly increasing urban populations and the rise of coastal mega-cities provides some opportunity to apply this approach. For example, every year in Hong Kong, an 'extra' 270 million m^3 of freshwater is available for human consumption because authorities have created a network to supply 8 out of 10 Hong Kongers with seawater for toilet flushing. Industrial water requirements vary from low-quality cooling water to the ultra-pure water needed in electronic circuit manufacturing. In this sense, the industrial sector has already gone a long way towards matching quality to use. Its success points to the potential efficiencies that could be realized both in industry and elsewhere. The Public Utilities Board of Singapore claims that their successful NEWater system accounts for 40% of the nation's current needs by recycling water mainly for industry but also for supply to households in dry times. According to the International Energy Agency, energy production consumed 66 billion m^3 of water in 2010 and is on a steep upward trajectory towards an increase of 85% by 2035. The water resources required can be of various qualities for the various stages of energy production, from extraction of resources (hydraulic fracturing for example uses water mixed with sand and chemicals) through to refining, distribution and electricity generation itself (water is often sourced from the sea to cool coastal nuclear plants). Yet many reports aimed at influencing policy, such as the 2010 *Water for Energy* report of the World Energy Council or the *Water 2012 World Energy Outlook* of the International Energy Agency, hardly mention the range of water qualities usable in energy production, preferring to focus almost exclusively on the demand for freshwater.

Water is essential to all development. The economic, social and environmental benefits of mobilizing waters of appropriate qualities would seem obvious. Yet, surprisingly, the current state of research is still unable to comprehensively identify the value of using appropriate qualities of water for each need. A study in 2013 looked at the three aspects of wastewater – generation, treatment and reuse – in 181 countries and found that only 55 had data on all three, while 69 had some data and 57 had no data at all (Sato et al., 2013). Other studies from UNEP on water quality guidelines for ecosystems, as well as by IWRA and IWA on water quality guidelines for different uses, have started to improve knowledge in this area. However, we sorely need a much more comprehensive understanding of the effective use of waters of different qualities as a resource.

There is a massive untapped potential in matching water qualities to uses in different sectors. To realize the full benefits of this potential, several paradigm shifts are required from the current focus on freshwater as primary supply for all purposes: societal paradigms for consumption; governance paradigms for policy and regulation; and professional paradigms for water management. Existing freshwater resources are limited, and different water qualities are a reality. Recognizing, understanding and effectively using waters of all qualities as potential resources will help us solve the water-scarcity problem.

References

Lavrnić, S., Zapater-Pereyra, M., & Mancini, M. L. (2017). Water scarcity and wastewater reuse standards in Southern Europe: Focus on agriculture. *Water Air Soil Pollution*, *228*, 251.

Luo, T., Young, R., & Reig, P. 2015. *Aqueduct projected water stress country rankings* (Technical Note). Retrieved September 10, 2017 from World Resources Institute website: www.wri.org/publication/aqueduct-projected-water-stresscountry-rankings

MacDonald, A. M., Bonsor, H. C., Ahmed, K. M., Burgess, W. G., Basharat, M., Calow, R. C., … Yadav, S. K. (2016). Groundwater quality and depletion in the Indo-Gangetic Basin mapped from in situ observations. *Nature Geoscience*, *9*, 762–766.

Sato, T., Qadir, M., Yamamoto, S., Endo, T., & Zahoor, A. (2013). Global, regional, and country level need for data on wastewater generation, treatment and use. *Agricultural Water Management*, *130*(2013), 1–13.

Wichelns, D., & Drechsel, P. (2011). Meeting the challenge of wastewater irrigation: Economics, finance, business opportunities and methodological constraints. *Water International*, *36*(4), 415–419.

The wicked problem of soil and water quality

Rabi H. Mohtar

The prospect of using the business-as-usual model for primary resource management over the next 20–30 years looks grim, with a projected 40% gap between aggregate demand and supply of water. Agriculture, already the largest user, would require more than 100% of the available water supply in order to meet the 50% or greater expected increase in food demand. This is compounded by the fact that no one predicts business-as-usual for precipitation, with projections centring on a reduction of 10–30% on average, with significant uncertainty and variability, both temporally and spatially. This vision of the future is well known. What is not well recognized is that in addition we will no longer have the same quantity of land available for agriculture, and that the quality of the soil is projected to decline. We are left with a very wicked problem; one that has no solution.

Most nonpoint-source pollution is due to agricultural production: as agricultural activity increases and intensifies, this pollution will worsen, unless serious efforts are made now to decouple agricultural production and soil and water pollution. At the same time, we must recognize that the soil, a naturally organized medium, is at the core, at the very heart, of water and food security. It is where water, nutrients and air circulate. It is where root distribution and respiration happen. It filters water and reduces or eliminates the cost of treatment of our surface water supply.

The main function of the soil is to continue to provide sufficient food for the world's growing population. Several factors affect the ability of soil to accomplish its function. These include but are not limited to proper soil management practices and land use such as tillage, soil additives and cropping system; and soil and water management practices such as irrigation using fresh and alternative sources of brackish, grey and wastewaters. So the question becomes: how can we evaluate the impacts on soil function of these factors in order to identify the best design and mitigation measures? To answer this question, we need to appreciate that the unique organization of the soil medium makes it possible for soil to perform several functions, and we must adequately and accurately represent that organization in modelling approaches.

We also need to recognize that soil security and water security are critical to achieving the post-2015 Sustainable Development Goals agenda. Implementation of that agenda impacts water, food, climate and energy nexus goals. The wicked problem in nonpoint-source pollution arises with the challenge to produce more food under growing constraints of water, land, and climate – where the risk of nonpoint-source pollution increases. Many communities will face a tight trade-off between increasing land and water activity and maintaining soil, water and ecosystem health: this includes avoiding nutrient pollution and maintaining biodiversity and ecosystem services.

To make the optimal trade-off between increasing food production and improving or maintaining soil and water quality requires that we:

(1) know the value of both human health and eco-system health – this will allow us to quantify the effect of agricultural production on soil, water quality and the quality of human life and develop the right policies and guidelines for this complex system, inclusive of economic as well as human and ecosystem indicators;
(2) understand soil dynamics and their role in filtering chemicals out of the water supply – this will allow us to explain and quantify the linkages between soil properties and its behaviour;
(3) understand the multi-scale, hierarchical hydrologic system based on soil structure – this will allow us to transfer knowledge across scales from the pedostructure to landscape scales, recognizing that interventions at the local scale can be linked to the higher scale and vice versa. Temporally, processes are often observed and modelled at the short-term scale, but estimates are also needed for the long term. The linkage and integration of the formulations at different scales has not yet been adequately addressed.

Only after all the above is done can we make the optimal trade-offs and solve this wicked problem. To reiterate: understanding the role of water and food security requires recognition of the soil as a natural and hierarchically organized medium. Quantitative approaches to studying the long-term impacts of agro-environmental practices on soil functioning are therefore urgently needed. Last but not least: valuing water, soil and water quality is vital to the sustainability of food production systems.

The wicked problem of pharmaceuticals in our waters

Gabriel Eckstein

Professor of Law and director of the Law Program in Natural Resources Systems, and the Graduate Faculty in Water Management and Hydrological Science, Texas A&M University. gabrieleckstein@law.tamu.edu

Pharmaceuticals and their wastes plague our water. Research conducted over the past two decades demonstrates that antibiotics, mood stabilizers, sex hormones, blood thinners, beta blockers and numerous over-the-counter medications have entered our rivers and lakes, groundwater and soils, wastewater treatment plants, and even public drinking water supplies. A 2013 study in the United States found at least 25 different active pharmaceutical ingredients in the waste stream of 50 large wastewater treatment plants nationwide. A 2016 global study conducted through a comprehensive literature review revealed that 631 different pharmaceutical substances were present in 71 countries covering all continents, some at levels exceeding predicted no-effect concentrations (aus Der Beek et al., 2016).

More troubling, however, are studies indicating that unintended exposure to these substances, whether individually or in synergistic combinations, may have adverse human health impacts. While inconclusive, they suggest that chronic exposure could contribute to testicular and breast cancers, reproductive abnormalities, early puberty and reduced sperm count (Eckstein, 2015). Unfortunately, most of the studies are based on acute exposures, while research on short-term exposures to varying levels of individual pharmaceuticals or pharmaceutical combinations is lacking.

In contrast to human exposure, many aquatic species are exposed to pharmaceutical pollutants throughout their lifetime and over multiple generations. As a result, studies on the health impacts of such exposure on aquatic species are considerably more categorical. The evidence shows that low-level, chronic exposure to certain pharmaceuticals or their residues, including antidepressants and oestrogen, has led to adverse biological effects, including the feminization of males, masculinization of females, impaired reproductive capacity, abnormal sexual development and impaired ability to heal (Eckstein, 2015).

While certain pharmacological substances, like oestrogen, can occur naturally in the environment, most are anthropogenic. Pharmaceutical pollutants can reach the environment from multiple sources. The more obvious include the improper handling of pharmaceutical products by pharmaceutical manufacturers, health care facilities, pharmacies, veterinary services and individual households. Less obvious sources include human and animal faeces and urine, bath water, coroner office wastes, leachates from landfill containing pharmaceutical wastes, and irrigation water from reclaimed wastewater (Eckstein, 2015).

Although human excretion is regarded as the main source of pharmaceutical pollutants in the environment, the relative contribution of each source to environmental loadings is still unknown. Nevertheless, the presence of these drugs in the environment should not be surprising. In 2016, the global community spent USD 1.13 trillion on prescription medicines (based on wholesale prices); Americans alone accounted for 40% of those expenditures (Berkrot, 2016). Moreover, most drugs are designed to persist while in the body in order to reach the organ or area targeted for pharmacological benefit. That persistence translates into resistance to natural degradation in the environment. One study indicated that 95% of pharmaceuticals assessed in Europe are not readily biodegradable (European Environment Agency, 2010).

Treating for these pollutants and wastes after they enter the sewage or wastewater system or the environment is costly and out of the financial reach of most municipalities and wastewater and drinking water treatment operators. As a result, communities, health care specialists and environmental professionals around the world are increasingly troubled about the ability of municipalities to ensure safe freshwater for their residents and for the surrounding natural environment.

Few governments around the world officially recognize the multitude of concerns that have been raised by experts in the field regarding exposure to pharmaceutical pollutants. And even fewer have developed broad management, mitigation or disposal-prevention strategies to address the challenges posed by the growing presence of these contaminants. Currently, no nation has implemented laws designed comprehensively to respond to the concerns associated with pharmaceuticals in the environment. A primary challenge of targeting pharmaceutical wastes and pollutants, especially in the aquatic environment, is that the problem typically crosses multiple statutory and institutional jurisdictions and authorities. Moreover, in those countries where some efforts have been made, most have targeted pharmaceuticals after they have entered the environment. Few have focused on preventing the improper disposal of pharmaceutical products.

In the United States, various federal and state agencies have attempted to apply disparate environmental statutes, medical-facility regulations, drug-collection laws, and even criminal laws to pharmaceuticals wastes and pollutants in the waste stream. Depending on where it may be found, pharmaceutical waste can be subject to the Resources Conservation and Recovery Act, Clean Water Act, or Safe Drinking Water Act, or the jurisdiction of local or state police, or the US Drug Enforcement Agency. Likewise, in Europe, nations and agencies have sought to tackle the concerns under the rubric of the Water Framework Directive, Drinking Water Directive, legislation focusing on managing chemicals and medicinal products, and other mechanisms. The European Commission is presently tasked with developing a strategic approach to water pollution by pharmaceuticals and to propose mechanisms to reduce their release into the water. That effort, however, has been delayed.

The presence of pharmaceutical wastes and pollutants in our water and broader natural environment is a wicked problem. The challenges are as much political and legal as they are technical. Certainly, how we respond could affect the availability of much-needed medications. Nevertheless, whether we confront this threat sooner than later could impact countless human lives and have unprecedented consequences for our environment.

References

aus Der Beek, T., Weber, F.-A., Bergmann, A., Hickmann, S., Ebert, I., Hein, A., & Küster, A. (2016). Pharmaceuticals in the environment—Global occurrences and perspectives. *Environmental Toxicology and Chemistry*, *35*(4), 823–835.

Berkrot, B. (2016, December 5). Global prescription drug spend seen at $1.5 trillion in 2021: Report. *Reuters*. Retrieved from https://www.reuters.com/article/us-health-pharmaceuticals-spending/global-prescription-drug-spend-seen-at-1-5-trillion-in-2021-report-idUSKBN13V0CB

Eckstein, G. (2015). Drugs on tap: Managing pharmaceuticals in our nation's waters. *New York University Environmental Law Journal*, *23*, 36–90. http://ssrn.com/abstract=2628047

European Environment Agency (2010). *Pharmaceuticals in the environment* (Technical report No 1/2010). Retrieved from https://www.eea.europa.eu/publications/pharmaceuticals-in-the-environment-result-of-an-eea-workshop/file

Monitoring and law enforcement: a wicked problem facing integrated water quality management

Jia Shaofeng

Key Laboratory of Water Cycle and Related Land Surface Processes, Institute of Geographic Sciences and Natural Resources Research, Chinese Academy of Sciences, Beijing, People's Republic of China

Threats to water quality are a more serious challenge to water resources security than diminishing supply

The total quantity of water available on Earth has remained more or less the same for millions of years. At a country level, average water resources seldom change more than 20% within one century. China's three assessments of total surface water supply, done in 1958, in 1980 and in the 2000s, all yielded 2.71–2.78 trillion km^3. The hydrological cycle works day in and day out, year by year, century by century.

Anthropogenic pollution is another matter.

Integrated water quality management is necessary but difficult

The diversity of sources and types of pollutants, technologies involved, shareholders, and policies and regulations, render integrated water quality management both necessary and difficult. The number-one difficulty may be identifying polluters, especially where they are clustered or are discharging surreptitiously. For regulators, it's a very big burden to supervise a large number of polluters, especially when the latter find it cheaper to bribe regulatory staff than to stop discharging wastes. In developing countries, the public finds it very difficult to get detailed information on water quality, because the detection of many pollutants requires complicated tests that only professionals can do using expensive equipment, and because the institutions responsible for monitoring water quality often do not want to publish the relevant information to escape their responsibility – a dereliction of duty. It is urgently necessary to enhance pollution source monitoring and open access to the monitoring results.

New technologies provide good opportunities

The rapid development of technologies, such as advanced and low-cost sensors for water quality, automatic data transmission and processing, and the Internet, have made full-time and comprehensive coverage of water quality monitoring feasible, as well as timely feedback. A good example is the National Urban Black and Malodorous Water Supervision and Administration Platform of China. This platform has one part for public supervision which can be accessed through the Internet and mobile websites. Anyone can check and report if black and malodorous water has been recorded within this system, if a remediation plan has been arranged, and if it has been rehabilitated as planned. A responsible person must give feedback within a stipulated time.

The enforcement of regulations is most important for developing countries

Nearly everyone recognizes the need for environmental protection. So almost every country, including the economically least developed, has formulated and promulgated environmental protection laws. But the problem is enforcement. Often, especially but not only in developing countries, environmental law exists only on paper, not in real life. Effective enforcement remains a wicked problem.

In China, a recent effective measure is that the central government sends high-ranking environmental teams to investigate whether local governments have strictly implemented national environmental legislation and policies. This has resulted in the punishment of some high-ranking local government officials. The punishments, including criticism or serious warnings, go into their dossiers and can have career implications.

Punishments should be severe enough to deter polluters

In principle, violators of water quality protection laws should be required to pay all direct and indirect costs of pollution, plus a penalty. China's environmental protection law provides that those who suffer direct losses may demand compensation, but there is no provision to recover indirect losses. Furthermore, until recently only those directly affected by pollution had standing to sue polluters. The latest revision (2014) of the environmental protection law of China has extended the right of litigation to environmental NGOs.

Persistent organic pollutants are a challenge for both developing and developed countries

Although the above mainly concerns developing countries, China in particular, there are wicked water quality problems facing economies at all levels of development. Organic pollution and the befouled water it produces, as well as most heavy metal pollution, are largely in the past for developed economies, but persistent organic pollutants (POPs), especially environmental hormones and those produced by pharmaceutical and personal care products, are proving hard to control in both developing and developed countries. POPs are widely used in agriculture as pesticides, in industry as heat-exchange media, intermediates and by-products, and in everyday life as personal care products. POPs disturb the reproduction and development of living beings, resulting in reproductive failure and developmental deformities, even cancer and extinction. Most environmental hormones are highly toxic, persistent in the environment, and bio-accumulate in the food chain, at the top of which are human beings. POPs are also mostly hydrophilic – they mix easily with water and are transported with it.

Because of the diversity and dispersion of the sources of POPs, they are difficult to control. Limitations on producers, such as bans on producing and selling of some products, may be an effective way to limit POPs, but they are most effective when there are good substitutes. Finding those substitutes or, in many cases, changing lifestyles and modes of production, will indeed be a wicked problem over the next two decades.

How to balance economic development and water protection: a wicked problem

Feng Hu
Development & Projects, China Water Risk
fh@chinawaterrisk.org

In becoming the world's factory, China has paid a heavy cost in water quality. To ensure continued growth, China is undergoing an economic transition towards a higher value-added and consumption-led economy. How well it balances economic development and water protection in the next two decades will be important for the country and have implications for the global economy. This is likely to be a wicked problem, not only for China but also for many other developing countries.

Over the past two decades, rapid economic growth and urbanization, coupled with fragmented environmental governance (Zhang and Feng, 2015) and barriers in policy implementation at levels from central to local (Kostka, 2014), have led to serious deterioration of water quality across China. According to the latest government report, despite overall improvement, surface water quality worsened in a few major rivers, such as the Hai and the Liao. Groundwater pollution also persists (China Water Risk, 2017).

The extent and scale of pollution led the government to declare a War on Pollution in early 2014. Since then, detailed action plans to tackle air, water and soil pollution have been set, and the amended Environmental Protection Law has been in force since the beginning of 2015. Various actions to rein in water pollution have deadlines by 2020. Clearly, the current five-year plan period (2016–2020) is key.

Managing water quality is essentially about managing the economy and the impacts of an increasingly urbanized population. Industries and cities are two main point sources of wastewater, accounting for 27.1% and 72.7%, respectively, of the total discharge of 73.53 billion tonnes in 2015. Of the industrial sources, 45.5% was from the top four polluting industries: chemicals; paper and pulp; textiles; and coal mining and washing (Ministry of Environmental Protection, 2017). Geographically, 40% of wastewater discharge came from five provinces, which also have the largest economies, representing 40% of the national GDP and 31.8% of China's population (National Bureau of Statistics of China, 2016).

For China, slower GDP growth has become the 'new normal'. The economy is shifting towards more added value from industries and services: in 2016, tertiary industry already accounted for over half (51.6%) of the GDP (National Bureau of Statistics of China, 2017). Traditional manufacturing such as textiles faces a double whammy, not only economically but also environmentally, to meet higher compliance requirements. If China wants to become a well-developed society by 2020, it has to increase the share of industries with higher added value, such as advanced machinery and IT. This is why there is the promotion of Strategic Emerging Industries and the Made in China 2025 strategy. Compared to traditional ones, most of these new industries are less resource-intensive and also less polluting.

Meanwhile, population growth slowed to 0.5% in 2016, less than half of the world average of 1.2% and also slower than countries like India (1.1%), the US (0.7%), Indonesia (1.1%) and Brazil (0.8%) (http://data.worldbank.org). Over 57% of the population are now living in cities (National Bureau of Statistics of China, 2017). By 2020, this is expected to rise to 60%. Pressure on municipal wastewater treatment will likely increase.

With various factors at play, it won't be an easy task to balance economic growth and pollution control in a rather short period, even 20 years. There will be trade-offs. To find a fine balance, a holistic approach is necessary in water policy making. This means that water quality management needs to be embedded in regional planning. Not only should water quality be integrated with water resource management, it should also be considered together with industrial transformation, food security, urban planning and climate impacts. The integration should be cross-regional, especially for upstream and downstream regions that share water systems.

We see such holistic thinking in the development plans of key regions such as the Yangtze River Economic Belt and the Pearl River Basin. For instance, a clear goal has been set to fully integrate and coordinate the socio-economic development of the three regions of the Yangtze River Economic Belt (Upper Reaches, Middle Reaches and the Yangtze River Delta) by 2030, to ensure overall water quality and ecological protection along the river.

According to the newly amended Water Pollution Prevention & Control Law, the environmental protection department within the State Council will act as the responsible party for setting water quality monitoring standards and coordinating unified monitoring stations and information disclosure. This will hopefully resolve the discrepancies between various standards and data produced by different ministries.

The new amendments also promote a national pollution permit system and the 'river chief' system to facilitate more integrated water management. According to the State Council's implementation plan, all point-pollution sources will be required to obtain such a permit by 2020. The first targeted industries are thermal power, and paper and pulp. By the end of 2017, all the companies in key industries identified in the Action Plan for Water Pollution Prevention and Control were to be covered. The 'river chief' system holds the highest-ranked government official accountable at each administrative level. This will help improve cross-department cooperation and facilitate more holistic decision making.

Other innovative mechanisms are also progressing. For instance, eco-compensation schemes are being promoted in key river basins between upstream and downstream (State Council of P.R. China, 2016). We have already seen progress on the Dongjiang River, an important water source for several cities in the Pearl River Delta. Two provinces, Guangdong and Jiangxi, signed an agreement in October 2016 to equally contribute RMB 100 million every year to protect the water quality of the river they share (Xinhua News, 2016).

These are positive steps. A holistic and integrated approach to manage the economy together with water is a must-do for China. Only by doing so can China ensure its long-term prosperity and water security. In this process, not only the government but also researchers and the private sector can play important roles. We can all make smarter decisions today to make tomorrow's water problems less wicked.

Note

1. For example, in 2017 the question was 'What scientific term or concept ought to be more widely known?' The 2013 question may be the closest to our concern here: 'What "should" we be worried about?' https://www.edge.org/annual-questions (viewed 14 February 2017).

References

China Water Risk. (2017). *2016 State of environment report review*. Retrieved from http://chinawaterrisk.org/resources/analysis-reviews/2016-state-of-environment-report-review/

Kostka, G. (2014). *Barriers to the implementation of environmental policies at the local level in China*. Retrieved from https://ssrn.com/abstract=2487614

Ministry of Environmental Protection of China. (2017). *National environmental statistical report (2015)*. Retrieved from http://www.mep.gov.cn/gzfw_13107/hjtj/hjtjnb/201702/P020170223595802837498.pdf

National Bureau of Statistics of China. (2016). *National data from the National Bureau of Statistics of China*. Retrieved from http://data.stats.gov.cn/index.htm

National Bureau of Statistics of China. (2017). *Statistical Communiqué of the People's Republic of China on the 2016 National Economic and Social Development*. Retrieved from http://www.stats.gov.cn/english/PressRelease/201702/t20170228_1467503.html

State Council of P.R. China. (2016). *Opinions on establishing eco-compensation systems*. Retrieved from http://www.gov.cn/zhengce/content/2016-05/13/content_5073049.htm

Xinhua News. (2016). *Jiangxi and Guangdong initiated eco-compensation pilot on Dongjiang River*. Retrieved from http://news.xinhuanet.com/local/2016-10/19/c_1119751405.htm

Zhang, B., & Cao, C. (2015). Four gaps in China's new environmental law: implementation and accountability will remain challenging, especially at the local level, warn Bo Zhang and Cong Cao. *Nature*, *517*, 433–435.

Water quality management from source to sea: from global commitments to coordinated implementation

Birgitta Liss Lymer, Joshua Weinberg and Torkil Jønch Clausen

ABSTRACT
This article reviews links and gaps related to water quality in the 2030 Agenda for Sustainable Development. It uses these as a starting point to analyze challenges in water quality management at national and basin levels from a 'source to sea' perspective. Experiences in the Danube River, Black Sea, East Asian Seas and the Baltic highlight key issues to be addressed during the implementation of the agenda to achieve water quality objectives in fresh, coastal and marine waters. It assesses priorities for supplementary actions to be supported to improve prospects for achieving targets in these areas.

Introduction

The need to control the cumulative effects of multiple sources and causes of deteriorating water quality in a given water body, from what can be a very large geographical area, represents one of the most cumbersome aspects of water quality management. Pollutants that enter water bodies can be deposited in sediments or transported long distances by rivers and currents, resulting in problems that can be local, regional or even global in nature (Granit et al., 2017). While some pollutants are transported as gas or aerosols through the atmosphere and deposited at remote sites by rainfall, others accumulate in the food chain and travel long distances in contaminated fish or seabirds. Alterations of water and sediment flows by activities that take place far upstream in a river system can influence water quality parameters such as water temperature, salinity and turbidity. While many of the impacts of poor water quality are felt locally, reduction of water pollution often require measures at national or transboundary levels.

Experiences in regions such as the Baltic and the Black Sea show that it is possible to reverse negative water quality trends even in water systems with large multi-country catchments, but that it requires decades of collaboration and investments in the range of hundreds of millions of dollars (Granit et al., 2017). While there are positive examples of pollution reduction in some regions, many trends are negative. The use of fertilizers and pesticides has increased exponentially since the beginning of the twentieth century (Tilman, 2001). Currently, 12.7 million tonnes of plastic waste enter the oceans each year (Jambeck et al., 2015), much of it via our waterways. Although there have been some positive trends (Pacyna et al., 2007; UNEP/GPA, 2016), heavy metals and persistent organic pollutants still pose major challenges in most regions. More than 70% of industrial wastes are dumped untreated into waters in many developing countries (WWAP, 2009) and only 40% of the global population is connected to sewage that is treated in any way before being discharged into the environment (Baum, Luh, & Bartram, 2013). New compounds that our wastewater treatment systems may not be equipped to handle, like pharmaceuticals, are also being added to the mix.

As a result, one-third of all river stretches in Latin America, Africa and Asia are affected by severe pathogen pollution (UNEP, 2016), and water quality has been severely affected in more than 80% of the world's transboundary basins (UNEP & UNEP-DHI, 2016). Further downstream, this has contributed to increasing eutrophication in fresh and coastal waters (Seitzinger et al., 2002; Smith, 2003), exponential increases in hypoxic areas and algal blooms (Diaz & Rosenberg, 2008; Dolah, 2000), and even changes in the chemistry of the open ocean (Doney, 2010). On the other side, salinity intrusion due to overexploitation of coastal aquifers, in some cases in combination with sea-level rise, threatens water security in coastal areas and small islands (McGranahan, Balk, & Anderson, 2007; Nurse et al., 2001).

In 2015, the 2030 Agenda for a Sustainable Development (UNGA, 2015) was adopted by the 193 member states of the United Nations to guide efforts towards a more sustainable future. An integral part of the agenda concerns the need for sustainable use of the planet's natural resources achieve and sustain inclusive and sustainable economic growth for all. The importance of good water quality for both human health and ecosystems is recognized and addressed in explicit terms throughout the Sustainable Development Goals (SDGs), not only in relation to water and sanitation, but also in relation to health, cities, consumption and production patterns, and oceans. Challenges in implementing the agenda will include addressing the complex links that span the different goals and targets, such as in the case of water quality.

The development of the 2030 Agenda for Sustainable Development (UNGA, 2015) was a highly participatory process. It was led by countries, and it benefited from broad participation from other stakeholders. The result, i.e. the formulation of the SDGs and their targets and indicators, could therefore be assumed to reflect the discussions that take place during national-level policy formation and planning, although in a simplified and aggregated form. This article reviews links and gaps as they relate to water quality in the SDGs, their targets and their proposed indicators and uses these as a starting point to analyze challenges in water quality management at national and basin levels.

Methods

The article is based on the understanding, as described by Granit et al. (2017), that flows of water, sediment, pollutants, materials, biota and ecosystem services connect ecosystems from source to sea.

A combination of analytical methods were applied. The research attempted to characterize the relation between water quality issues in fresh and marine waters and the 2030 Agenda and connect the findings to recent efforts to address water quality degradation in coastal and marine waters.

To identify potential gaps in how the SDGs address water quality, a method for basic content analysis (Elo & Kyngäs, 2008; Krippendorff, 2004) was developed to analyze the relation between 'implied' and 'explicit' links to water quality in the different SDG targets. The formulation of the SDGs and their targets was adopted by 193 countries at the United Nations Sustainable Development Summit in September, 2015. To enable the monitoring of progress on their implementation, the development of an indicator framework was initiated shortly thereafter, led by the UN Inter-Agency and Expert Group on Sustainable Development Goal Indicators. This framework consists of a list of indicators, which is complemented by metadata for each indicator specifying concepts and definitions, methodology, data sources, data availability and data providers. The metadata for several indicators is still under development.

First, the sample for analysis was restricted to those SDG targets with implied links to water quality, that is, those targets where UN-Water (2016) had identified links to the water quality target (Target 6.3), in terms of whether or not achieving a particular target has synergies with the water quality target (in terms of mutually reinforcing, positive interdependencies) or represents a potential conflict with the water quality target.

Second, the documentation available to outline the scope of those SDG targets with implied links to water quality was coded to identify the extent to which they include explicit links, i.e. whether or not explicit aims to reduce water pollution or improve water quality are included in the formulation of the targets (UNGA, 2015), their indicators (IAEG-SDG, 2017) or the metadata currently available for those indicators (IAEG-SDG, 2016).

Third, implied and explicit links to water quality were compared to analyze gaps and identify areas in need of attention during planning and review of SDG implementation.

Lessons from regional experiences to improve water quality in the Black Sea, East Asian Seas and the Baltic, as documented in literature reviews by Liss Lymer and Zavadsky (2017), Tengberg (2017), and Nõmmann and Granit (2017), were then drawn on to further analyze the efforts required to address the integrated and indivisible nature of the 2030 Agenda as it relates to water quality during its implementation.

Analysis: water quality in the 2030 Agenda

The SDGs are the first globally agreed target for improved fresh and marine water quality for all countries. The Millennium Development Goals, which preceded the SDGs, referred broadly to principles of environmental sustainability, but did not include any specific targets related to water quality. The Global Programme of Action for the Protection of the Marine Environment from Land-Based Activities aims to deal with a number of pollution sources that harm the marine environment (UNEP, 1995) and is the only global intergovernmental mechanism that addresses the links between terrestrial, freshwater, coastal and marine ecosystems. Multilateral environmental agreements (MEAs) relevant to water quality management tend to focus on issues related to specific pollutants (e.g. the Stockholm Convention on Persistent Organic Pollutants and the Minamata Convention on Mercury), the handling of pollutants across borders or at sea (e.g. the Basel, MARPOL and London Conventions), or broader aspects of transboundary environmental cooperation, which may or may not include aspects related to water quality (e.g. the UNECE Water Convention and the Espoo Convention). The

SDGs complement the existing global governance framework by making specific reference to aims by countries to both improve freshwater quality and reduce marine pollution. In addition, the need for improved water quality spans the social, environmental and economic dimensions of several other SDGs.

The formulation of the SDGs and their targets is broad, offering room to consider a wide range of relevant links. In March 2017, 244 indicators covering the different SDG targets were agreed to (UNSC, 2017). These are intended to provide the means to assess the progress on related targets. Even though a particular indicator does not cover the full extent of its related target, there is a risk that it will be used to define how the targets should be interpreted for the sake of practicality.

As an example, there are many implied links between good water quality and other social, environmental and economic objectives in the SDGs. UN-Water (2016) identified links in terms of both synergies and potential conflicts between water quality targets and all other SDGs. Synergies between the water quality target (Target 6.3) and other environmental targets are strong. For example, it can be argued that the rate of success in protecting water-related ecosystems (Target 6.6), inland freshwater ecosystems (Target 15.1), marine and coastal ecosystems (Target 14.2), and threatened species (Target 15.5) is closely linked to efforts to improve water quality (Target 6.3). Likewise, success in reducing child mortality (Target 3.2) depends on access to clean drinking water (Targets 6.1 and 6.3). Potential conflicts were identified primarily in relation to targets that aim to increase production (e.g. food production and industrialization), access to basic services (e.g. water and sanitation and energy), and economic growth. Progress towards these targets may increase the risk of pollution and water quality degradation, unless adequate measures are put in place to minimize such risks.

Explicit references to water pollution occur in the targets to reduce marine pollution (Target 14.1) and deaths and illnesses from water pollution and contamination (Targets 3.3 and 3.9). Although several other (implied) links are apparent, the SDG targets and indicators are not necessarily designed to make explicit reference to all implied links in the formulation used to define them. Nevertheless, gaps in how cross-cutting issues are reflected in the formulation of targets and their indicators risk contributing to compartmentalization of the implementation of an otherwise integrated agenda.

Further review of how improved water quality would enable or depend on different targets in the 2030 Agenda and how such links are captured in the formulation of the different goals, targets and proposed indicators reveals a number of areas that merit special attention.

The capacity to monitor water quality

Improved data and monitoring capacity are needed to understand and address water quality issues, particularly in developing regions and those experiencing rapid economic growth (UNEP, 2016). Reliable water quality monitoring and reporting is also needed to justify the required investments to reduce water pollution. The SDG target for water quality (6.3) recognizes that the ability of countries to report on progress in water quality is heavily dependent on their capacity to monitor water quality. To increase the likelihood of receiving comparable data across countries, the baseline selection of water quality parameters is limited to a number of determinants of water quality impairment: total dissolved solids, percentage dissolved oxygen, dissolved inorganic nitrogen and phosphorous, and *E. coli* (IAEG-SDG,

2016). They do not cover all the pollutants the SDG targets aim to reduce, i.e. hazardous chemicals, materials, wastes and marine pollution of all kinds. Parameters to track 'marine pollution of all kinds' are limited to coastal eutrophication index and floating plastic debris density. In the case of freshwater quality monitoring, a monitoring ladder is proposed through which countries are encouraged to expand their monitoring programme as national capacities and requirements increase (IAEG-SDG, 2016). The selection of relevant parameters and assignment of different priorities for different water types and uses will be a critical process to observe.

Tracing the sources and causes of water quality degradation

Applying water pollution control at the appropriate level, be it district, basin, national or transboundary, requires knowledge of the pollutant flows from the source to the recipient. To prevent pollution, rather than treating its symptoms, the causes and sources of deteriorating water quality in a particular water body have to be identified. In addition to water quality monitoring, this requires pollution source assessments, ideally in combination with modelling applications to assess management alternatives. The EU Water Framework Directive, for example, requires member states to identify significant point and diffuse sources of pollution for water bodies in each river basin district, to assess their impacts and to design measures to regulate and control point-source emissions and control or prevent the input of diffuse sources of pollutants to achieve good status of all water bodies (EU, 2000). The United States' Clean Water Act (USC, 1972) requires that total maximum daily loads of pollutants be assigned for water quality standards in impaired waters. Knowledge of the different pollution sources is then used to allocate necessary reductions in one or more pollutant sources.

The SDG indicators of water quality are designed to assess the progress in reaching their respective SDG targets. The SDG indicators that include an explicit reference to water pollution or water quality focus on the mortality rate attributed to unsafe water (3.9.2); the proportion of wastewater safely treated (6.3.1); the proportion of bodies of water with good ambient water quality (6.3.2); the number of parties to international multilateral environmental agreements on hazardous waste and chemicals that meet their commitment in transmitting information (12.4.1); and the index of coastal eutrophication and floating plastic debris density (14.1.1). There are no indicators defined, however, to assess the actual release of pollutants into water bodies from different sources. Such data are crucial to guide the management measures that are needed to improve water quality. In this sense they represent an end rather than a means to good water quality management. As constructed, the indicators have limitations in guiding water quality management efforts on their own, in particular with respect to the capacity to trace the source and flow of pollutants, which is necessary to enable progress towards improved water quality.

The accountability and engagement of water-polluting sectors and other key stakeholders

Ensuring effective regulation and accountability systems of pollution from agriculture, industry and other relevant sectors is a fundamental component of water quality and overall environmental quality management. There are a number of ways this is commonly done, including laws, regulations, emission permits, environmental impact assessments, strategic environmental assessments, and enforcement of the polluter-pays principle. Another method, often used in combination with other regulatory instruments, is incentivising, promoting or requiring best practices and best available technologies in agriculture or other industries.

The SDGs do not assign accountability to water-polluting sectors in the sense of including targets for reduced water pollution from any particular sector, such as food or industry. Several SDGs that deal with areas whose contribution is critical to improve water quality include no explicit reference to water pollution. Targets 6.3 and 14.1 can be seen as all-encompassing, aiming to 'improve water quality by reducing pollution, eliminating dumping and minimizing release of hazardous chemical and materials, halving the proportion of untreated wastewater and substantially increasing recycling and safe reuse globally' and to 'reduce marine pollution of all kinds'. However, the proposed indicators of progress towards this target focus solely on the 'proportion of wastewater safely treated', the 'proportion of bodies of water with ambient good water quality' and the 'index of coastal eutrophication and floating plastic debris density'.

As implied in the formulation of the target, improved water quality would also benefit greatly from pollution prevention at the source. This is reflected in Target 12.4, to 'achieve environmentally sound management of chemicals ... and reduce their release to air, water and soil'. Target 12.4 is however the only SDG target outside of the 'water goal' (SDG 6) that explicitly aims to reduce water-pollutant emissions. The targets and proposed indicators of the sustainability of other common water-pollution sources tend to focus on other parameters. In the case of sustainable industries (Target 9.4), the focus is on CO_2 emissions. The focus of reducing the environmental impact of cities (Target 11.6) is on air pollution and solid waste. Efforts to decouple economic growth from environmental degradation (Target 8.4) are proposed to be measured against material footprint and material consumption. The proposed indicator for Target 2.4, on sustainable agriculture, focuses on maintaining the natural resource base to ensure sufficient productivity, but does not refer to runoff pollution, for example. This could lead to a national plan where the authorities responsible for oversight of the main drivers of pollution are not assigned responsibility to report their actions to reduce pollution of water bodies.

Such gaps point to one of the challenges in water quality management. Where resources are limited and the institutional capacity to regulate and enforce pollution reduction is weak, efforts and investments to increase the environmental sustainability of a particular sector or industry may be directed towards issues that are deemed of highest priority. In the cases where poor water quality is the result of cumulative effects of multiple pollution sources in a large geographical area, this may be at the expense of efforts to reduce water pollution from multiple sources of little individual but large combined impact.

Discussion: How to move from an agenda to action? Managing water quality 'from source to sea'

Overall, the SDGs provide a measurable, aspirational set of goals aiming to steer global progress towards sustainable development. Together, they form an agenda that is 'integrated and indivisible' (UNGA, 2015). Countries will now have to design strategies and action plans to achieve these goals. An important challenge in such a process will be to understand and address interlinkages between the delivery of different SDGs and targets as they relate to the needs and priorities of different sectors, stakeholders and ecosystems. In the case of water quality, this entails balancing actions to achieve goals for sustainable economic growth, food security and inclusive industrialization with those to improve water quality and reduce marine pollution for human health and ecosystem well-being.

These are challenges that many countries and regions have been battling for decades. In 2017, Granit et al. identified some of the main stumbling blocks for effective resource and water quality management across connected land, river, coastal and ocean systems (source-to-sea systems) as: the fragmented, often sectoral, governance and management frameworks that make strategic planning across connected systems cumbersome; the challenges in engaging upstream stakeholders, who may or may not experience the impacts of deteriorating downstream water quality, in pollution-reduction activities; and the collaboration, investments and long timeframes required to achieve positive outcomes.

The gaps found in the 2030 Agenda also point to some of these issues. Pollution source prevention often requires measures that depend on sectoral engagement, but such engagements are not always clearly incentivized by the targets and indicators set out in the SDGs. The way the SDG water-quality-related indicators, for example, are currently designed, they do not stimulate specific action to trace pollutant flows through the environment, from the source to the recipient (Berggren & Liss Lymer, 2016). Another fundamental concern is to ensure a sense of priority and accountability of polluting sectors, such as agriculture and industry, towards implementing measures and reporting on progress in reducing water pollution. Here, the tendency to focus on 'one issue at a time', such as CO_2 emissions in industry, could lead to slow or no action on water pollution control. Unless productive sectors, such as agriculture, take responsibility for the targets related to improved water quality in fresh and marine waters, it is unlikely that those targets will be achieved. However, the potential conflicts between increasing food and industrial production and improving water quality are not captured in the current formulation of the indicators, which were designed to track countries' progress towards more sustainable production in these areas.

Even where there is good understanding of water quality issues and strong political commitment to remediate them, a number of common pitfalls arise. Experiences in large-scale regional initiatives exemplify some of the challenges facing regions in addressing these water quality challenges across multiple scales.

The importance of water quality monitoring and data as means for justification of large-scale interventions can be noted in relation to the cooperation between Danube River and Black Sea countries. In 1990 about 40,000 km^2 of the Black Sea's North-West Shelf was effectively considered a dead zone. In the mid-1990s, both Black Sea and Danube River countries formally committed to protect the Black Sea from pollution. Since then, eutrophication and hypoxia have decreased in frequency and severity following annual reductions of 25,000 metric tonnes of nitrogen and 4000 metric tonnes of phosphorus pollution (Liss Lymer & Zavadsky, 2017). This resulted from both policy and regulatory reforms and large investments in nutrient-reducing water treatment and better farming practices. In the Danube Basin, nitrogen emissions decreased by 20% and phosphorus by almost 50% over 15 years (ICPDR 2007). This case featured several favourable conditions for transformation: an upstream community was more affluent than the downstream area and proactively engaged in reducing pollution to the sea; the requirements of recent EU legislation contributed to commitment to nutrient reduction in many of the countries; partners had access to large external investments and support for environmental restoration; and those investments coincided with a dramatic drop in agricultural production and resulting runoff due to market forces (Liss Lymer & Zavadsky, 2017). However, the programme had difficulty establishing an effective monitoring system in the Black Sea to detect changes in its water quality (Liss Lymer & Zavadsky, 2017). Without clear information on the effectiveness of the measures already undertaken, it is hard to justify continued investments in pollution reduction or to incentivize continued action by upstream authorities.

The Baltic Sea case provides an example of a long-term regional effort to trace the causes and sources of water quality degradation and take remedial action. Collaboration within and between Baltic countries to monitor water quality, identify and quantify pollution-load sources, conduct research and set national nutrient-reduction targets, coupled with investment programmes, have successfully led to drastic declines in nutrient inputs (Nõmmann & Granit, 2017). Yet, the concentration of nutrients in the Baltic has not declined at the same rate (Andersen et al., 2015). This is because the Baltic is a slow system with a long residence time, and also due to the delayed release of phosphorous from bottom sediments that entered the system in past years. To achieve good water quality in the Baltic and to reach the targets that countries have agreed on, voluntary actions outside the legal requirements for wastewater management are required (Andersen et al., 2015), but most of the municipalities in the Baltic Sea region are either unaware of the extent of the problem or lack the resources to implement actions beyond the mandated ones. There is some asymmetry between different stakeholders in terms of their resources, interest, and capacity to contribute to reducing pollution to improve the state of the Baltic. Coastal municipalities are more aware of the critical state of the Baltic than those inland, and are trying influence upstream authorities to prioritize nutrient runoff prevention even more (Dahlgren et al., 2015). Several of the countries that represent the highest nutrient input to the Baltic are also the countries where the general population values the Baltic least and is least willing to pay to improve the state of the Baltic (BalticStern, 2013).

The differences in awareness and, supposedly, willingness to engage and invest in water quality improvements, between those directly affected (coastal municipalities) and those located at a distance from negative impacts on coastal water quality (upstream municipalities) point towards the third area needing special attention: the accountability and engagement of water-polluting sectors. Despite noting several primary priority sources of pollution upstream of the Black Sea, the Black Sea Ecosystem Recovery Project was unable to focus on agriculture, industry and transport because it could not effectively engage or influence government authorities in those areas. Even though best agricultural practices and the use of phosphate-free detergents were promoted, the lack of engagement from relevant ministries meant that few changes were made in the practices of farming and other industries that impair water quality (Liss Lymer & Zavadsky, 2017). In the Seas of East Asia, Partnerships in Environmental Management for the Seas of East Asia has successfully supported the implementation of integrated coastal management to improve coastal and environmental quality (Tengberg, 2017). In 2015, it was estimated that countries had scaled up integrated coastal management programmes to cover 14% of the region's 234,000 km coastline. Nonetheless, Partnership sites face common upstream/downstream dilemmas, in which upland local government units have to invest in activities that will largely benefit coastal local government units (Tengberg, 2017). Incentivizing upstream communities and holding polluting industries accountable are the primary challenges. In these cases, willingness to pay for schemes for compensation for actions taken by upstream actors is not necessarily matched by an ability to oversee regulation of the industrial and agricultural sectors that act outside their jurisdiction.

In many places, it is the coastal zone where one finds a consolidation of wealth, power and also pollution loads (Blackburn & Pelling, 2014). Their value can provide impetus for the downstream communities to try to improve water quality upstream, before it reaches their communities. In the case of the coastal city of Xiamen in China's Fujian Province, compensation of roughly USD 8 million per year is provided to the upstream municipalities of Longyan and Zhangzhou for measures to reduce pollution of the Jiulong River (Lundqvist, Bjørnsen, Weinberg, Liss Lymer, & Qinhua, 2013). This is generally effective to improve 'green investment' in treatment and control of industrial pollution, but agricultural runoff has been more difficult to address through compensation policies.

These few examples show some actions taken that have positively contributed to water and coastal quality management 'from source to sea'. Each features some basic enabling conditions that stimulate the developments that took place: an engaged upstream actor in the Danube River basin authority and its riparian countries were committed to supporting improved environmental quality in the Black Sea; in the Baltic Sea, a high value is placed on environmental quality and restoration, and citizens of many countries are willing to pay and support political commitment to the cause; and in Fujian, the coastal city is an economic power with high willingness to pay for investments and compensation for measures to improve water quality upstream. These situations also benefited from access to substantial funding for pollution reduction, either from within the region or from external donors. Even so, each still faces major challenges, as noted in this section, particularly in controlling diffuse pollution sources in upstream communities. Reducing pollution can be even more cumbersome in places where the upstream development options, be it for industrial, agricultural or hydropower development, for example, are prioritized; where there is limited knowledge of pollution loads and the negative impacts associated with water pollution; and/or where financing from outside parties is not available. These tend to also be places where coastal communities are most vulnerable to losses of environmental quality in their livelihoods. A fundamental additional challenge faced in these situations is to assess and integrate the value provided by the water and environmental quality in the coastal zone into regional planning and development that better incorporates impacts on human livelihoods (beyond GDP), the potential future cost of restoration, and the benefits of coastal-zone development.

Conclusions

The SDGs are an achievement in their capacity to align and raise collective ambitions for global development, but they are a result of negotiation. As such, they represent what a large and diverse group of countries and stakeholders has been able to agree on as key markers of development progress. The challenge ahead is to move from a series of dots in the form of ambitious goals and targets, towards connecting these dots through national plans of action. The spaces between the dots do not need to be viewed as gaps in a global agenda but can be seen as priority issues that need to be addressed in those national action plans. These action plans will only be truly useful and result in movement towards fresh, coastal and marine water quality objectives if they are able to address the cross-sectoral priorities needed to implement pollution source prevention more effectively than the laws, regulations, strategies and plans that are already in place.

This article has shown that deeper inspection of the suggested indicators for each target only widens the gaps between the goals and targets set out and the frameworks suggested for reaching them. For example, the only water-quality items that countries need to report, according to the currently suggested indicators, are the proportion of wastewater safely treated and the proportion of bodies of water with good water quality. For marine waters, the indicator focuses on indexes of coastal eutrophication and amount of floating plastic debris. Apart from investing in wastewater treatment, the only second clear action for countries to take is to monitor fresh and marine water quality. There are no distinct targets aiming to prevent water pollution from key sectors, such as agriculture and industry. The selected indicators quickly steer into familiar patterns to cover what is actually under the jurisdiction of the authorities responsible for water and environment.

As a result, the SDGs as such may not provide guidance to better address the more complex issues of source prevention. Countries need to look independently at how such issues can be best addressed in each particular context. Experience with action plans to achieve the Millennium Development Goals suggests that, for practical purposes, the indicators tend to 'become the targets' in policy processes (Berggren & Liss Lymer, 2016). If this is again the case, source prevention, nutrient and water reuse, and other opportunities could be lost if priority actions are decided sector by sector.

The coordinated and effective management of pollution flows from their source and into fresh and marine waters could thus be considered a priority area for supplementary actions to be supported in the UN system or by other actors explicitly committed to ensuring that the SDGs are met. A logical starting point for this could focus on ensuring that management of land-based activities is effectively considered and coordinated in the strategies set out to achieve the SDG targets set out in 6.3 and 14.1. This could include alternative entry points for intervention dependent on whether basic enabling conditions for engaging upstream actors in pollution prevention and control exist in the region. If downstream actors lack both the technical capacity and the resources to influence polluters upstream, more direct financial and technical support may be warranted. In this way, the aspirations set out in the SDGs can be used as a real opportunity to support implementation of national strategies to improve water and environmental quality 'from source to sea'.

Disclosure statement

No potential conflict of interest was reported by the authors.

References

Andersen, J. H., Carstensen, J., Conley, D. J., Dromph, K., Fleming-Lehtinen, V., Gustafsson, B. G., ... Murray, C. (2015). Long-term temporal and spatial trends in eutrophication status of the Baltic Sea. *Biological Reviews, 92*(1), 135–149.

BalticStern. (2013). The baltic sea: our common treasure. Economics of saving the sea. Gothenburg, Sweden Retrieved from https://www.havochvatten.se/download/18.2a9b232013c3e8ee03e46f5/1363249834195/rapport-hav-2013-04-baltic-sea-our-common-treasure-english.pdf

Baum, R., Luh, J., & Bartram, J. (2013). Sanitation: A global estimate of sewerage connections without treatment and the resulting impact on MDG progress. *Environmental Science & Technology, 47*, 1994–2000.

Berggren, J., & Liss Lymer, B. (2016). *Source to sea: linkages in the 2030 Agenda for Sustainable Development* (*Report 2016:22*).

Blackburn, S., & Pelling, M. (2014). Coastal megacities: Risks and opportunities. *Glob Change, 82*, 8–11.

Dahlgren, S., Källström, N., Lind, F., Morin, M., Mrozowski, T., Seppä, T., & Wallin, M. (2015). *Restoring waters in the Baltic Sea Region – A strategy for municipalities and local governments to capture economic and environmental benefits*. The Boston Consulting Group. Retrieved from http://image-src.bcg.com/Restoring-Waters-Baltic-Sea-Region-Feb-2015-Nordics_tcm22-29045.pdf

Diaz, R. J., & Rosenberg, R. (2008). Spreading dead zones and consequences for marine ecosystems. *Science, 321*, 926–929.

Dolah, F. M. V. (2000). Marine algal toxins: Origins, health effects, and their increased occurrence. *Environmental Health Perspectives, 108*, 133.

Doney, S. C. (2010). The growing human footprint on coastal and open-ocean biogeochemistry. *Science, 328*, 1512–1516.

Elo, S., & Kyngäs, H. (2008). The qualitative content analysis process. *Journal of Advanced Nursing, 62*, 107–115.

EU. (2000). Establishing a framework for community action in the field of water policy. Directive 2000/60/EC of the European Parliament and of the Council. *Official Journal of the European Union, L, 327*, 1–72.

Granit, J., Liss Lymer, B., Olsen, S. B., Tengberg, A., Nõmmann, S., & Clausen, T. J. (2017). A conceptual framework for governing and managing key flows in a source-to-sea continuum. *Water Policy, 19*(4), 673–691.

IAEG-SDG. (2016). Compilation of Metadata for the Proposed Global Indicators for the Review of the 2030 Agenda for Sustainable Development received as of 4 March 2016.

IAEG-SDG. (2017). Revised list of global Sustainable Development Goal indicators. Retrieved from https://unstats.un.org/sdgs/indicators/Official%20Revised%20List%20of%20global%20SDG%20indicators.pdf

ICPDR. (2007). Joint Danube Survey 2: Water pollution in the Danube River Basin. Retrieved from https://www.icpdr.org/jds/files/JDS_FS_Pollution_EN.pdf

Jambeck, J. R., Geyer, R., Wilcox, C., Siegler, T. R., Perryman, M., Andrady, A., ... Law, K. L. (2015). Plastic waste inputs from land into the ocean. *Science, 347*, 768–771.

Krippendorff, K. (2004). *Content Analysis: An introduction to its methodology* (2nd ed.). USA: Language Arts & Disciplines. SAGE Publications. ISBN 0-7619-1544-3.

Liss Lymer, B., & Zavadsky, I. (2017). (2017). Danube River and Black Sea Collaboration. Case study supplementary material in Granit et al. *Water Policy, 19*, 2.

Lundqvist, J., Bjørnsen, P. K., Weinberg, J., Liss Lymer, B., & Qinhua, F. (2013). Bridging divides between land, water, sea and coast. In A. Jägerskog, T. J. Clausen, K. Lexén, & T. Holmgren (Eds.), *Cooperation for a Water Wise World – partnerships for sustainable development* (pp. 55–61). Stockholm: Stockholm International Water Institute.

McGranahan, G., Balk, D., & Anderson, B. (2007). The rising tide: Assessing the risks of climate change and human settlements in low elevation coastal zones. *Environment and Urbanization, 19*(1), 17–37.

Nõmmann, S., & Granit, J. (2017). The Baltic Sea Region. Case study supplementary material for Granit et al., 2017. *Water Policy, 19*, 2.

Nurse, L. A., Sem, G., Hay, J., Suarez, A., Wong, P. P., Briguglio, L., & Ragoonaden, S. (2001). Small island states. In J. J. McCarthy, O. F. Canziani, N. A. Leary, D. J. Dokken, & K. S. White (Eds.), *Climate change 2001: Impacts, adaptation, and vulnerability. Contribution of Working Group II to the Third Assessment Report of the Intergovernmental Panel on Climate Change* (pp. 844–875). Cambridge: Cambridge University Press.

Pacyna, E. G., Pacyna, J. M., Fudala, J., Strzelecka-Jastrzab, E., Hlawiczka, S., Panasiuk, D., … Friedrich, R. (2007). Current and future emissions of selected heavy metals to the atmosphere from anthropogenic sources in Europe. *Atmospheric Environment, 41*(38), 8557–8566.

Seitzinger, S. P., Kroeze, C., Bouwman, A. F., Caraco, N., Dentener, F., & Styles, R. V. (2002). Global patterns of dissolved inorganic and particulate nitrogen inputs to coastal systems: Recent conditions and future projections. *Estuaries, 25*, 640–655.

Smith, V. H. (2003). Eutrophication of freshwater and coastal marine ecosystems a global problem. *Environmental Science and Pollution Research, 10*, 126–139.

Tengberg, A. (2017). Collaboration in the Seas of East Asia - Partnerships in Environmental Management of the Seas of East Asia (PEMSEA). Case study supplementary material in Granit et al., 2017. *Water Policy, 19*, 2.

Tilman, D. (2001). Forecasting agriculturally driven global environmental change. *Science, 292*, 281–284.

UNEP. (1995). *Washington declaration on protection of the marine environment from land-based activities*. Retrieved from https://wedocs.unep.org/bitstream/handle/20.500.11822/13421/WashingtonDeclaration.pdf?sequence=1&isAllowed=y

UNEP. (2016). A snapshot of the World's Water Quality; Towards a global assessment. (978-92-807-3555-0). Retrieved from https://uneplive.unep.org/media/docs/assessments/unep_wwqa_report_web.pdf

UNEP, & UNEP-DHI. (2016). Transboundary River Basins: Status and trends. Nairobi, Kenya Retrieved from http://twap-rivers.org/assets/GEF_TWAPRB_FullTechnicalReport_compressed.pdf

UNEP/GPA. (2016). The State of the Marine Environment: Trends and processes. Retrieved from http://www.unep.org/gpa/documents/publications/TheStateMarineEnvironmentTrendsProcesses.pdf

UNGA. (2015). *Transforming our world: The 2030 Agenda for Sustainable Development* (A/RES/70/1)

UNSC. (2017). *Report of the Commission on its forty-eighth session*. Retrieved from https://unstats.un.org/unsd/statcom/48th-session/documents/Report-on-the-48th-Session-of-the-Statistical-Commission-E.pdf

UN-Water. (2016). Water and sanitation interlinkages across the 2030 Agenda for Sustainable Development. Geneva Retrieved from http://www.unwater.org/fileadmin/user_upload/unwater_new/docs/UN-Water_Analytical_Brief_Water%20and%20sanitation%20interlinkages%20across%20the%202030%20Agenda%20for%20Sustainable%20Development.pdf

USC. (1972). *An Act to amend the Federal Water Pollution Control Act*. Retrieved from https://www.epa.gov/sites/production/files/2017-08/documents/federal-water-pollution-control-act-508full.pdf

WWAP. (2009). *The United Nations World Water Development Report 3: Water in a Changing World. World Water Assessment Programme*. Paris; London. Retrieved from http://unesdoc.unesco.org/images/0018/001819/181993e.pdf

Adaptive or aspirational? Governance of diffuse water pollution affecting Australia's Great Barrier Reef

Poh-Ling Tan and Fran Humphries

ABSTRACT

The natural attributes of Australia's Great Barrier Reef, a UNESCO world heritage site listed for its natural beauty and biological diversity, are rapidly declining due to major threats from diffuse water pollution and climate change. The environmental, social, political and legal conditions that have enabled or blocked successful management of diffuse water pollution are analyzed. We find that the management approach has transitioned towards resilience-focused adaptive management of impacts from outside the marine park. Despite key enablers of adaptive governance, deep-seated political ideology is a major barrier to transformational adaptive governance to improve reef water quality.

Introduction

The Great Barrier Reef (hereafter, the Reef), the world's largest coral reef ecosystem, lies north-east of Queensland, Australia. Stretching over 2300 km and receiving runoff from 35 major catchments, it covers an area of almost 350,000 km^2, about the size of Japan.

Australia moved to protect the natural assets of the Reef in 1975, aiming to provide for 'the long term protection and conservation of the environment, biodiversity and heritage values of the Great Barrier Reef Region' (Great Barrier Reef Marine Park Act 1975 (Cth) §2A). It established a marine park over the region and created the Great Barrier Reef Marine Park Authority as an independent statutory body responsible for the day-to-day management of the park. UNESCO listed the Reef on the World Heritage Register for its 'outstanding universal value', as the first of 46 marine World Heritage areas.

The Reef contributes significantly to the Australian economy. In 2015–16, it contributed AUD 6.4 billion to the Australian economy, including AUD 5.7 billion from tourism, AUD 199 million from commercial fishing and aquaculture, AUD 415 million from recreation, and AUD 182 million from scientific research. Beyond this, the Reef's total economic, social and icon value is estimated at AUD 56 billion (Deloitte Access Economics, 2017). Management critically requires addressing the strong agricultural interests in the Reef catchments. Grazing is the dominant agricultural land use;

sugarcane and horticultural crops grow in the coastal floodplain; and grain crops and irrigated cotton grow inland of the Fitzroy region.

Despite protection by the Commonwealth and Queensland governments, an assessment showed a decline of more than 56% of the Reef's World Heritage attributes (GBRMPA, 2014). These attributes include loss of seagrass in areas affected by cyclones and river floods; decline in coral reef cover on mid-shelf and offshore reefs; decline of populations of dugong, an endemic species of mammal; regular outbreaks of crown of thorns starfish, which feed on species of corals; and increased frequency of massive coral bleaching events, including in early 2016 (GBRMPA, 2014; Hughes et al., 2017).

The four highest risks to water quality and the health of the Reef are climate change, land-based runoff, coastal land-use change, and aspects of direct use, such as fishing, shipping and port activities (GBRMPA, 2014; Reef Taskforce, 2016). These risks led the World Heritage Committee to put Australia on notice that the Reef may be listed as a World Heritage in Danger (World Heritage Committee [WHC], 2015a).

Management of water quality issues over the past 30 years has mainly concentrated on point-source pollution. Since scientists have identified diffuse water pollution as an immediate and system-wide threat to the resilience of the Reef, that focus is shifting. Diffuse water pollution emanates from many sources, and its impacts tend to be cumulative, significant and episodic, for example from extreme floods bringing sudden high pollution loads. Both point-source and diffuse pollution impact ecological and heritage values, and their relative contributions are not known precisely, but the former is considered to be better managed in the Reef, and perhaps less alarming overall (GBRMPA, 2014; Reef Taskforce, 2016).

The overarching research question of this article is: How effective, equitable and resilient is Australia's approach to designing and implementing policy and law for diffuse water pollution in the Reef? The analytical framework focuses on three major sub-questions. First, what are the major theoretical frameworks underpinning the governance of the Reef? Second, what are the major issues and current trends relating to the Reef? Third, what conditions have enabled or blocked its success, including environmental, social, political and legal? In relation to the history of the management of the Reef and identification of issues confronting decision makers we undertook literature searches on theoretical frameworks for marine park management; considered relevant public policy and regulatory documentation for the case-study region; and examined other background materials, such as media and political commentary on policy challenges and responses.

Key theoretical frameworks

We rely on resilience theory in our analysis of the Reef's governance. Resilience is the ability of a system to absorb shock and withstand disturbances while retaining its function (Holling, 1978; Walker & Salt, 2006). When subject to disturbance, the ecological responses of organisms are unpredictable, but the greater the resilience, the more able the system is to recover. When critical thresholds and tipping points are reached, the system may drastically transform (Walker & Salt, 2006; GRMPA, 2014).

Derived from resilience theory, adaptive management acknowledges that environmental management actions need to respond to changes in the social–environmental system; and in a feedback loop, any changes that management actions set off require monitoring and further management response (Gunderson, Holling, & Light, 1995; Lee, 1993). Extending that concept, adaptive *governance* incorporates the social and political context in which adaptive management takes place (Gunderson & Light, 2006). The focus is on bridges and barriers to implementing adaptive management (Gunderson et al., 1995; Lee, 1993). Such barriers may include deeply held values about how a geographical area should be developed, vested interests driving people to militate for or against proposed management actions, short-term political incentives that frame issues as win-or-lose battles, or a history of conflict that precludes dialogue (Lewicki, Gray, & Elliott, 2003). Transformational adaptive governance requires opening debates to different perspectives, allowing the questioning of assumptions and reconsideration of beliefs and values, and new approaches to managing uncertainty (Pahl-Wostl, 2009).

To maintain the resilience of marine environments, governments have created reserves (Evans, Bax, & Smith, 2016). Using a representative area model, countries have investigated a range of restrictions, from partial (e.g., fishing gear) to high ('no-take') and almost total ('no-entry') exclusions in the reserves or 'protected areas' (Devilliers et al., 2015). In 1992 the United Nations Convention on Biological Diversity (1992, Article 8) required countries to 'establish a system of protected areas'. However, these static exclusionary representative areas were criticized for being incapable of responding to more indirect and diffuse threats, including poor water quality and climate change (McLeod et al., 2012).

In 2003 the Fifth World Parks Congress explicitly rejected the top-down, exclusionary management model of the previous century (Dudley, 2014). International effort shifted towards recognizing integrated coastal zone management using adaptive governance for protected areas. McCay and Jones's (2011) study of 20 marine parks worldwide identified five categories of incentives for improving marine park governance:

- Participative – encouraging a wide range of stakeholders to collaborate and co-manage, providing coordination of top-down and bottom-up approaches;
- Legal – providing a clear framework for use, roles and responsibilities of parties;
- Interpretative – communicating the rationale and outcomes;
- Knowledge – improving scientific, local and traditional forms of knowledge, and independent advice;
- Economic – gaining support of local stakeholders to overcome resistance.

The Great Barrier Reef Marine Park, created under a static representative area planning model, has transitioned towards adaptive management to tackle cumulative threats from outside the marine park and increase resilience to these threats. There has been a shift in understanding of the key pressures affecting the Reef, consciously adopting a risk management approach, which assesses 'not only pressures from direct use, but also pressures, such as climate change, coming from outside our jurisdiction' (GBRMPA, 2016, p. 3). The remainder of this article explores the steps that the Commonwealth and Queensland governments have taken to implement adaptive resilience-based management to address diffuse water pollution in the Reef.

Current trends and major issues

We highlight the water quality issues caused by two key threats to Reef health identified by the Australian Government and the WHC – climate change and land-based runoff (GBRMPA, 2014; WHC, 2015a). The remaining two threats – direct use and coastal development, including mine expansions, port development and increased shipping through the Reef – are generally considered to be point sources of water pollution, and thus beyond this article's scope.

The 2017 Scientific Consensus statement (Waterhouse et al., 2017, p. 7), a review of the significant advances in scientific knowledge and understanding of the Reef, states:

> Key [Reef] ecosystems continue to be in poor condition. This is largely due to the collective impact of land run-off associated with past and ongoing catchment development, coastal development activities, extreme weather events and climate change impacts such as the 2016 and 2017 coral bleaching events.
>
> Current initiatives will not meet the water quality targets. To accelerate the change in on-ground management, improvements to governance, program design, delivery and evaluation systems are urgently needed. This will require greater incorporation of social and economic factors, better targeting and prioritisation, exploration of alternative management options and increased support and resources.

Climate change impacts

The direct and indirect impacts of climate change are relevant to water quality. Direct impacts derive from rising ocean temperatures and acidification, whereas indirect impacts derive from extreme weather events – droughts, flooding, decreases in salinity and damage from cyclones. The 2017 Scientific Consensus statement confirms that 'climate change is predicted to increase the intensity of extreme weather events, which are significant in driving impacts to coastal and marine ecosystems' and recommends 'developing a coherent climate adaptation strategy for the Great Barrier Reef catchment' (Waterhouse et al., 2017, p. 9).

While research suggests that marine reserves influence the resilience of ecosystems to local climate-related disturbances (e.g. severe weather events), it does not address global-scale events, e.g. rising temperatures (Olds et al., 2014).

> The projected, far-reaching, impacts from climate change are likely to ultimately over shadow the effects of other factors and uses. Addressing the other factors, which can be more directly managed, will improve the resilience of the Region's values to any future impacts of a changing climate. (GBRMPA, 2014, Chapter 6, para. 6.7.5)

Under the United Nations Framework Convention on Climate Change, the Commonwealth committed to a 2030 target of reducing emissions to 26–28% below 2005 levels (Department of Environment and Heritage, 2017).

Runoff from land-based activities

A Reef Taskforce report found that

> Agricultural land uses are the main source of nitrogen, sediment and pesticides to the Reef and its ecosystems. Other land uses, such as industrial, mining, port development,

> dredging and urban development contribute relatively small loads of pollutants to the Reef but can be locally significant. These industries are generally more heavily regulated than agriculture. (Reef Taskforce, 2016, p. 16).

Although the overall extent of agricultural land has remained stable in recent years (GBRMPA, 2014), scientific studies of the Reef attribute the impact of agricultural runoff primarily to land clearing, cane farming and grazing, with main pollutants being:

- nutrients such as nitrogen and phosphorus, used in sugarcane farms, causing outbreaks of crown of thorns starfish and most inordinate algal growth;
- sediments from land clearing and grazing, impacting seagrass growth and inshore coral reefs;
- pesticides, also impacting seagrass and coral growth.

The 2017 Scientific Consensus Report provides more specificity than ever on the types, levels and sources of pollutants, highlighting problems of sedimentation and a quantified anthropogenic increase in some areas (Bartley et al., 2017).

Complex jurisdictional and regulatory environment

Australia's jurisdictional arrangements mean that local-government, Queensland and Commonwealth policies, legislation and regulation apply to the Reef region.[1] There are 26 Commonwealth and state legislative instruments for the management of the Reef (Jacobs, 2014); some of the primary legislation is shown in Figure 1. There are also 17 local governments fringing the reef (from Bundaberg to Cooktown), each with policies, bylaws and initiatives that affect waste, land and water management.[2]

Point-source water pollution measures are not well integrated with regulations regarding diffuse-source pollution (Waschka & Gardner, 2016). Regulating diffuse water pollution is challenging because it is difficult to identify, measure and attribute, including at the single-property scale. Its impacts tend to be cumulative in nature, and the episodic variability of the pollution load is spiked by flood events, bringing sudden high pollution loads (Waschka & Gardner, 2016).

The Commonwealth government has jurisdiction over the World Heritage Area (the broadest area) and the Commonwealth's Marine Park (the same as the World Heritage Area except for 13 coastal exclusion areas). It regulates causes of diffuse water pollution principally through:

- Great Barrier Reef Marine Park (Aquaculture) Regulations 2000 under the Great Barrier Reef Marine Park Act 1995 (Cth), regulating the discharge of aquaculture water that impacts the marine park, extending its reach into the wider Reef region outside of the marine park and into streams that discharge into the Reef region; and
- Environment Protection and Biodiversity Conservation Act 1999 (Cth). A new dam in 2003 on the Dawson River proposed by the Queensland government triggered consideration by the Commonwealth Minister for the Environment and Heritage of both the direct *and the indirect* impacts of the construction and

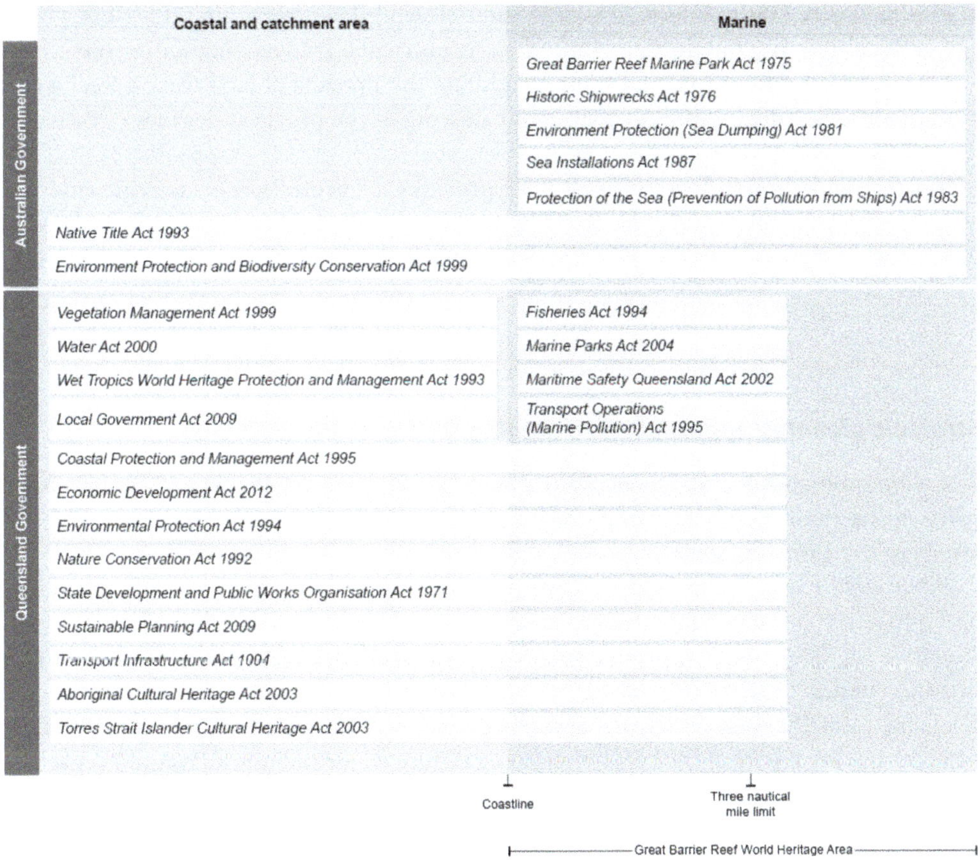

Figure 1. Primary Commonwealth and Queensland legislation used to manage the Great Barrier Reef (Commonwealth Government 2015a, p. 19).

operation of the dam on listed threatened species and certain ecological communities. The indirect impact was the runoff from agricultural development, enabled by the construction of the new dam, into Reef catchments (McGrath, 2008; Waschka & Gardner, 2012).

Queensland has power to regulate causes of diffuse water pollution through:

- Environmental Planning Act 1994 (EPA): imposes a general environmental duty on each person not to carry out any activity that causes environmental harm unless the person takes all reasonable and practicable measures to prevent or minimize the harm (§ 319(1)). Despite this broad power, this act has been used primarily to regulate point-source pollution, not diffuse water pollution (Waschka & Gardner, 2012).
- Environmental Protection (Water Quality) Policy 2009: provides for setting water quality targets and guidelines. It does so for sewage management, urban stormwater waste management, etc., and has been extended to the agricultural sector in Reef catchments.

- Vegetation Management Act 1999: land-clearing laws are contentious but play a critical role in containing pollution from nutrients. Implementation of these laws is a critical element in beginning to address the impacts of land use on Reef water quality by reducing the extensive land clearing of the previous decades (GBRMPA, 2014).
- Regional Planning Interests Act 2014: provides for protection of strategic environmental areas (including the Reef) by restricting resource or regulated activities in areas of regional interest. The government has not used these powers in relation to the Reef.

Strategic planning approach to managing diffuse water pollution

The Commonwealth and Queensland governments have undertaken a multitude of plans and actions, either separately or jointly, from 1975 to 2016 to manage a range of issues in the Reef (Figure 2).

Since 1992 Australia has adopted two national frameworks for developing guidelines and processes for managing water quality.[2] This contrasts with the National Water Initiative of 2004, which predominantly provides for reform in addressing water quantity issues. Under the Caring for our Country framework, and specific to the Reef, the Reef Plans have set overarching goals for halting and reversing the decline of water quality entering the reef by reducing diffuse water pollution (Figure 2). Each of the Reef plans from 2003 to 2013 (currently under review) has adopted different approaches to managing water quality.

Reef Plan 2003 devolved responsibility to six regional natural resource management (NRM) bodies for the Reef. This government-imposed regionalism was seen as heralding an integrated, devolved model for policy development and implementation. A risk-based management approach identified 10 nutrient management zones, with the three highest priorities being the Wet Tropics, Mackay/Whitsunday and Burdekin. But the NRM bodies considered all of their catchments as high-risk (Peterson, Walker, Maher, Hoverman, & Eberhard, 2010). Evaluation showed an impressive range and level of funded activity (Peterson et al., 2010), which was however ultimately ineffective in solving the issue (Waschka & Gardner, 2012).

Reef Plan 2009 and its replacement Reef Plan 2013 set clear targets, moved away from detailed-strategy action-based approaches, and added long-term water quality targets by 2020. These two well-funded programmes of work adopted a 'soft', best-management-practices, voluntary approach, which has been modestly successful. From 2009 to 2014, long-term average loads of nitrogen declined by 17% and sediments by 12% (Reef Report Card, 2014), but this was not fast enough to make a significant difference in ecosystem health. Take-up of stakeholders using best management practices on their land in 2016 fell far short of the target of 90% of land area by 2018. The area of land covered by best management practices was about 50% for cane and 10% for grazing (Reef Taskforce, 2016). Several commentators have criticized the reliance on soft approaches which have not been backed up by legislative instruments, calling for a stronger regulatory approach (Craig & Roberts, 2015; Waschka & Gardner, 2016).

1975	**Commonwealth passes GBR Marine Park Act (Cth)** Established a marine park and Authority for conservation and protection of GBR
1979	**Off-shore constitutional settlement (Emerald agreement)** • 1979 – title and powers over coastal waters (to 3 nautical miles offshore) vested in Queensland subject to the GBR Marine Park Act and a joint consultative arrangement between Queensland and commonwealth for management and preservation of GBR • GBR Intergovernmental Agreement 2009 – updated 1979 agreement to recognise challenges such as water quality that were not foreseen at the time of the original agreement and to clarify financial resources for management. • GBR Intergovernmental Agreement 2015 – updated previous agreement to reflect the shared vision outlined in Reef 2050.
1981	**Inscribed on UNESCO's World Heritage Register** Protection of natural beauty and natural habitats for the conservation of biological diversity.
2000	**Commonwealth proclaims GBR aquaculture regulations** Regulate the discharge of waste from aquaculture.
2003	**Commonwealth and Queensland develops Reef Water Quality Protection Plan - 'Reef Plan'** • Program of mainly voluntary activities for improving water quality. • 2003 goal was to halt and reverse decline in water quality entering Reef by 2013, $375 million invested from 2008 to 2013. • 2009 goal was shifted to 2013 so that by 2020 water quality from adjacent catchments would not threaten resilience of Reef. Specific reduction targets set for nutrients, sediment, pesticides. • 2013 confirmed goal, restated targets, $575 million expected to address water quality initiative from 2015 to 2020. • 2017 Reef Plan under review to incorporate Reef2050 and GBR Water Science Taskforce outcomes.
2007	**Reef Water Quality Partnership** Established between regional National Resource Management (NRM) organisations and the Commonwealth/ Queensland governments to enable coordinated, scientifically robust and collaborative target setting, monitoring and reporting arrangements.
2008	**First Scientific Consensus Statement** Scientists provided understanding of water quality issues. **GBR Reef Protection Amendment Act (Qld)** Regulatory measures imposed to reduce impact of agricultural activities on water quality. **Queensland Report Card** First report offered base-line data of 2009, evaluating progress towards Reef Plan's goals and targets.
2012	**UNESCO's WHC Mission Report** Noted that future conservation of GBR world heritage area is at cross roads, recommended several decisive and immediate actions.
2013	**Reef Plan Protection Legislation suspended, Best Practice Management introduced** Liberal -National Party (conservative political party) puts into place voluntary instead of legislative measures to address water pollution causes. **Second Scientific Consensus Statement** Leading independent scientists agreed that diffuse source pollution from agriculture is main source of excess nutrients and sediment, and that water quality in the GBR is declining.
2014	**GBR Outlook Report** Produced by GBRMPA, stated that overall outlook for Reef is poor, worsened since 2009 and expected to deteriorate in the future. **GBR Regional and Coastal Zone Strategic Assessment** Response to UNESCO WHC's recommendation for comprehensive analysis of issues affecting OUV.
2015	**Sustainable Ports Act (Qld)** Capital dredging within and adjoining GBR WHA prohibited; restricts port development to four major existing port areas. **Commonwealth develops Reef Long-term Sustainability Plan -** Protects water quality values to maintain World Heritage listing by providing overarching framework from 2015 to 2050. **'Reef 2050'** 2050 Reef Plan endorses the 2013 Reef Plan adopts end-of-catchment pollution load reduction targets, plus proposes targets for urban, industrial and port activities..
2015/ 2016	**GBR Water Science Taskforce** • Provided advice to Queensland on how to achieve water quality targets and priorities over 5 years for investment of $90 million. • Queensland accepts the recommendations in 2017.

Figure 2. Key actions taken for protection of the Reef.

Sources: Queensland Government, Reef 2050 Water Quality Improvement Plan (http://www.reefplan.qld.gov.au/about/history/); Commonwealth Government (2015a); Reef Taskforce (2016); Waschka and Gardner (2016).

In 2013, considering the state of conservation of the Reef World Heritage Area, the WHC and the International Union for Conservation of Nature (IUCN) expressed concern at the lack of progress on reducing threats to the area, including poor water quality. They recommended that Australia address fully the direct, indirect and cumulative impacts of developments in and adjacent to the property and lead to concrete measures to ensure the overall conservation of the OUV [outstanding universal value] of the Reef (World Heritage Committee, 2013). GBRMPA's Outlook Report 2014 raised further concerns *that even with recent initiatives the overall outlook for the Reef had worsened since 2009 and was expected to further deteriorate.* Following this grim report and other evidence of decline, the WHC put Australia on notice that the Reef could be listed as World Heritage in Danger.

Reef 2050, a long-term water quality improvement plan, was formulated in 2015 in response to the notice (Commonwealth Government, 2015a). It provides agreed leadership at Commonwealth and state levels, and builds actions, targets, objectives and outcomes for the four key risks to the Reef. Australia set up a Reef Taskforce to make recommendations on how to achieve Reef 2050's water quality targets. It acknowledged that over the next 5–10 years transformational change in reef governance is required in order to reach the 2025 water quality targets of up to 80% nutrient and 50% sediment reduction (Reef Taskforce, 2016).

Although the WHC decided not list the Reef as World Heritage in Danger in 2015, it advised that it would closely monitor how Australia addresses these threats to the Reef and requested by December 2019 'an overall state of conservation report ... demonstrating effective and sustained protection' of the Reef's listed values (2015a, p. 69).

Conditions that have enabled or hindered success of diffuse water quality governance

We now examine the range of environmental, social, political and legal conditions that have affected diffuse water quality governance affecting the Reef.

Enablers

Enabler 1: high level of science consensus assists policy development and funding allocation

The 2017 Scientific Consensus Statement demonstrates significant agreement in the scientific community about the causes, risks and extent of poor water quality in the Reef. This impressive level of consensus was built up over the years by several panels providing independent multidisciplinary scientific advice to the Commonwealth and Queensland governments.

There is consensus on key threats to water quality, the relative 'weight' of these threats and the geographical areas at greatest risk. For example, the ranking of the relative risk of degraded water quality between the regions in the Great Barrier Reef (from highest risk to lowest) is: Wet Tropics, Fitzroy, Burdekin, Mackay Whitsunday, Burnett Mary, Cape York (Brodie et al., 2013). Findings are communicated for a general audience (Figure 3).

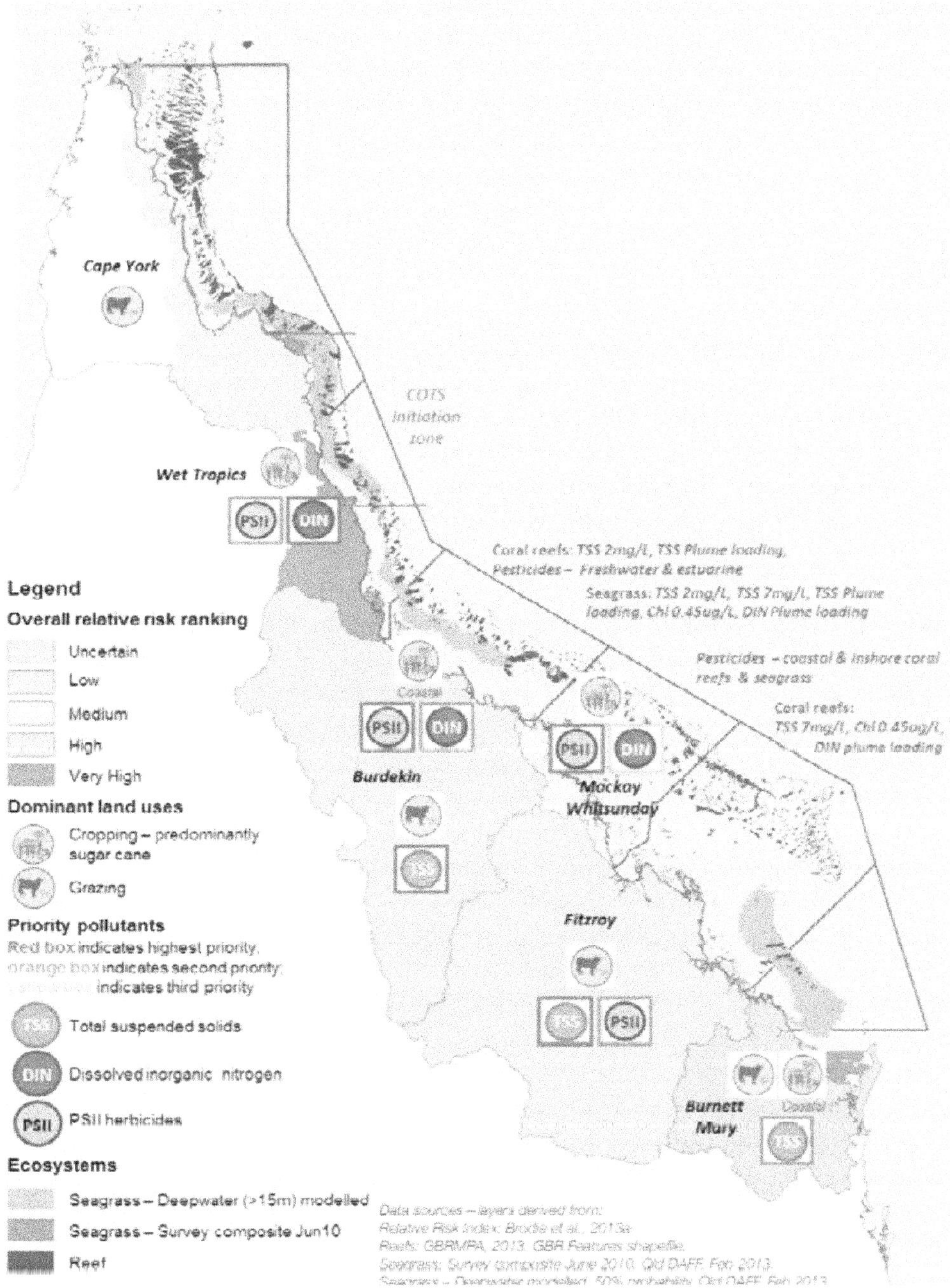

Figure 3. Relative risk of degraded water quality in the Great Barrier Reef.
Source: Reef Water Quality Protection Plan Investment Strategy 2013–2018.

The Reef Taskforce (2016) acknowledged the strong level of science already available, but noted a lack of alignment of research to priority needs. It recommended to 'better align science and fund development of new ideas and solutions' (p. 66). While it is accepted there is a high level of funding from governments, with AUD 1.28 billion for

Reef 2050 actions over the next five years, including 44% directed to water quality (WHC 2017), it would appear that much of the funds invested in water quality may be directed to tackle point-source pollution. For example, almost AUD 620 million is allocated to upgrade sewage treatment plants at the three largest urban centres along the Reef coast (Reef 2050 2015), while diffuse water quality may lack similar investment. Robertson and Slezak (2016) argue that governments need to boost funding, as it would cost AUD 8.2 billion over 10 years to reach the 2025 water quality targets.

The taskforce also recommended that the water quality targets in the Reef 2050 Water Quality Improvement Plan (up to 80% reduction in nitrogen runoff and up to 50% reduction in sediment runoff in key catchments) be reviewed and set at the basin scale. This strong scientific consensus on the 'problem' means that governments could make swift policy responses and funding allocations – if there were a comparably strong political consensus. Having scientists aware of policy implications also means that findings are expeditiously translated into action; for example the 2017 Scientific Consensus states that 'In parallel to the update of the Scientific Consensus Statement in 2017, new catchment-based pollutant load reduction targets were developed for the Reef 2050 Plan' (Brodie et al., 2017).

Enabler 2: cooperation and strong buy-in from governments and stakeholders

Cooperative management of the Reef, formalized in 1979, is long-standing and robust. Independent reports note a high level of cooperation between the Commonwealth and Queensland governments over marine port management, especially at the operational level (Jacobs, 2014). Government cooperation has resulted in relatively rapid agreement on, and responses to, the iterations of the reef water quality plans. For example, under Reef Plan 2009 almost all government departments responsible for implementing actions have reallocated resources to fulfil their obligations (Lloyd Consulting, 2011).

Cooperative management is supplemented by strong buy-in from stakeholder groups. The Reefocus Summit in October 2008 was a pivotal event in stakeholder engagement, with attendance by Queensland and Commonwealth ministers, leading scientists, and representatives from a range of sectors, including industry, NRM bodies, agencies, local government, and environment groups. The resulting Reef Plan 2009 adopted a more strategic and adaptive approach, incorporated and recognized industry and community initiatives, and provided new policy and regulatory frameworks. Work focused on the highest-priority areas, and implementation plans were developed for each area. Lead entities were identified to drive implementation of actions, working with nominated 'supporters' to deliver outcomes. There was awareness that long-term action needed collaboration (Howard Partners, 2005).

Enabler 3: co-management with Traditional Owners

Collaborative problem solving for water quality issues is seen as a key mechanism for building bridges between the Reef's governing bodies and Traditional Owner groups (Nursey-Bray & Rist, 2009; Zurba et al., 2012). Over 70 Traditional Owner groups maintain cultural values in the Reef region (GBRMPA, 2014). Elsewhere there are perceptions that Indigenous claims of native title over land and water resources compete against current users or owners of the resources or development proposals (Robertson, 2017), adding to already complex management decisions. However, with respect to the Reef, the ongoing links with Traditional Owners is a specific attribute contributing to the values of the Reef as World Heritage property (GBRMPA, 2014) and could be a significant driver for tackling water quality, specifically through funding of increased numbers of Indigenous rangers for land-care activities to minimize soil erosion. Deloitte Access Economics (2017, p. 9) emphasize that cultural ecosystem services, currently not assigned a quantified value, provide further justification of increased funding to protect the Reef.

There are diverse co-management arrangements, ranging from a system of Indigenous Protected Areas run by either Traditional Owners with support from the Commonwealth, or looser arrangements to co-manage resources (Nursey-Bray & Rist, 2009). For example, in 2005 Girringun Aboriginal Corporation, representing a group of nine Traditional Owners, signed an agreement for the management of traditional hunting of protected species in the greater Hinchinbrook Island area. An Aboriginal Ranger programme in 2010 included a water quality monitoring programme in the Tully basin. This programme built trust between groups and led to a 10-year Indigenous Protected Area in 2013 over the Girringun area for co-management of traditional land and sea-estates.

Enabler 4: UNESCO listing and global/local networks build social awareness

When considering whether to list the Reef as World Heritage in Danger, the WHC (2015b, p. 11) stated that 'the establishment of [Reef 2050] through a multi-stakeholder process at the scale of the property is in itself a major technical and policy achievement'. The potential listing prompted social awareness in a broad range of stakeholders of the key threats to the reef, including water quality decline. For example, a survey of 230 residents of the Reef region found that the environmental non-use values, such as the health of coral reefs, are 'well recognized' by its residents, and are important contributors to their well-being and quality of life (Larson, Stoeckl, Farr, & Esparon, 2015, p. 172). Both use and non-use values related to the natural environment were reported as more important to local residents than commercial activities such as mining, agriculture and commercial fishing. Social awareness of the uniqueness of the Reef has built partnerships across diverse stakeholder groups, including the commercial and conservation sectors. Stakeholders have funded evaluations to make a strong economic case for protection of the Reef. For example, the latest evaluation of the Reef's value was supported by a range of experts from various universities and funded by a leading bank, the Reef Foundation and the Great Barrier Reef Marine Park Authority (Deloitte Access Economics, 2017). Reef report cards (see below) are also a collaboration between stakeholder groups.

Enabler 5: systematic monitoring and evaluation

To build resilience, management frameworks need to be adaptive, innovative and responsive to new information as it becomes available (Evans et al., 2016). On the whole, there are extensive monitoring programmes in the Reef, although there are gaps in some regional areas and 'insufficient subregional or fine scale [farm] monitoring' (Reef Taskforce, 2016, p. 67). There are a number of evaluation mechanisms for the Reef, including:

- *Report Cards 2009–2016* assess the results of Reef Plan actions for the health of the reef, with annual data from the 2009 baseline position. The large volume of reported data creates a yearly time lag in Report Card publication, which has a follow on effect for timely adaptive management. Although noting that water quality was poor, reporting until 2015 appeared to gloss over the slow progress of improvement. Report Card 2015 introduced a five-point (A, B, C, D, E) scoring system to make it easier to track progress towards the targets. As a report card system is familiar to most members of the community, it is considered a better way to communicate complicated scientific findings. While the 2015 report card included a D for water quality (Queensland Government, 2015), the 2016 report card found moderate improvements. For example, thanks to research findings on the causal effect of sedimentation, and catchment modelling to prioritize locations for investment, programmes achieved an annual reduction of 4.1% in the Fitzroy region. This occurred mainly as a result of graziers protecting stream banks by limiting cattle access in locations where large amounts of sediment are washed into the Reef lagoon (Queensland Government, 2016).
- Under the Reef Plans, *independent audits* were carried out in 2005 and also 2010 (Lloyd Consulting, 2011). The 2010 audit gathered information and views of Queensland and Commonwealth governments, industry, NRM bodies and community groups, and adopted a participatory approach, which contributed to social learning.
- *Outlook reports* are created by the Great Barrier Reef Marine Park Authority every five years as a statutory requirement under the Great Barrier Reef Marine Park Act 1975 (§54). These examine the Reef's health, pressures, and outlook. The most current report is from 2014.

While there is a high level of Reef monitoring and evaluation compared with other areas in Australia, the Reef Taskforce (2016, pp. 69–71) recommended 'additional long-term and finer scale [end-of-farm] catchment monitoring, modelling and reporting for improved decision-making and adaptive management'.

Hurdles

Hurdle 1: complicated governance arrangements

The Reef's complicated jurisdictional and governance arrangements for water quality in general (Figure 4) are a major challenge for addressing ecosystem threats to the Reef caused by diffuse water pollution. In an attempt to provide coordination, the Reef Ministerial Forum oversees implementation of Reef 2050 reforms (including water quality), supported by a Standing Committee of Officials and an Executive Steering Committee, and guided by a Reef Advisory Committee and an Independent Expert Panel. The Reef Marine Park Authority is responsible for the day-to-day management of the Commonwealth sections of the Marine Park Area but has representatives on the various committees, panels and working groups for managing impacts outside the Marine Park. The Queensland government's Department of Environment and Heritage Protection is responsible for day-to-day management of the state sections of the Marine

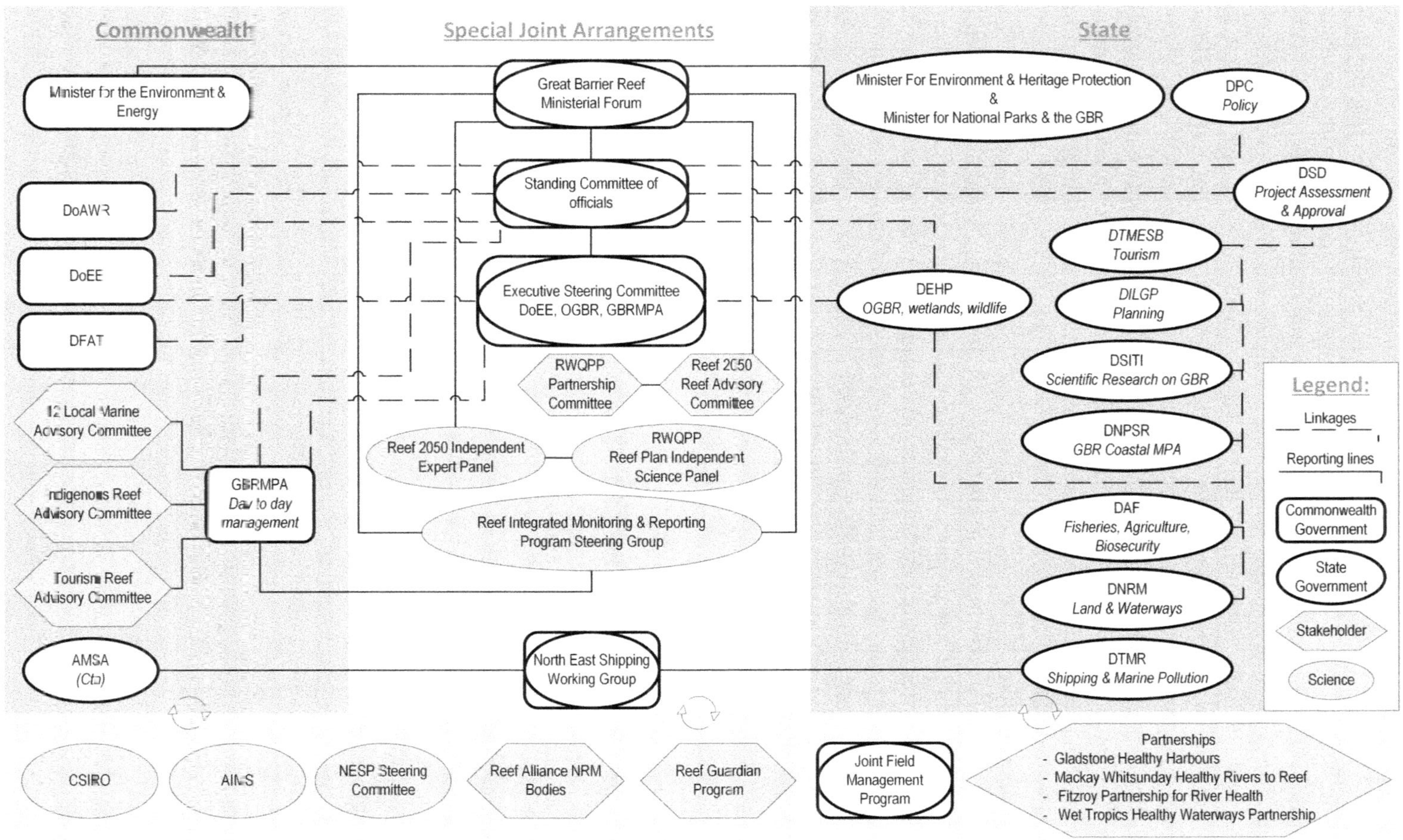

Figure 4. Governance structure for water quality as of December 2017. Functions of government departments relating to water quality are in italics. See the Appendix for definitions of the abbreviations.

Park. The Office of the Great Barrier Reef promotes coordination across Queensland government and ministerial portfolios whose functions impact water quality.

The Reef Taskforce (2016) observed that water quality governance affecting the Reef continues to be fragmented, complex and poorly coordinated across the three levels of government, multiple committees, advisory bodies and funding bodies. A primary failing in Reef-wide water quality governance is the lack of agreement about key catchment-scale delivery mechanisms (Reef Taskforce, 2016). Despite best attempts to streamline governance arrangements, the complexity of functions and responsibilities undermines engagement with, and feedback from, landowners and stakeholders about water quality initiatives, which are essential for resilience-based adaptive management. The Reef Taskforce (2016, p. 77) recommended implementing a 'simplified and more effective governance structure across Queensland and Commonwealth governments to deliver better joint arrangements in funding and decision-making, and more efficient delivery arrangements'.

However, Australia's complex jurisdictional and governance structure is a deeply entrenched problem that has plagued Australian management across a range of areas, including natural resource and environmental management (Productivity Commission, 2016). There appears to be little political will to reform Australia's governance structure. In 2013 the Commonwealth government promised a White Paper on the Reform of the Federation, which was subsequently abandoned (Borrello, 2016). International attention and the unfavourable 2017 Scientific Consensus Statement are likely to put Australia under pressure to urgently reform water quality governance structures. Measures will need to include a refreshed National Water Initiative that addresses quality issues.

Hurdle 2: ideological and socio-political factors

Ideological paradigms have been barriers to effectively managing diffuse water pollution in the Reef.

We noted earlier that Queensland has regulatory powers over the causes of diffuse water pollution. Instead, successive state governments have demonstrated a see-saw approach to regulation – introducing regulatory measures while exhibiting a reluctance to proceed with the measures. For example the Great Barrier Reef Protection Amendment Act 2009 (Cth) included regulatory measures to reduce the impact of agricultural activities on the quality of water entering the Reef. But these were put on hold. Another measure, the 2009 amendments to Queensland's Environmental Protection Act 1994 and Chemical Usage (Agricultural and Veterinary) Control Act 1988, regulated nitrogen, phosphorus and chemical application in three high-priority catchment areas. These required an environmental risk management plan for commercial cane growing or cattle grazing on farms larger than 2000 hectares in the three areas (Waschka & Gardner, 2012). However, with a change in state government, these regulatory mechanisms were suspended in 2012, and the Queensland government relied on voluntary best management practices for dealing with water quality (Kroon, Thorburn, Schaffelke, & Whitten, 2016).

An audit of the two Reef water quality programmes found that the right balance had not been reached between regulatory and voluntary approaches (Queensland Audit Office, 2015). The audit's final recommendations included 'implementing staged regulations to reduce water pollution throughout the Reef regions'. In response, the

Queensland government reorganized reporting arrangements for the Reef Secretariat and formed the Reef Taskforce to consider the audit's recommendations. Queensland has agreed in principle to all of its recommendations and has commenced implementation (Commonwealth Government, 2016).

The second set of ideological and socio-political issues relate to land clearing and demonstrate how difficult it is to effect land-use change. The Australian Labor Party (ALP) government passed the Vegetation Management Act 1999 to put in place controls on land clearing. Amendments in 2013 by the conservative Liberal National Party government, a coalition with rural party interests, loosened land clearing controls by repealing regulations that applied to clearing high-value regrowth on freehold and Indigenous lands; allowing broad-scale clearing for 'high-value' intensive agricultural production; and promoting self-assessment of areas that contain remnant or high-value regrowth. Satellite imagery showed that clearing in catchments that drain onto the Reef has increased dramatically, and constituted 35% of total clearing across Queensland in 2013–14 (Taylor, 2016). The loss of native vegetation cover in such regions is one of the major drivers of the water quality deterioration in the Reef's lagoon.

In 2016, with yet another change in government, the incoming ALP government introduced the Vegetation Management (Reinstatement) and Other Legislation Amendment Bill 2016. The bill attempted to reinstate the protection of high-value regrowth on freehold and Indigenous land; remove provisions which permit clearing applications for high-value agriculture and irrigated high-value agriculture; broaden the protection of regrowth vegetation in watercourse areas to the Burnett Mary, Eastern Cape York and Fitzroy Great Barrier Reef catchments; and reinstate compliance provisions for the reverse onus of proof and remove the 'mistake of fact' defence for vegetation-clearing offences. Explanatory notes to the bill expressly refer to the objective of reversing the decline in the reef's health by maintaining vegetation cover in Reef catchments and that this is a key action of the Reef 2050 Plan. Proposed amendments to this Bill were not passed, as the minority ALP government needed the votes of two independent members, who opposed controls on farmers' rights.

The competing political policies of the three levels of government can undermine water quality reform. For example, the Commonwealth's northern Australia policy seeks to develop northern Australia by facilitating investment and reforming regulation to attract more agribusiness, mining and other business sectors (Commonwealth Government, 2015b; Humphries et al, 2017). The Reef Taskforce (2016, p. 35) observed that this has the 'potential to undo some of the positive changes made to land management and water quality outcomes'.

Hurdle 3: climate change issues continue to be controversial, hindering a holistic approach

Climate change is the biggest long-term, system-wide threat to coral reefs globally (Reef 2050, Queensland Government, 2015). Reef 2050, despite being a long-term plan, does not incorporate any direct action on climate change (Hughes, Day, & Brodie, 2015) but claims to provide the 'best insurance for the [Reef] against climate change by reducing direct and indirect threats' (Reef 2050, Queensland Government, 2015, p. 22). Australia is committed to reducing its emissions to 26-28% below 2005 levels by 2030, and there is Commonwealth funding for mitigation programmes. The Queensland government is committed to reducing emissions by taking up renewable energy technology, but has not been able to deliver enhanced practical regulatory controls on land clearing.

Recently, 46 scientists published results from aerial and in-water surveys of the Reef combined with satellite-derived sea surface temperatures in 2016. They found that 'during 2015–2016, record temperatures triggered a pan-tropical episode of coral bleaching, the third global-scale event since mass bleaching was first documented in the 1980s' (Hughes, et al, 2017, p. 373). The study also found that water quality and fishing pressure had minimal effect on the unprecedented bleaching in 2016, suggesting that local protection of reefs affords little or no resistance to extreme heat (p. 375). They found however that improved water quality, given enough time, may improve the prospects for recovery (p. 376). Paradoxically, the increased extreme flooding events and other effects of climate change have been shown to have a negative impact on Reef water quality.

Again in 2017, the Reef experienced bleaching, a result of warming seas. Aerial surveys show while 2016's bleaching occurred mostly in the northern Reef stretches, in 2017 bleaching was found along 300 km of the Reef's middle section due to warming seas, killing almost half of the coral (Knaus & Evershed, 2017; Slezak, 2017). These events led UNESCO to observe that despite the good work under Reef 2050, 'progress towards achieving water quality targets has been slow, and the most immediate water quality targets … are not expected to be achieved within the foreseen timeframe', warning that climate change remained the most serious threat to the future of the Reef (WHC, 2017, p. 24).

Although tackling climate change effectively lies beyond the ability of any one country, Australia's ineffective and piecemeal approach has not set a good example. Australia had close to the worst ranking in the Climate Change Performance Index 2017, which compares and measures countries' efforts and policies on tackling climate change. An expert report found a wide gap between national and state policies. It concluded that Commonwealth policies are 'unambitious and uninspired', with states managing 'to some extent to take independent action' (Burch, Marten, Bals, Dertinger, & Uhlich, 2017, p. 6). With leaders of peak bodies such as the National Farmers Federation accepting that agricultural industries need to prepare for climate change by both mitigation and adaptation strategies (National Farmers Federation 2016), it is time for an opening of rational debate and the reconsideration of these beliefs and values.

Queensland adopts a co-regulatory approach

The Reef Taskforce (2016, p. 62) recognized that government should continue to progress towards using approaches that recognize industry best practices, while finding that 'regulation … is important to ensure continuous improvement in water quality'. Such regulation includes setting and progressively reducing catchment pollution load limits in legislation to meet water quality targets (recommendation 5.1) and establishing a water quality offset framework that can apply across industries (recommendation 5.6).

Co-regulation (a new term coined for the tailored approach to include 'soft' and market-based instruments) will provide government recognition of producers accredited against a recognized best management practices programme. Those accredited will demonstrate compliance with a minimum practice regulatory standard, and therefore have greater access to government grant schemes or other incentives. Under a carrot-and-stick approach, those not in compliance will be required to write an up-front environmental risk management plan. Existing minimum standards for point-source polluters will also be strengthened (Reef Taskforce, 2016).

The current government has re-established a compliance programme against previous regulatory standards, with an initial focus on the Wet Tropics and Burdekin catchments (Reef Taskforce, 2016). Tougher regulations are planned, setting pollution load limits for reef catchments that contribute the highest pollutant loads to the Reef.

Markets for offsets are being considered. The Paddock to Reef Monitoring, Modelling and Reporting programme under Reef Plan 2009 models pollutant loads, which will allow targets to be established for each of the 35 catchments draining into the reef. This was considered a 'world-class leading-edge' programme (Lloyds, 2011, p. 53). Once load limits are set, water quality offsets may be used to ensure compliance of existing and new development activities. Research is currently underway to develop offset models and schemes for pollutants and sediments (Queensland Government, 2017).

Conclusion

The Great Barrier Reef is valued and protected for its cultural identity, economic returns and international significance. Australia's early experience in managing the Reef Marine Park relied on static reserve management to achieve conservation outcomes, which, despite its being a world-leading arrangement at the time, could not address external threats such as water quality. However, recognizing that the majority of threats to water quality come from outside the marine park, management has shifted to activities beyond the reserve areas. The listing of the Reef in 1981 under the World Heritage Convention provided the impetus for change. Its outstanding universal values includes natural beauty and a highly evolved and biologically diverse ecosystem that is of the most significant for biodiversity conservation.

The Great Barrier Reef Intergovernmental Agreement 2009 formalized the Commonwealth and Queensland's partnership in protecting and managing the Reef region, resulting in joint management on many issues within the Great Barrier Reef Marine Park, the adjacent Queensland Coast Marine Park and Queensland's island national parks (GBRMPA, 2014). However, pinpointing the legislative and policy reform necessary to effectively tackle diffuse water pollution is as difficult as pinpointing the origin of diffuse water pollution.

There is evidence that the Reef's health has been declining over the last three decades. Ambitious water quality targets set by Reef Plans since 2003, sometimes seen as aspirational or 'entirely unrealistic' (Craig & Roberts, 2015, p. 59), have not been achieved. The highest risks to its resilience come from climate change, land-based runoff, coastal land-use change and direct use such as port activities. Sophisticated monitoring of water quality consistently shows that impacts of agricultural activities – nutrients and pesticides from runoff and sediments from land use – are still not well managed. Strong consensus by independent scientists, corresponding with international pressure from UNESCO, has put pressure on the Commonwealth and Queensland governments to respond with increased funding and a raft of measures.

While elsewhere problems with management of marine parks relate to issues of top-down governance, resistance from communities, and lack of public participation in policy development (McCay & Jones, 2011; Sayce et al., 2013), these have not been the Reef's key issues. Since 2007, top-down governance in the Reef has been matched with regional input and a strong multi-stakeholder approach with emphasis on co-management with Traditional Owners, demonstrating exemplary attempts at a collaborative and equitable approach. But these measures were not sufficient to address diffuse water pollution, and the WHC put Australia on notice in 2014 that the Reef could be deemed World Heritage in Danger.

Switching from initial five-year plans focused on improving water quality, Australia in 2015 formulated Reef 2050, a *holistic* long-term plan for the next 35 years. The latest Reef plan has a science-based adaptive management approach that focuses on reducing diffuse water pollution and building resilience. Flexibility and feedback loops through rigorous monitoring and evaluation are among the enabling factors of adaptive governance highlighted above.

Australia is moving towards supplementing voluntary best practices for managing water quality with strengthened regulations, pollution-load limits and the introduction of economic instruments such as nutrient-offset mechanisms. This approach attempts to incorporate participative, legal, interpretative, knowledge and economic incentives required to improve resilience-based adaptive management of protected areas. As Australia adopts measures to address land-based runoff, water quality threats from climate change appear to be escalating, countering gains in resilience. Unprecedented coral bleaching events in 2016 and 2017 killed almost half of the Reef's coral, placing the Reef yet again in a precarious position.

Despite the willingness of State and Commonwealth governments to work together to protect the values of the Reef, we find three major barriers to implementing *transformational* adaptive governance to improve reef water quality. First, within Queensland, despite evidence that land clearing is a significant cause of diffuse water pollution, ideological combative politics explains the government's inability to strengthen land-clearing legislation. In the US, regulatory reform of diffuse water pollution has met similar hostility from the powerful lobbies of agriculture and development and dysfunctional partisan politics (Craig & Roberts, 2015). Since Queensland elections in late 2017 returned a majority ALP government, land-clearing legislation has become a high priority. It is important that beyond these changes, the political debate over land clearing needs to transcend short-term vested interests framing the issues as 'win-lose'. However, it is beyond the scope of this article to fully map out the political tensions behind state and national government policy responses on diffuse pollution to protect the reef.

Second, the Commonwealth government has been slow to take strong action and international leadership on tackling climate change, despite its demonstrated effect on increasing runoff and reef pollution. The Reef Taskforce (2016, p. 14) emphasizes that

> while efforts to reduce global climate change are underway, the focus must be on reducing all pressures [including poor water quality] to improve the Reef's resilience. This will give the Reef a greater capacity to bounce back from climate change related disturbances and survive well into the future.

In other words, while the Reef needs strong Australian leadership and international cooperation to reduce carbon emissions, taking strong action to improve Reef resilience will have immediate gains for withstanding the effects of climate change.

Third, despite scientific consensus about the causes, extent, and risks of poor water quality in the Reef, governments have had difficulties with taking timely responses and aligning research to priority needs. This may in part be the result of the complex Reef governance structure with its emphasis on strong stakeholder participation, where governments must take into account a range of interests in their decision making. The WHC recognized this multi-stakeholder process as a major technical and policy achievement. It does, however, reduce the timeliness of action.

When considering the question, 'How effective, equitable and resilient is Australia's approach to designing and implementing policy and law for diffuse water pollution in the Reef?' it is clear that the approach is not yet effective. The recent 2017 Consensus Statement has unequivocally concluded that current initiatives will not meet Australia's water quality targets. This article demonstrates that in the recent past, soft approaches were adopted rather than legislative instruments. Australia is pursuing an invigorated approach towards managing diffuse water pollution. First, by supporting best practice management of agricultural land, it aims to reinforce developing a culture of stewardship beyond meeting minimum standards. Second, while systematic monitoring and evaluation provides data for adaptive management, overall targets are seen as aspirational. These targets should be tightened for Reef catchment which are 'very high and high risk' (see Figure 4). Strengthened catchment-based pollutant load limits, tougher regulations and enforcement of these regulations would build a stronger compliance culture. Queensland also has a number of legislative instruments, not currently being used, that can be relied on. Scientists are now able to identify fine-grained information on where to restore catchments thus enabling focused management efforts. Continual improvements to Australia's adaptive management approach provides some hope that governments and stakeholders will turn the tide on diffuse water pollution affecting the Reef.

Notes

1. There are three levels of government in Australia: federal (Commonwealth), state and local. The Commonwealth Constitution of 1901 gives authority over land and water within state boundaries to state government (Kildea & Williams, 2010; Horne, 2017).
2. These are the National Water Quality Management Strategy, which gives guidelines for fresh and marine water quality, and the National Action Plan for Salinity and Water Quality. The latter ran from 2000 to 2008 before becoming the Caring for our Country initiative, one arm of which extended to the Reef (Waschka & Gardner, 2016).

Disclosure statement

No potential conflict of interest was reported by the authors.

References

Bartley, R., Waters, D., Turner, R., Kroon, F., Wilkinson, S., Garzon-Garcia, A., ... Waterhouse, J., (2017). *Scientific Consensus Statement 2017. A synthesis of the science of land based water quality impacts on the Great Barrier Reef, Chapter 2: Sources of sediment, nutrients, pesticides and other pollutants to the Great Barrier Reef.* Retrieved from http://www.reefplan.qld.gov.au/about/assets/2017-scientific-consensus-statement-summary-chap02.pdf

Borrello, E. (2016, April 28). Malcolm Turnbull scraps federation white paper after $5 million work. *ABC News*. Retrieved from http://www.abc.net.au/news/2016-04-28/malcolm-turnbulls-$5-million-tax-white-paper-scrapped/7367204

Brodie, J., Baird, M., Waterhouse, J., Mongin, M., Skerratt, J., Robillot, C., . . . Warne, M. (2017). *Development of basin-specific ecologically relevant water quality targets for the Great Barrier Reef*, TropWATER Report number 17/38. Australia: James Cook University, published by the State of Queensland. Retrieved from https://www.reefplan.qld.gov.au/about/assets/gbr-water-quality-targets-june2017.pdf

Brodie, J., Waterhouse, J., Schaffelke, B., Kroon, F., Thorburn, P., Rolfe, J., ... McKenzie, M. (2013). Scientific consensus statement: Land use impacts on great barrier reef water quality and ecosystem condition. Retrieved from http://www.reefplan.qld.gov.au/about/assets/scientific-consensus-statement-2013.pdf

Burch, J., Marten, F., Bals, C., Dertinger, A., & Uhlich, T. (2017). *Climate change performance index: Results 2017, Climate change network*. Berlin. Retrieved from https://germanwatch.org/en/ccpi

Committee, World Heritage. (2013). *Decision: 37 COM 7B.10*. Great Barrier Reef (Australia) N 154. Retrieved from http://whc.unesco.org/en/decisions/4959

Commonwealth Government. (2015a). *Reef 2050 Long-Term Sustainability Plan* (Commonwealth of Australia). Retrieved from http://www.environment.gov.au/marine/gbr/publications/reef-2050-long-term-sustainability-plan

Commonwealth Government. (2015b). *Our north, our future: White Paper on developing Northern Australia* (Australian Government 2015). Retrieved from http://northernaustralia.gov.au/page/publications

Commonwealth Government. (2016). *Reef 2050 2016 Annual Report and implementation Strategy*. Retrieved from http://www.environment.gov.au/marine/gbr/publications/reef-2050-plan-annual-report-implementation-strategy

Convention on Biological Diversity. (1992, June 5). *Opened for signature*, 1760 U.N.T.S. 79. Retrieved from https://www.cbd.int/doc/legal/cbd-en.pdf

Craig, R. K., & Roberts, A. M. (2015). When will governments regulate nonpoint source pollution? A comparative perspective. *Boston College Environmental Affairs Law Review*, *42*, 1–64.

Deloitte Access Economics. (2017). *At what price? The economic, social and icon value of the Great Barrier Reef*. Retrieved from https://www2.deloitte.com/content/dam/Deloitte/au/Documents/Economics/deloitte-au-economics-great-barrier-reef-230617.pdf

Department of Environment and Heritage. (2017). *Review of climate change policies*. Retrieved from http://www.environment.gov.au/system/files/resources/18690271-59ac-43c8-aee1-92d930141f54/files/2017-review-of-climate-change-policies.pdf

Devilliers, R., Pressey, R. L., Grech, A., Kittinger, J. N., Edgar, G. J., Ward, T., & Watson, R. (2015). Reinventing residual reserves in the sea: Are we favouring ease of establishment over need for protection? *Aquatic Conservation: Marine and Freshwater Ecosystems*, *25*(4), 480–504.

Dudley, N., Groves, C., Redford, K. H., & Stolton, S. (2014). Where now for protected areas? Setting the stage for the 2014 World Parks Congress. *Oryx*, *48*(04), 496–503.

Evans, K., Bax, N., & Smith, D. C. (2016). *Australia state of the environment 2016: marine environment*. Canberra: Independent Report to the Australian Government Minister for the Environment and Energy, Australian Government Department of the Environment and Energy. Retrieved from https://soe.environment.gov.au/download/reports

GBRMPA. (2014). *Great Barrier Reef Outlook Report 2014*, Townsville: Great Barrier Reef Marine Park Authority. Retrieved from http://www.gbrmpa.gov.au/cdn/2014/GBRMPA-Outlook-Report-2014/

GBRMPA. (2016). *Annual Report 2015-16*. Townsville: Great Barrier Reef Marine Park Authority. Retrieved from http://www.gbrmpa.gov.au/about-us/corporate-information/annual-report/annual-report-2015-16

Gunderson, L., & Light, S. S. (2006). Adaptive management and adaptive governance in the everglades ecosystem. *Policy Sciences*, *39*(4), 323–334.

Gunderson, L. H., Holling, C. S., & Light, S. S. (Eds.). (1995). *Barriers and bridges to the renewal of ecosystems and institutions*. New York: Columbia University Press.
Holling, C. S. (1978). *Adaptive environmental assessment and management*. Chichester, UK: John Wiley & Sons.
Horne, J. (2017). The politics of water reform and environmental sustainability in the Murray–Darling Basin. *Water International, 42*(8), 1000–1021.
Howard Partners. (2005). *Audit of the Reef water quality protection plan: Audit report*. Retrieved from http://www.reefplan.qld.gov.au/about/assets/pmp-report-310306.pdf
Hughes, T., et al. (2017). Global warming and recurrent mass bleaching of corals. *Nature, 543*, 373–377.
Hughes, T., Day, J., & Brodie, J. (2015). Securing the future of the great barrier reef. *Nature Climate Change, 5*, 508–511.
Humphries, F., Anton, D. K., Tan, P. L., Akhtarkhavari, A., Butler, C., & England, E. (2017). Ecological governance and the development plan for Northern Australia. *Australian Environment Review, 32*(2), 46–50.
Jacobs. (2014). *Independent review: Institutional and legal mechanisms that provide coordinated planning, protection and management of the Great Barrier Reef World Heritage Area* (Review Report). Retrieved from http://www.environment.gov.au/system/files/resources/8b9fd5af-99ce-4a18-9ca5-6bb1829fd9e1/files/gbr-independent-review.pdf
Kildea, P., & Williams, G. (2010). The Constitution and the management of water in Australia's rivers. *Sydney Law Review, 32*, 595–616.
Knaus, C., & Evershed, N. (2017, April 10). Great Barrier Reef at 'terminal stage': Scientists despair at latest coral bleaching data. *The Guardian*. Retrieved from https://www.theguardian.com/environment/2017/apr/10/great-barrier-reef-terminal-stage-australia-scientists-despair-latest-coral-bleaching-data
Kroon, F. J., Thorburn, P., Schaffelke, B., & Whitten, S. (2016). Towards protecting the Great Barrier Reef from land-based pollution. *Global Change Biology, 22*, 1985–2002.
Larson, S., Stoeckl, N., Farr, M., & Esparon, M. (2015). The role the great barrier reef plays in resident wellbeing and implications for its management. *AMBIO, 44*, 166–177.
Lee, K. N. (1993). *Compass and gyroscope: Integrating science and politics for the environment*. Washington, DC: Island Press.
Lewicki, R. J., Gray, B., & Elliott, M. (2003). *Making sense of intractable environmental conflicts: Frames and cases*. Washington, DC: Island Press.
Lloyd Consulting. (2011). *Reef Water Quality Protection Plan audit report 2010* (Consultant Report). Retrieved from http://www.reefplan.qld.gov.au/measuring-success/assets/reef-plan-audit-report.pdf
McCay, B. J., & Jones, P. J. S. (2011). Marine protected areas and the governance of marine ecosystems and fisheries. *Conservation Biology, 25*(6), 1130–1133.
McGrath, C. (2008). Flying foxes, dams and whales: Using federal environmental laws in the public interest. *Environmental and Planning Law Journal, 25*, 324.
McLeod, E., Green, A., Game, E., Anthony, K., Cinner, J., Heron, S. F., ... Woodroffe, C. (2012). Integrating climate and ocean change vulnerability into conservation planning. *Coastal Management, 40*(6), 651–672.
National Farmers Federation. (2016). *Climate change policy*. Retrieved from http://www.nff.org.au/file preview.html?dboid=5476
Nursey-Bray, M., & Rist, P. (2009). Co-management and protected area management: Achieving effective management of a contested site, lessons from the Great Barrier Reef World Heritage Area (GBRWHA). *Marine Policy, 33*, 118–127.
Olds, A. D., Pitt, K. A., Maxwell, P. S., Babcock, R. C., Rissik, D., & Connolly, R. M. (2014). Marine reserves help coastal ecosystems cope with extreme weather. *Global Change Biology, 20*(10), 3050–3058.
Pahl-Worstl, C. (2009). A conceptual framework for analysing adaptive capacity and multi-level learning processes in resource governance regimes. *Global Environmental Change, 19*, 354–365.

Peterson, A., Walker, M., Maher, M., Hoverman, S., & Eberhard, R. (2010). New regionalism and planning for water quality improvement in the Great Barrier Reef, Australia. *Geographical Research*, *48*(3), 297–313.

Productivity Commission. (2016). *Marine fisheries and aquaculture* (Productivity Commission Inquiry Report No. 81, 19 December 2016). Retrieved from http://www.pc.gov.au/inquiries/completed/fisheries-aquaculture/report/fisheries-aquaculture.pdf

Queensland Audit Office. (2015). *Managing water quality in Great Barrier Reef catchments* (Report 20: 2014-15). Retrieved from https://www.qao.qld.gov.au/reports-parliament/managing-water-quality-great-barrier-reef-catchments

Queensland Government (2015). *Reef Report Card 2015*. Retrieved from http://www.reefplan.qld.gov.au/measuring-success/report-cards/2015/assets/report-card-2015-scoring-system.pdf

Queensland Government (2016). *Reef Report Card 2016*. Retrieved from https://www.reefplan.qld.gov.au/measuring-success/report-cards/2016/assets/report-card-2016-detailed-results.pdf

Queensland Government. (2017). *Enhancing regulations to ensure clean water for a healthy Great Barrier Reef and a prosperous Queensland* (Discussion Paper). Retrieved from http://www.ehp.qld.gov.au/assets/documents/reef/gbr-discussion-paper.pdf

Reef Taskforce. (2016). *Clean water for a healthy reef: Final report*. Department of Environment and Heritage Protection. Canberra: Office of the Great Barrier Reef. Retrieved from https://research.jcu.edu.au/tropwater/news-and-events/great-barrier-reef-water-science-taskforce-delivers-its-final-report/image/image_view_fullscreen

Robertson, J. (2017, February 7). Indigenous owners threaten legal action unless Adani abandons land access deal. *The Guardian*. Retrieved from https://www.theguardian.com/australia-news/2017/feb/07/indigenous-owners-threaten-legal-action-unless-adani-abandons-land-access-deal?CMP=Share_iOSApp_Other

Robertson, J., & Slezak, M. (2016, December 2). Great Barrier Reef Report: We have to do better on water quality, says Australia. *The Guardian*. Retrieved from https://www.theguardian.com/environment/2016/dec/02/great-barrier-reef-progress-report-we-have-to-do-better-on-water-quality-says-australia?CMP=Share_iOSApp_Other

Sayce, K., Shuman, C., Connor, D., Reisewitz, A., Pope, E., Miller-Henson, M., … Owens, B. (2013). Beyond traditional stakeholder engagement: Public participation roles in California's statewide marine protected area planning process. *Ocean & Coastal Management*, *74*, 57–66.

Slezak, M. (2017, May 25). Great Barrier Reef 2050 plan no longer achievable due to climate change, experts say. *The Guardian*. Retrieved from https://www.theguardian.com/environment/2017/may/25/great-barrier-reef-2050-plan-no-longer-achievable-due-to-climate-change-experts-say

Taylor, M. L. (2016, February 22). Queensland land clearing is undermining Australia's environmental progress, *The Conversation*. Retrieved from http://theconversation.com/queensland-land-clearing-is-undermining-australias-environmental-progress-54882

Walker, B., & Salt, D. (2006). *Resilience thinking: Sustaining ecosystems and people in a changing world* (1st ed.). Washington, DC: Island Press.

Waschka, M., & Gardner, A. (2012). Using regulation to tackle the challenge of diffuse water pollution and its impact on the Great Barrier Reef. *The Australasian Journal of Natural Resources Law and Policy*, *15*(2), 109–149.

Waschka, M., & Gardner, A. (2016). Diffuse source pollution and water quality law for the Great Barrier Reef: Why the reticence to regulate? In J. Gray, C. Holley, & R. Rayfuse (Eds.), *Trans-Jurisdictional water law and governance*. London, UK: Routledge.

Waterhouse, J., Schaffelke, B., Bartley, R., Eberhard, R., Brodie, J., Star, M., ... Kroon, F. (2017). *Scientific consensus statement: Land use impacts on Great Barrier Reef water quality and ecosystem condition*. Retrieved from http://www.reefplan.qld.gov.au/about/assets/2017-scientific-consensus-statement-summary.pdf

World Heritage Committee. (2013). Decision: 37 COM 7B.10, Great Barrier Reef (Australia) N 154. Retrieved from http://whc.unesco.org/en/decisions/4959

World Heritage Committee. (2015a). *Great Barrier Reef* (39th Sess. Decision 39 COM 7B.7).

World Heritage Committee. (2015b). *State of conservation of properties.* United Nations Educational, Scientific and Cultural Organization Convention Concerning the Protection of the World Cultural and Natural Heritage, 39th Sess. WHC-15/39.COM/7B.Add)
World Heritage Committee. (2017). *State of conservation of World Heritage properties inscribed on the World Heritage List* (41st Sess. WHC/17/41.COM/7B.Add). United Nations Educational, Scientific and Cultural Organization Convention Concerning the Protection of the World Cultural and Natural Heritage
Zurba, M., Ross, H., Izurieta, A., Rist, P., Bock, E., & Berkes, F. (2012). Building co-management as a process: Problem solving through partnerships in Aboriginal country, Australia. *Environmental Management*, *49*, 1130–1142.

Appendix: Table of abbreviations

AIMS	Australian Institute of Marine Science (Cth)
AMSA	Australian Maritime Safety Authority (Cth)
CSIRO	Commonwealth Scientific and Industrial Research Organisation (Cth)
DAF	Department of Agriculture and Fisheries (Qld)
DEHP	Department of Environment and Heritage Protection (Qld)
DFAT	Department of Foreign Affairs and Trade (Cth)
DILGP	Department of Infrastructure, Local Government and Planning (Qld)
DNPSR	Department of National Parks, Sport and Racing (Qld)
DNRM	Department of Natural Resources and Mines (Qld)
DoAWR	Department of Agriculture and Water Resources (Cth)
DoEE	Department of Environment and Energy (Cth)
DPC	Department of Premier and Cabinet (Cth)
DSD	Department of State Development (Qld)
DSITI	Department of Science, Information Technology and Innovation (Qld)
DTMR	Department of Transport and Main Roads (Qld)
DTMESB	Department of Tourism, Major Events, Small Business and the Commonwealth Games (Qld)
GBRMPA	Great Barrier Reef Marine Park Authority
IPA	Indigenous Protected Areas
IUCN	International Union for Conservation of Nature
MPA	Marine Protected Area
NESP	National Environmental Science Program
NRM	Natural Resource Management
OGBR	Office of the Great Barrier Reef
RWQPP	Reef Water Quality Protection Plan
WHC	World Heritage Committee

The social discourses on market-based instruments to manage non-point-source water pollution in the Oldman River basin, southern Alberta

Gillian Kerr and Henning Bjornlund

ABSTRACT

The Canadian province of Alberta has incorporated market-based instruments into recent policy to manage non-point-source pollution. Investigating context-specific social discourses through the Q-method provides a timely understanding of why these instruments have not been well implemented in southern Alberta, and may assist in developing their potential. This article identifies four distinct discourses, named Incentive Orienteers, Rural Advocates, Honest Brokers and Progressive Producers.

Introduction

Ensuring adequate water quality in rivers and other water bodies – for ecosystems, human consumption and economic use – is becoming an increasingly urgent issue across the world (United Nations, 2011). Two main categories of pollution influence water quality: point-source and non-point-source. Point-source pollution is relatively easy to manage, as there is a specific source of the pollution, such as a sewage treatment plant, which authorities can effectively monitor and, therefore, regulate (Russell & Clark, 2006), whereas non-point-source pollution is caused by disparate sources, which it is far more difficult to monitor and therefore to regulate (Oldman River Basin Water Quality Initiative, 2002). Non-point-source pollution is the predominant water quality issue in developed countries, and it has increased management challenges for governments (Dai, Sun, Zhang, & Guo, 2015; Graham, Schempp, & Troell, 2011).

Globally, policy experts have evaluated mechanisms to control and manage non-point-source pollution. Many authors have promoted market-based instruments, such as tradable pollution rights, as effective tools, because they incentivize individual actors to change behaviour on their land to minimize or avoid issues with non-point-source pollution (e.g., Shortle & Horan, 2001). Further, these instruments are considered the most economically efficient way to regulate pollution, as it is cheaper for some polluters to reduce pollution than others (Russell & Clark, 2006). However, their application in these situations has been limited, and success is rare, which is partly due to implementation challenges, such as having adequate monitoring to demonstrate additionality (Russell & Clark, 2006; Stirrett, Rolfe, & Shewchuk, 2012).

Agriculture is a key contributor to non-point-source pollution. Water runoff from agricultural land carries pollutants into rivers, such as pesticides, fertilizer and manure, and cattle with access to rivers urinate and defecate directly into the water. These sources of pollution are hard to measure, but their cumulative downstream effects can be significant (Oldman River Basin Water Quality Initiative, 2002). Best management practices have been used to manage these issues, such as buffer strips to trap nutrients and chemicals before they enter water courses; laser grading of land to prevent feedlot runoff draining into water courses; fencing to prevent cattle accessing the water course, and off-site watering points to provide alternative water supply (Agriculture and Agri-Food Canada, 2004; Fitch, Adams, & O'Shaughnessy, 2003). Governments have looked to market-based instruments to enhance the use of best management practices. However, there are a number of challenges associated with evaluating these practices, which include establishing water quality testing to assess actual positive change from best practices; how to design monitoring to ensure that the promised management practice is implemented; and how to assure that a landowner will continue with the best management practices for the length of the program. With respect to the use of taxpayers' money to provide incentives for changed practices, it is often debated whether it is reasonable to pay landowners for activities that they should undertake as part of responsible citizenship (Baird, Belcher, & Quinn, 2014; Pannell, 2008). These are critical issues for the success of market-based programs.

The government of Alberta has promoted the use of market-based instruments for water management through legislation and policy, such as the Environmental Protection and Enhancement Act (2000), Water Act (2000) and Wetland Policy (Government of Alberta, 2013). However, there is a gap between the support for and the use of market-based instruments, which has resulted in limited on-the-ground implementation. In relation to agriculture, this implementation gap is likely to be associated with the challenges of using market-based instruments in the water-pollution context and the perspectives of critical stakeholders, particularly the farming community and their level of trust in the government of Alberta and their understanding of market-based instruments. The objective of this research is to reveal the social discourses that exist on the acceptance and use of market-based instruments to manage non-point-source water pollution from livestock and manure management in the Oldman River basin. A case study in Alberta, Canada, provides an ideal context to investigate the social discourses associated with market-based instruments, which is of value to market-based instrument implementers more broadly. The study uses Q methodology to both qualitatively and quantitatively reveal stakeholder perspectives on these instruments.

Social discourse and market-based instruments

Traditional or neoclassical economics have heavily influenced contemporary water policy and management processes, which assume that humans make economically rational decisions, leading to predictable behaviours (Ingram, Schneider, & Deleon, 2007). This has led to the development of heuristics and models that rely on generalizations of human behaviour and decision making (Howlett, 2009). However, evidence suggests that people do not behave in purely economically rational ways, and in fact community, values and preferences influence their decisions and behaviour (Brown, 2006). Modern policy design often disregards these crucial influences (Pannell, 2008). Empirical studies have shown that policy tools, thought to be efficient and appealing based on economic rationality, have failed due to a lack of understanding of the interests of target populations (e.g., Pannell, 2008; Valbuena, Bregt, McAlpine, Verburg, & Seabrook, 2010). The integration of key community perspectives is likely to increase the success of policy tools (Brown, Durning, & Selden, 2008). There is a need to reveal the distinctive social discourses of the community to inform future governance actions and policy design and implementation, including the use of market-based instruments.

There is a long tradition in rural and agro-environmental studies to employ behavioural approaches to study the links between farming perspectives, values and behaviour (Burton & Wilson, 2006). Many behavioural economic models make restrictive assumptions that farmers are short-term and rational wealth optimizers (Hodge, 2001). However, other researchers are critical of this approach. Maybery, Crase, and Gullifer (2005) argued that farmers may show strong preferences for non-economic factors, which include environmental protection, community values and land stewardship for future generations. Later work by Howard, Bjornlund, and Xu (2012) confirmed this.

While recent rural research has focused extensively on farmers, this article uses the Q-method to reveal community perspectives from knowledgeable stakeholders about non-point-source pollution, water quality management and the use of market-based instruments. The Q-method, developed in the 1930s to understand and reveal human subjectivity (Brown, 1980; Stephenson, 1953), has been recognized as a useful approach to capture and categorize the subjective and latent community discourses that are missed by other methods (Addams & Proops, 2000). The main strength of the Q-method is its ability to identify and explore these shared perceptions and discourses in a more structured way than can be achieved with other analyses (Barry & Proops, 1999). This method has recently become more popular in environmental management research for issues such as natural restoration and sustainability (Colorado Institute of Public Policy, 2006), agricultural practices (Davies & Hodge, 2012; Pereira, Fairweather, Woodford, & Nuthall, 2016), land use (Coke & Brown, 1976), water management (Montgomery, Bjornlund, Xu, & Edwards, 2016; Strickert et al., 2016) and grizzly bear management (Rutherford, Gibeau, Clark, & Chamberlain, 2009).

This article uses the Q-method to reveal the social discourses that exist on the use of market-based instruments for the management of non-point-source water pollution in the Oldman River basin, and to contribute new information and analysis of the market-based-instrument implementation gap.

The Alberta context

'Context' means the historical, political, economic and cultural conditions in which environmental management policies are embedded (Swainson, De Loë, & Kreutzwiser, 2011). In the late 1800s, European immigrants, mainly farmers or ranchers, settled

Alberta. Today, livestock and crop production are major components of the agricultural industry and account for the largest number of cow-calf and intensive livestock operations in Canada (Johnson et al., 2003). The agricultural economy has contributed greatly to the growth and development of the province; however, the focus on livestock production, particularly in southern Alberta, has degraded water quality. The governance of both the agricultural industry and environmental quality is primarily the responsibility of government of Alberta, with participation of industry groups such as the Alberta Irrigation Project Association and watershed groups such as the Oldman River Basin Watershed Planning and Advisory Council.

The province of Alberta has experienced substantial economic and population growth during the past three decades. This has had wide-reaching environmental implications, particularly in terms of competition for resources, conflicts over land use, and environmental degradation. The provincial government's pro-development agenda has resulted in increased water management challenges (Government of Alberta, 2009; Kennedy, 2010).

In response to these environmental challenges, the government of Alberta has been examining its environmental governance and policy approaches, which have traditionally taken a command-and-control approach. The province introduced the Alberta Water for Life Strategy in 2003 (Government of Alberta, 2003) and the broader Land Use Framework in 2007 (Government of Alberta, 2008) to address the cumulative impacts of development on water and land. These policy commitments represent a shift from the government's being responsible for environmental outcomes to policy that shares governance responsibilities with non-state actors. Market-based approaches became part of this shared approach, which fit into the conservative, resource-centred government of the time (Clare & Krogman, 2013).

Market-based instruments have a number of theoretical strengths, such as increasing the flexibility of non-state actors having to meet targets; enhancing technology shifts; increasing the ability to overcome information asymmetry; and increasing economic efficiency or cost savings (Stavins, 2003). One of the challenges associated with a change in governance approach is the need to understand and integrate how the community will respond to specific policy design and implementation (Clarkson & Woods, 2010: Howlett, 2009). Despite market-based instruments having been formally promoted and legally enabled by the government of Alberta, and endorsed by environmental and agricultural industry groups (e.g., Agri-Environmental Partnership, 2012; Kennedy, 2010), few have been implemented for their agro-environmental outcomes. The limited use of market-based instruments has resulted in an implementation gap between the government of Alberta's support and commitment to integrating market-based instruments and their limited use on the ground.

In southern Alberta, numerous studies in the past decade have examined various aspects of farmer/producer preferences and choices (Baird et al., 2014: Bjornlund, 2010; Bjornlund, Nicol, & Klein, 2007). Most of these studies have had an economic focus and have examined preferences for specific applications or concepts: for example, types and costs of best management practices or perspectives on water trading. Recent studies have broadened the economic typology approach to explore how non-economic factors influence irrigators' policy preferences for water sharing, including market-based instruments (Bjornlund, Parrack, & De Loë, 2013; Bjornlund, Zhao, & Xu, 2014;

Lafreniere, Deshpande, & Bjornlund, 2015; Zuo, Wheeler, Bjornlund, Edwards, & Xu, 2015) and how they influence landowners' perspectives on the use of market-based instruments to increase the provision of ecosystem services (Howard et al., 2012). However, no studies have looked at the concept of market-based instruments for agricultural environmental improvements, either from the perspective of the agricultural actors and their local community or from a policy maker's perspective.

This article provides a critical examination of the social discourses around market-based instruments for the management of non-point-source pollution to protect water

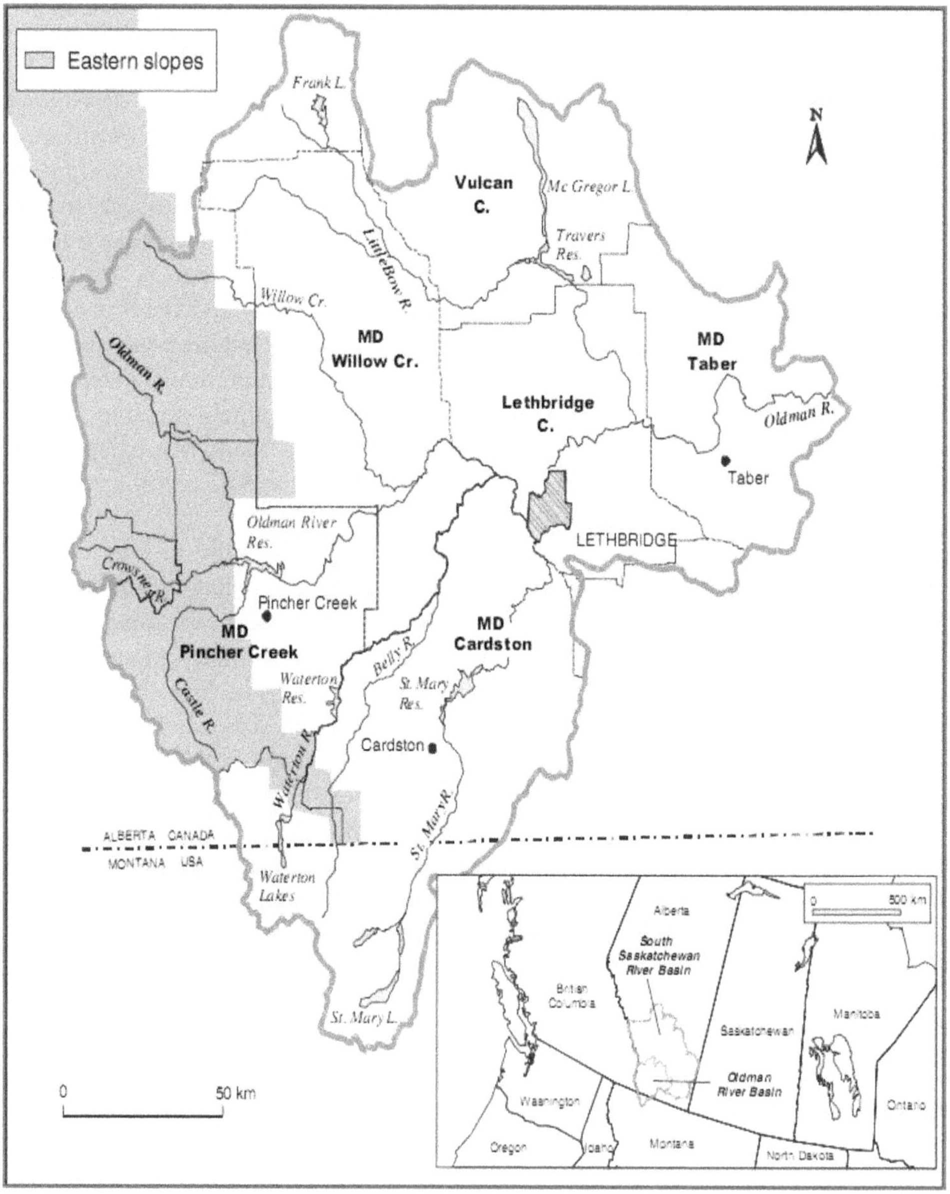

Figure 1. The Oldman River basin, Alberta, Canada (adapted from Ivey, de Loë, Kreutzwiser, & Ferreyra, 2006).

quality to help understand issues of acceptance in an affected community. Specifically, this research looks at discourses associated with non-point-source pollution and livestock and manure management and the potential use of market-based instruments.

Methods

Study area

The Oldman River basin is a major watershed in southern Alberta, and the river is surrounded by agricultural land (Figure 1). Water quality monitoring in some of the tributaries of the Oldman River has shown total phosphorous and total nitrogen concentrations exceeding the Alberta Surface Water Quality Guidelines and levels of faecal coliforms that occasionally exceed provincial guidelines for contact recreation and irrigation (Little, Kalischuk, Gross, & Sheedy, 2010). One study found that 31 of the 33 main-stem rivers ranged from 'moderately impacted' to 'degraded' (Government of Alberta, 2006).

This article focuses on two areas with known non-point-source issues and elevated nitrogen and phosphorous levels from agricultural practices (Alberta Agriculture, Food and Rural Development, 2012): the Battersea Drain Area in the county of Lethbridge, which has the highest concentration of intensive livestock operations in Alberta (Olsen, Miller, Rodvang, & Yanke, 2005; Rodvang, 2002), and the municipal district of Willow Creek, where several waterways have water quality issues associated with agricultural practices (Depoe, 2006; SEAWA, 2010). The two regions border each other but differ agriculturally and culturally.

The Battersea Drain Area is a homogeous agricultural area dominated by intensive livestock operations. Many farmers are first- or second-generation Dutch farmers, and there is a close social fabric based on a common heritage and way of life. Battersea has been part of a long-term water quality monitoring program associated with the Oldman Watershed Planning and Advisory Council, which is part of testing and assessing the use of best management practices to improve water quality.

Willow Creek is more heterogeneous, with a wide range of agricultural practices and industries, which produce a variety of crops. The mixed cropping and livestock operations vary in size, and many include a number of cow calf operations, which are often established on or close to waterways. It is a rural area with five small towns, which in recent years has experienced a steady influx of urban dwellers.

While the two communities differ substantially in their agricultural practices and community make-up, the rural south of Alberta is well known for its historically strong social fabric.

Q-methodology

The Q-method is a factor-based, analytic approach that identifies the principal dimensions of respondents' attitudinal perspectives by extracting factors relating to scores assigned to a set of statements (Watts & Stenner, 2012). There are generally six steps: concourse; Q-set; participant selection; Q-sort; factor analysis; and factor interpretation.

The first step gathered a body of opinion statements (Q-statements) called a 'concourse', which represented all the relevant perspectives on the subject area (Stephenson, 1978), namely, the use of market-based instruments to manage non-point-source pollution from agricultural manure runoff. As recommended in the Q-literature (Barry & Proops, 1999; Watts & Stenner, 2012), we purposefully assembled the concourse based on observations at local workshops and 26 semi-structured interviews. A significant design consideration in the Q-method is the choice of statements in the Q-set. The Q-set statements should be sufficiently broad so that they capture all potential viewpoints on the research subject, in this case the government of Alberta's interest in market-based instruments for water quality issues associated with manure management (Davies, Blackstock, & Rauschmayer, 2005). Almost all of the Q-statements were 'verbatim' from the interviews, which is recommended to limit bias (McKeown & Thomas, 2013). We selected interviewees to be representative of the range of relevant organizations and perspectives, (Watts & Stenner, 2012) using a snowball sampling technique that starts with a few key informants and asks them to identify other key informants (Lewis-Beck, Bryman, & Futing Liao, 2004). Most of the interviewees would later participate in the Q-sort (Table 1). Tables 1 and 2 provide demographic information on the Q-sort participants. Demographic information was collected as part of the Q-sort process, but it does not contribute significantly to the interpretation of the discourses produced by the Q-sort.

The second step developed an initial Q-set of approximately 140 statements from the concourse, which were then divided into themes (following Brown, 1980) by first assessing the concourse statements and then reviewing environmental market-based-instrument studies and the government of Alberta's water management approaches. Four themes emerged from this process to guide the number of statements for the Q sort: specific market-based instruments; non-market-based instrument policy tools; concerns about market-based instruments; and broader concerns about water management solutions. This process produced a final Q-set of 36 statements (Table 2) representing the four themes. To allow participants in the Q-sort to more fully express their perspectives, statements were 'rich' with multiple meanings (Brown, 1980).

The third step identified the participants (P-set) who would conduct the Q-sort. The participants chosen all lived in the Oldman River basin, were subject-matter experts and represented the diversity of views and perspectives on market-based instruments as a tool for dealing with non-point-source pollution.

The fourth step was a ranking procedure, which Brown (1980, p. 7) calls 'the technical means whereby data are obtained for factoring'. Participants were provided a set of 36 cards (one for each of the Q-set statements and randomly numbered) and a large sorting sheet with a ranking continuum from least representative (−4) to most representative (+4). All participants received a brief description of the research focus and 'Conditions of Instruction' for the Q-sort to ensure consistent understanding of the Q-sort task. The participants then had to:

Table 1. Number and distribution of participants.

Sector representation[a]	Sub-sector of representation	Number of desired interviewees	Interviewees	P-set for Q-sort
Agriculture[b]	Farmers: ranching	2	1	1
	Farmers: intensive livestock	2	2	2
	Farmers: mixed	1	1	1
	Agricultural service groups	2	2	2
	Agricultural organizations	1	1	1
Government of Alberta	Ministry of Environment	5	5	3
	Alberta Agriculture and Forestry	3	3	2
	Natural Resource and Conservation Board	1	Declined	Declined
Other government	Municipal government councillors	2	1	2
	Municipal government staff associated with water and agriculture	2	1	2
Rural community	Rural community leaders (non-government)	2	2	2
Indigenous community	Local Blackfoot First Nations	1	1	1
NGO	Environmental	2	2	2
	Oldman Watershed Council	2	2	1
Alberta academics	Universities of Lethbridge	2	2	2
Media	Local newspaper	1	Declined	0
Total		31	26	24

[a]This column represents the primary occupation; however, participants defined themselves by a minimum of two and maximum of four other sectors. Interviewees and Q-participants were asked to respond naturally versus representing a single sector.
[b]Two farmers were also municipal councillors.

- read all the cards and identify the two cards that most represented their perspective (+4) and the two that least represented their perspective (−4) and place them on the sheet; and
- sort the remaining statements into three piles: those that represented their perspectives, those that did not and those towards which they had a neutral response; and place them on the sheet in accordance with their ranking.

Figure 2 provides an example of a completed Q-sort, which illustrates the 'forced' and quasi-normal distribution.

The researchers administered the Q sort on 23 May 2012 in Fort McLeod, a central town in southern Alberta. Eight participants could not attend; therefore, supplemental sort days were set, on 30 May and 6 June 2012, in their respective communities.

At the end of the Q-sort, participants were asked to fill out a short post-sort survey to provide additional information on their age, sex, primary career, other community roles and why each participant chose their four most and least representative statements. There was also an open-ended question for general feedback.

The fifth step was a factor analysis of each of the Q-sorts using PQMethod 2.12 software, which is specifically designed for the Q-method (Schmolck & Atkinson, 2012). This process produced a correlation matrix and first generated factors using centroid analysis and then rotated using varimax to provide the most statistically significant factors. The number of factors for each sort was determined using a combination of eigenvalues (>1) and a loading criterion, which requires at least two participants to have loaded significantly onto each factor (Table 3).

Table 2. Factor and discourse characteristics from the Q-sort statistical analysis.

	Factor			
Factor characteristic	Incentive Orienteers	Rural Advocates	Honest Brokers	Progressive Producers
How discourse is distinct from other factors understanding	Recognition of urgency of the issue Need to work practically with industry to resolve issues	Need for	understanding of rural perspective	Need for brokers to enhance trust and
Affiliated sector (s)[a]	Farmer, broader ag industry, environmental NGO, provincial and municipal government, academic	Farmer and rural community leader	Provincial government and academia	Intensive livestock farmers
Number of defining variables (participants)	11	5	3	2
Gender of variables (number of men and women participants)	8 men 3 women	3 men 2 women	1 man 2 women	2 men
Age range of variables (years)	26–65	36–65	36–54	56–65
Observation about where participants live	Predominantly born in the area	Four live in Willow Creek	City of Lethbridge	Both in Battersea Drain Area
Observation about participants' relevant experience	Have participated for a long time in farming and water management	Have farmed or have family that farm	Have worked on environmental water issues in academia, government of Alberta, or as volunteers for a long time	Have farming, council or watershed leadership experience
Variance explained	21%	13%	6%	6%

[a]Includes how participants listed their own affiliations

Results: factor interpretation and understanding the social discourse (Step 6)

The factor analysis uncovered four unique and statistically significant factors. Table 2 presents the factors and the associated factor scores for each of the 36 statements. Analysis of the data permitted the development of narratives to provide a better, practical understanding of the discourses inherent in the four factors. As such, descriptive names were given to the four discourses: Incentive Orienteers (Factor 1), Rural Advocates (Factor 2), Honest Brokers (Factor 3), and Progressive Producers (Factor 4). Table 3 summarizes the characteristics of the four factors and discourses.

A critical observation across the four discourses is that none is against the use of market-based instruments; however, the narratives expressed in the discourses show differing perspectives on the process of arriving at a solution for non-point-source pollution. Further, each placed a different emphasis on the types of regulatory and market-based tools that the government of Alberta should use.

Table 3 . Q-statements, with associated factor scores for each discourse.

Statement	Factors			
	IO[a]	RA[b]	HB[c]	PP[d]
1 If the government is going to use MBIs[e] for water quality they need to know what the equity issues will be. Who benefits and who pays?	+3[f]	+1	+3	+2
2 Government should place more trust in the local people and in what local knowledge can offer to the water quality solutions.	−2	+3	+2	+2
3 Taxes won't be accepted to change farmer behaviour for water quality.	+3	−3	0	+1
4 Watershed groups are making a difference on these environmental issues and should be tightly tied into what the government is doing on MBIs.	0	+3	−1	−2
5 Farmers need technical help and knowledge, not just money to make good water quality changes.	−1	−1	+1	−2
6 Regulations are fairly blunt instruments and so MBIs have a place for certain issues.	+4	0	+1	+4
7 Education is the most important to see if info can change behaviour but if it costs money then there should be an incentive.	+1	+1	−1	0
8 An effective solution would be to provide payments annually for best management practices. Start with payments lower in the beginning and increase once they have proven they are working.	+3	−1	−4	0
9 The Natural Resource and Conservation Board does a lot of work with agricultural industry, particularly on manure but as a regulator, they do not use MBIs.	−1	−3	−3	+3
10 There are a range of things that can be done from economic incentives to voluntary measures, but they take time for the community to hear about them, absorb them, then adapt and do the changes with the time and resources available to them.	−3	+3	+3	+2
11 There is a lot of confusion and fear around these MBIs and if they will make industry uncompetitive or shut them down.	+4	−2	−1	0
12 To get a landowner to buy into an incentive program they would need to know that what was being offered (dollars) was as much or more than it would cost them to do it.	+2	−2	−2	+1
13 I will not be convinced of the effectiveness of an MBI until I know that the change being paid for is going to make things better and not just pay someone to maintain the status quo.	+2	+1	+1	+2
14 There are a lot of good local initiatives that need to be looked at first, before the government adds a new tool or incentive.	0	0	+1	−1
15 Environmental farm plans could have led to better environmental management solutions but the funding ended as they were just getting accepted and adopted.	−2	+1	−3	+4
16 There is a culture of financial incentives for many objectives in Alberta and there is an appetite.	0	−2	−2	+1
17 If you want acceptance of MBIs you will need an honest broker that have a level of credibility and can deliver the information and open the doors to the community if you are going to have acceptance of these tools.	−1	−1	+4	+3
18 The biggest challenge with MBIs and government tools is persistence – will they last long enough to help meet the intended solution?	−3	+2	0	−3
19 Farmers already have enough financial incentives.	−1	−4	−1	−3
20 Water is very, very special in Alberta so any MBI tool will need to be positioned very carefully.	+3	+4	+3	+1
21 MBIs will need to fit into the other layers of governments and other programs so that it doesn't 'butt heads' with another program or set of rules.	−2	0	+3	0
22 In Willow Creek one of the issues is still stock watering in the creek so an incentive program for off-stream watering would be the biggest bang for the buck.	1	+1	−1	+1
23 Cattle should not be allowed to water in creeks.	−3	−3	+1	−3
24 When dealing with intensive livestock operations they should have a specific tool because they are so different than other farm types and have a bigger impact.	−3	0	0	−3
25 Money is only one way to connect people to the benefits society wants.	−4	−2	+4	−2
26 It takes 3–5 years to gain the trust of the people and their respect. You have to have their best interest in mind.	0	+3	−3	−4
27 The best incentive that government could give in Feedlot Ally would be to support and broker a manure compost facility or such thing so that producers could get rid of their manure and potential buyers could access it.	−4	−1	−2	0
28 What is the path of least resistance that gets you to your goal? If there is a pool of funds that the public generally agrees that can be dedicated, then what is the best way?	+1	0	0	−1
29 The best way to do it is to ask the community 'you tell me what you need and think' and have a consultation about it, like Water for Life.	+1	−1	+2	0
30 If fiscally possible, subsidies are a desirable way to go because not controversial.	0	−4	0	−4

(*Continued*)

Table 3 . (Continued).

Statement	Factors			
	IO[a]	RA[b]	HB[c]	PP[d]
31 Have to be careful as people may start to expect the money to do things that they would have done voluntarily before.	+1	+2	0	−1
32 If you are talking about carbon offsets to the third world I don't really think the community would agree.	+1	−3	+2	−1
33 We would all want to make sure there are proper checks and balances to make sure that the person is doing what they said they would do on their land and making the difference.	+2	+2	+2	−1
34 The farmers are providing valuable services on the land. They are being stewards to keep the land intact and water clean and should be rewarded for that.	−2	+1	−3	−2
35 I believe penalties are the wrong way as some people get upset and don't want to work with the government.	0	+2	−2	+3
36 The trend seems to be that most innovative approaches are being done by arms-length groups to government or outside government completely.	+2	0	−4	+3

[a]Incentive Orienteers, [b]Rural Advocates, [c]Honest Brokers, [c]Progressive Producers, [d]market-based instruments, [e]factor score.

Incentive Orienteers

Incentive Orienteers strive to investigate the best water-quality incentives for the agricultural industry. They understand the industry and believe that regulations ('sticks') are too blunt to use with agriculture; hence, market-based instruments have a place to help the government of Alberta pursue water-quality incentives (Statement 6, Rank +4). However, the government of Alberta's water management approaches have created a lack of trust in the government of Alberta. As one participant stated, 'many farmers have survived Government of Alberta programs' (post-Q-sort survey, Participant 3).

Incentive Orienteers do not strongly believe that it will take several years for the government to gain the trust of the community (26, 0). They suggest that the government of Alberta can start to build acceptance by addressing the broad confusion in the community around market-based instruments (11, +4) and the government of Alberta's rationale for using them (1, +3). Incentive Orienteers also recommend that the government of Alberta abandon its need to spend significant time examining all options (10, −3), to ensure that market-based instruments fit perfectly with other institutional arrangements (21, −2), and that the government of Alberta get something implemented. These recommendations could be useful advice for the government of Alberta, no matter what the beliefs are about market-based instruments, as they would provide basic and important information to the community.

Incentive Orienteers strongly disagree that money is 'only one way' to connect with farmers to have improved land-use management (25, −4), which reflects that they see monetary incentives as a central mechanism to convince farmers to make changes to improve water quality. However, they highlight specific tools that will or will not work: an annual payment program based on farmers' performance would be effective (8, +3); while taxes will not be accepted by the community as the way to change behaviour (3, +3). Therefore, this group recognizes that the type of market-based instrument implemented is critically important. Incentive Orienteers view market-based instruments as incentives ('carrots'), which are preferred to 'sticks'. However, they do not believe it is the right time to promote site-specific market-based instruments, such as manure composting facilities (27–4), until there is greater certainty about who pays and who benefits from the government of Alberta's market-based

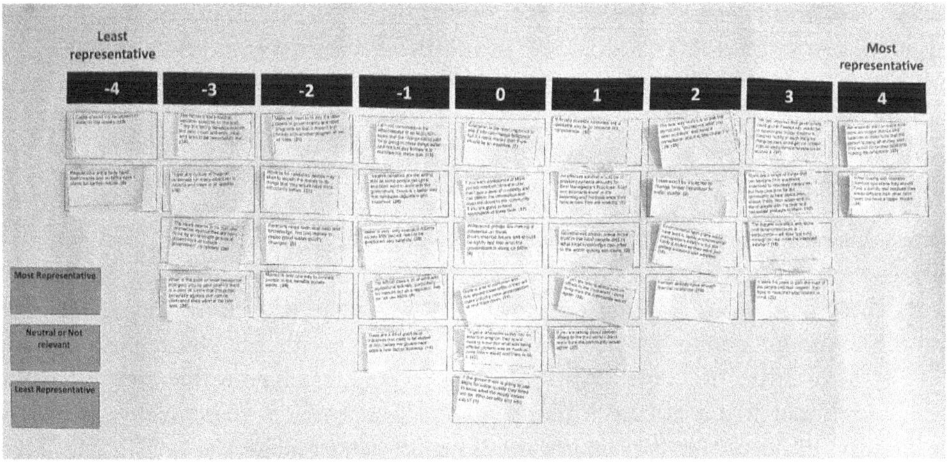

Figure 2. Example of a completed Q-sort.

instruments (1, +3) and farmers are convinced that the benefits are equal to or exceed the cost (12, +2).

Rural Advocates

Rural Advocates believe in the need to defend the rural way of life, particularly in their engagement with the government of Alberta. Rural Advocates believe that, more than any other factor, the government of Alberta's success is tied to building trust in the community. It will take three to five years, and a demonstration by the government that they understand the needs and issues of the community (26, +3), for a new government approach to be trusted. Therefore, they advise the government of Alberta to proceed cautiously when introducing any new market-based instrument tool (20, +4). Further, they believe strongly that governance solutions, such as market-based instruments, should be tied to the work of local watershed groups, which are making a difference in water management (4, +3).

Rural Advocates understand that manure has the largest impact on water quality, and they advocate tools to deal with manure, such as preventing livestock from accessing creeks and rivers as watering points (22, +4). Rural Advocates believe in positive approaches such as education and persuasion and carrots in the form of market-based instruments, as well as sticks (regulations) as policy approaches, versus the carrot-only approach of the Incentive Orienteers. The Rural Advocates see that there is a need to financially assist the agricultural community, but they express caution that monetary incentives may lead to an expectation that farmers will be paid for all the work that they would previously have done for free as part of a moral code (31, +2). They reject the assumption that taxes would not be accepted by the community as a tool to change farmer behaviour for water quality (3, −3), and disagree that farmers have enough financial incentives (19, −4). They stress that proper checks and balances should be in place to ensure that people do what they agree to do (33, +2) and that processes and policy tools to address non-point-source pollution need to be in place for long enough to achieve successful outcomes (18, +2). For example, the government of Alberta ended funding for the Environmental Farm Plan program just as it was gaining acceptance

(15, +1). The lack of persistent policy is a concern for Rural Advocates, as it can take 3–5 years to gain local trust (26, +3). This reinforces their belief that the government of Alberta should work through watershed groups already engaged with communities (4, +3).

Honest Brokers

Honest Brokers recognize that there is a need to enhance trust and understanding between the government of Alberta and the community to have successful water management. These participants have a long history of working on water management issues with the community, and the environment more broadly, either as government employees or as researchers. This experience of the Honest Brokers suggests that the government of Alberta should include honest brokers, such as local agricultural people trusted by the community, to develop and promote solutions to water-quality issues (17, +4). These trusted liaisons, such as agricultural extension specialists, have credibility in the community and can help the government of Alberta deliver its programs and policy tools, both generally and market-based instruments more specifically (as expressed in post-sort interviews with participants associated with this factor). This is because the agricultural industry and broader community need a translator to facilitate and explain the government's intentions with regard to programs and solutions. The Honest Broker's role is not to replace the government of Alberta, because the government is not innovative or effective (36, −4); rather, Honest Brokers have value that the government of Alberta should take advantage of when implementing measures to manage water quality, including market-based instruments.

Honest Brokers want the government of Alberta to understand that funding is only one of a number of ways to connect with the community (25, +4). They stress that market-based instruments must link with the programs, legislation and stakeholder activities that are already in place (21, +3), for example the Battersea Water Quality Program. Further, the Honest Brokers recognize that it is possible to integrate a range of governance tools into water quality management; however, new ones, like most of the market-based instruments, will take time for the community to understand and adopt (10, +3).

Progressive Producers

Progressive Producers believe in practical solutions to help the industry evolve environmentally, and this includes anything the government of Alberta intends to do in the Oldman River basin. This factor is characterized by intensive livestock operators, who have been actively involved in their community's water management solutions, including leadership in local government and water-quality programs. The Progressive Producers acknowledge that there are a few poor performers but say that most of the farmers are 'good' players who are committed to making the industry more sustainable (as expressed in post-sort interviews). However, the reality is that there are two long-term non-point-source pollution challenges facing the region: the number of livestock creating manure and the significant cost of transporting manure for disposal (post-Q-sort interviews).

Progressive Producers believe it is important to develop practical water-quality solutions that help farmers change their practices and that market-based instruments will work in the intensive livestock industry and help the government of Alberta achieve its goals sooner (comment made during post-sort interview). However, the government of Alberta needs to give these tools time to work; for example, the Environmental Farm Plans could have made a significant environmental difference and also built trust in the community, but the funding ran out just as the farmers were getting on board (15, +4). At the same time, the Progressive Producers do not believe it will take a lot of time to develop trust (26, −4). The government of Alberta can build on the success of the Natural Resource and Conservation Board, which has worked with the community to understand the issues and develop solutions (9, +3).

The Progressive Producers are concerned about the effectiveness of government subsidies (30, −4), as some farmers have come to expect them and, therefore, will continue to demand compensation for improvements when they have a responsibility to manage land properly. Further, they do not see the need to treat intensive livestock producers differently (24, −3); rather, it is necessary to involve the agricultural industry as a whole in solution building. There is a range of tools, from voluntary to prescriptive, and the industry can help identify solutions that will work, such as providing livestock watering stations away from creeks (22, +1).

Consensus statements

Three consensus statements resulted from the Q-sort. First, there is neutral agreement that the government of Alberta should look at local initiatives first before adding new tools (Statement 14; factor scores 0, 0, +1 and −1). Second, there is a shared neutral response that education is an important aspect of behavioural change (Statement 7; +1, +1, −1 and 0). Third, there is a concern that market-based instruments must actually create positive change rather than simply paying someone to maintain the status quo. The latter would not be acceptable to the community or to farmers (Statement 13; +2, +1, +1 and +2). The proximity of all the statements to neutral may mean that they were insignificant points on which Q-sorters were ambivalent, or statements toward which they have mixed feelings.

Discussion and conclusions

The objective of this article was to reveal the prevalent discourses on the use of market-based instruments as a tool to manage non-point-source pollution in southern Alberta. Four distinct discourses were identified (Incentive Orienteers, Honest Brokers, Rural Advocates and Progressive Producers), and each provides a unique perspective and advice that the government of Alberta needs to consider in its pursuit of integrating market-based instruments into water management.

The three consensus statements provide the government of Alberta a starting point to engage the community in water-quality solutions. The first statement (14) advises the government of Alberta to look at local initiatives first, before adding new tools. This provides the government of Alberta a strategic opportunity to build trust, particularly with the Rural Advocates and Incentive Orienteers, and to acknowledge the various good works underway in the Oldman River basin. While Statement 14 is not strongly positive or negative, it does represent a discussion point to start engagement. The other two consensus statements relate to solutions. First, there is a recognition that education is an important aspect of behavioural change (7, +1, +1, −1 and 0); and second, a concern is expressed that simply paying someone to maintain the status quo would not be acceptable to the community or to farmers (13, +2, +1, +1 and +2).

The results indicate a common understanding that agriculture contributes to non-point-source pollution having a negative impact on water quality and agreement that market-based instruments, carefully developed, can be part of the solution to non-point-source pollution. The discourses provide insights into why market-based instruments are not more widely introduced and used, such as a general confusion about what market-based instruments actually are, and questions about why the government of Alberta wants to use them. The discourses provide advice that the government of Alberta needs to be persistent in the Oldman River basin and spend time educating and building trust in the community. This is a way to ensure that the community is willing to try new tools, including market-based instruments. The results suggest that an effective way of achieving environmental quality improvements might be to work through established community groups, already known and respected in the community, such as watershed organizations and extension specialists. As Alberta continues to grow and develop, successful resolution of non-point-source pollution issues will be important for all parties.

Understanding community knowledge, perspectives and concerns can provide a government with important information on specific education needs or where there will be areas of conflict or consensus. The Q-method takes time and effort to design and implement; however, the benefits of understanding where political land mines might be – or testing assumptions about how local communities will respond to a new policy design that policy makers are unfamiliar with – could save time and political capital and build more trust in the process. This research demonstrates that it is possible to measure and analyze distinct perspectives held by groups of stakeholders on important issues and processes using the Q-method. It also reveals consensus areas that the government can use as initial discussion points in building understanding and trust with the community to inform water quality solutions.

These results are important for policy makers. The government of Alberta has committed to policy excellence (Hicks & Watson, 2007) and can use these results, as well as the Q-method more broadly, to enhance the environmental policy development process by proactively capturing, assessing and integrating key community viewpoints on an issue. Unlike other methods, the Q-method can reveal a great diversity and complexity of environmental beliefs, beyond responses that simply state that someone is for or against something (Addams & Proops, 2000). Government staff can use these local discourses to build credibility and trust with the community.

Whilst the Q-method provided new and rich insights into the distinct discourse on non-point-source pollution and water quality and market-based instruments as a solution, it did not tell us the percentage of the population represented by each factor. Further, the Q-method is time-intensive, may be difficult for respondents, and requires specific training for those administering it. Therefore, it is most appropriate when undertaking in-depth analysis on contentious or wicked issues, where solutions have eluded the government and where the method can reveal latent social discourses.

An extension of this research could be to apply the discourse-generation abilities of the Q-method to water policy development more broadly, by using it at each stage of the process to ground-truth a government's analysis, policy options and implementation preferences with that of a specific community. This might provide a more successful outcome, as well as saving time, money and resources in the development and implementation of environmental policy.

Disclosure statement

No potential conflict of interest was reported by the author.

ORCID

Gillian Kerr http://orcid.org/0000-0002-7300-9706
Henning Bjornlund http://orcid.org/0000-0003-3341-5635

References

Addams, H., & Proops, J. (Eds.). (2000). *Social discourse and environmental policy*. Cheltenham, U.K.; Northampton, MA: Edward Elgar.

Agriculture and Agri-Food Canada. (2004). *Using off-stream water sources as a beneficial management practice in riparian areas - a literature review*. Retrieved from http://www.agr.gc.ca/eng/science-and-innovation/agricultural-practices/soil-and-land/riparian-areas/using-off-stream-water-sources-as-a-beneficial-management-practice-in-riparian-areas-a-literature-review/?id=1220560749957

Agri-Environmental Partnership. (2012). *AEPA 2012 annual report*. Retrieved from http://www.agpartners.ca/aepa/Portals/0/AEPA_2012AnnualReview.pdf

Alberta Agriculture, Food and Rural Development. (2012). *Surface water quality risks for the agricultural areas of Alberta*. Retrieved from http://agriculture.alberta.ca/acis/maps/agricultural_land_resource_atlas_of_alberta/water/water_quality/surface_water_quality_risk_2001_big_map.png

Baird, J., Belcher, K., & Quinn, M. S. (2014). Context and capacity: The potential for performance-based agricultural water quality policy. *Canadian Water Resources Journal / Revue Canadienne Des Ressources Hydriques*, *39*(4), 421–436.

Barry, J., & Proops, J. (1999). Seeking sustainability discourses with Q methodology. *Ecological Economics*, *28*(3), 337–345.

Bjornlund, H. (2010). *The competition for water: Striking a balance among social, environmental and economic needs* (Commentary No 302). Toronto: Governance and Public Institution, C.D Howe Institute.

Bjornlund, H., Nicol, L., & Klein, K. (2007). Challenges in implementing economic instruments to manage irrigation water on farms in southern Alberta. *Agricultural Water Management*, *92*, 131–141.

Bjornlund, H., Parrack, C., & De Loë, R. (2013). Segmenting the urban and rural populations of southern alberta for improved understanding of policy preferences for water reallocation. *Society & Natural Resources*, *26*(11), 1330–1350.

Bjornlund, H., Zhao, X., & Xu, W. (2014). Variation in the perspective on sharing water - irrigators, their communities and the wider society. *International Journal of Water Governance*, *2*(1), 19–41.

Brown, S. R. (1980). *Political subjectivity: Applications of Q methodology in political science.* New Haven, CT: Yale University Press.
Brown, S. R. (2006). A match made in heaven: A marginalized methodology for studying the marginalized. *Quality & Quantity, 40,* 361–382.
Brown, S. R., Durning, D. W., & Selden, S. C. (2008). Q methodology. In K. Yang & G. J. Miller (Eds.), *Handbook of research methods in public administration* (2nd ed., pp. 721–763). New York: CRC Press/Taylor & Francis.
Burton, R. J. F., & Wilson, G. A. (2006). Injecting social psychology theory into conceptualisations of agricultural agency: Towards a post-productivist farmer self-identity? *Journal of Rural Studies, 22,* 95–115.
Clare, S., & Krogman, N. (2013). Bureaucratic slippage and environmental offset policies: The case of wetland management in Alberta. *Society & Natural Resources, 26*(6), 672–687.
Clarkson, S., & Woods, S. (2010). *A perilous imbalance: The globalization of Canadian law and governance.* British Columbia: University of British Columbia Press.
Coke, J. G., & Brown, S. R. (1976). Public attitudes about land use policy and their impact on state policy-makers. *Publius: The Journal of Federalism, 6*(1), 97–134. Retrieved from http://www.jstor.org/stable/3329607Colorado
Dai, H., Sun, T., Zhang, K., & Guo, W. (2015). Research on rural nonpoint source pollution in the process of urban-rural integration in the economically-developed area in china based on the improved STIRPAT Model. *Sustainability, 7,* 782–793.
Davies, B. B., Blackstock, K., & Rauschmayer, F. (2005). 'recruitment', 'composition', and 'mandate' issues in deliberative processes: Should we focus on arguments rather than individuals? *Environment and Planning C: Government and Policy, 23,* 599–615.
Davies, B. B., & Hodge, I. D. (2012). Shifting environmental perspectives in agriculture: Repeated Q analysis and the stability of preference structures. *Ecological Economics, 83,* 51–57.
Depoe, S. (2006). *Water quality monitoring of small streams in agricultural areas: Alberta environmentally sustainable agriculture water quality monitoring program* (2004 Annual Technical Report). Edmonton, Alberta, Canada: Alberta Agriculture, Food and Rural Development.
Environmental Protection and Enhancement Act, E-12 R.S.A. (2000).
Fitch, L. B., Adams, B., & O'Shaughnessy, K. (2003). *Caring for the green zone: Riparian areas and grazing management.* (3rd ed.). Lethbridge, Alberta: Cows and Fish Program. Retrieved from http://cowsandfish.org/pdfs/greenzone3rd.pdf
Government of Alberta. (2003). *Water for life: Alberta's sustainability strategy.* Retrieved from http://aep.alberta.ca/water/programs-and-services/water-for-life/strategy/documents/WaterForLife-Strategy-Nov2003.pdf
Government of Alberta. 2006) *Approved south Saskatchewan River Basin Management Plan.* Retrieved from http://environment.alberta.ca/documents/SSRB
Government of Alberta. (2008). *Land use framework.* Edmonton: Author. Retrieved from https://landuse.alberta.ca/PlanforAlberta/LanduseFramework/Pages/default.aspx
Government of Alberta. (2009). *Water for life: Action plan.* Retrieved from http://aep.alberta.ca/water/programs-and-services/water-for-life/strategy/downloadable-information-about-the-water-for-life-strategy.aspx
Government of Alberta. (2013). *Alberta wetland policy.* Edmonton: Author. Retrieved from http://aep.alberta.ca/water/programs-and-services/wetlands/alberta-wetland-policy.aspx
Graham, S., Schempp, A., & Troell, J. (2011). Regulating nonpoint source water pollution in a federal government: Four case studies. *Journal International Journal of Water Resources Development, 27*(1), 53–69.
Hicks, R., & Watson, P. (2007). *Policy capacity: Strengthening the public service's support to elected officials.* Edmonton, Alberta: Unpublished Internal Document.
Hodge, I. (2001). Beyond agri-environmental policy: Towards an alternative model of rural environmental governance. *Land Use Policy, 18,* 99–111.

Howard, K., Bjornlund, H., & Xu, W. (2012). Securing societal benefits through increased provision of ecosystem services using incentives. In D. Pepper & V. Brebbia (Eds.), *Water and society* (pp. 487–500). Southampton: WITPress.

Howlett, M. (2009). Governance modes, policy regimes and operational plans: A multi-level nested model of policy instrument choice and policy design. *Policy Sciences*, *42*, 73–89.

Ingram, H., Schneider, A. L., & Deleon, P. (2007). Social construction and policy design. In P. A. Sabatier (Ed.), *Theories of the policy process* (pp. 93–126). Boulder, CO: Westview Press.

Institute of Public Policy. (2006). *Water in 2025: Beliefs and values as a means for co-operation.* Fort Collins: Author.

Ivey, J. L., De Loë, R., Kreutzwiser, R., & Ferreyra, C. (2006). An institutional perspective on local capacity for source water protection. *Geoforum*, *37*, 944–957.

Johnson, J. Y. M., Thomas, J. E., Graham, T. A., Townshend, I., Byrne, J., Selinger, L. B., & Gannon, V. P. J. (2003). Prevalence of Escherichia coli O157: H7and Salmonella spp. in surface waters of southern Alberta and its relation to manure sources. *Canadian Journal of Microbiology*, *49*, 326–335.

Kennedy, M. (2010). *Towards a market-based approach for ecosystem services in Alberta* (Report prepared for the Institute for Agriculture, Forestry and the Environment). Retrieved from: https://ecoservicesnetwork.ca/building-blocks/

Lafreniere, K., Deshpande, S., & Bjornlund, H. (2015). Segmenting and targeting irrigators' preferences regarding proposed water transfers. *Society & Natural Resources*, *28*(4), 423–438.

Lewis-Beck, M. S., Bryman, A., & Futing Liao, T. (2004). *The SAGE encyclopedia of social science research methods.* Thousand Oaks, CA: SAGE Publications.

Little, J., Kalischuk, A., Gross, D., & Sheedy, C. (2010). *Assessment of water quality in Alberta's irrigation districts.* Alberta, Canada: Alberta Agriculture and Rural Development.

Maybery, D., Crase, L., & Gullifer, C. (2005). Categorising farming values as economic, conservation and lifestyle. *Journal of Economic Psychology*, *26*, 59–72.

McKeown, B., & Thomas, D. (2013). *Q methodology* (2nd ed.). Newbury Park, CA: Sage.

Montgomery, J., Bjornlund, H., Xu, W., & Edwards, J. (2016). A table for five: Stakeholder perceptions of water governance in Alberta. *Agricultural Water Management*, *174*, 11–21.

Oldman River Basin Water Quality Initiative. (2002). *Oldman river basin water quality initiative* (Partnering for our Future! Annual Report 2001/2002). Lethbridge, Alberta: Oldman River Basin Water Quality Initiative.

Olsen, B. M., Miller, J. J., Rodvang, J., & Yanke, L. J. (2005). Soil and groundwater quality under cattle feedlot in southern Alberta. *Water Quality Research Journal*, *40*(2), 131–144.

Pannell, D. J. (2008). Public benefits, private benefits, and policy mechanism choice for land-use change for environmental benefits. *Land Economics*, *84*(2), 225–240.

Pereira, M. A., Fairweather, J. R., Woodford, K. B., & Nuthall, P. L. (2016). Assessing the diversity of values and goals amongst Brazilian commercial-scale progressive beef farmers using Q-methodology. *Agricultural Systems*, *144*, 1–8.

Rodvang, S. J. (2002). *Groundwater quality in the Battersea Drainage Basin* (Factsheet 1B002-2002). Lethbridge, Alberta, Canada: Irrigation Branch, Alberta Agriculture, Food and Rural Development.

Russell, C. S., & Clark, D. C. (2006). Economic instruments and nonpoint source water pollution. In A. K. Biswas, C. Tortajada, B. Braga, & D. J. Rodriguez (Eds.), *Water quality management in the Americas* (pp. 17–45). doi:10.1007/3-540-30444-4_2

Rutherford, M. B., Gibeau, M. L., Clark, S. G., & Chamberlain, E. C. (2009). Interdisciplinary problem solving workshops for grizzly bear conservation in Banff National Park, Canada. *Policy Sciences*, *42*, 163–187.

Schmolck, P., & Atkinson, J. (2012). *PQ method: Release 2.25*. Retrieved from http://schmolck.userweb.mwn.de/qmethod/

Shortle, J. S., & Horan, R. D. (2001). The economics of nonpoint pollution control. *Journal of Economic Surveys*, *15*, 255–289.

South East Alberta Watershed Alliance (SEAWA). (2010). *The impact of agriculture on water quality in the south Saskatchewan River sub-basin: SEAWA state of the watershed report*

2010-7. Lethbridge: Author. Retrieved from http://seawa.ca/state-of-the-watershed/seawa-state-of-the-watershed-report/
Stavins, R. N. (2003). Experience with market-based environmental policy instruments. In K. G. Mäler & J. I. Vincent (Eds.), *Handbook of environmental economics* (pp. 355–435). Amsterdam, Netherlands: Elsevier Science.
Stephenson, W. (1953). *The study of behaviour: Q-technique and its methodology*. Chicago, IL: University of Chicago Press.
Stephenson, W. (1978). Concourse theory of communication. *Communication*, *3*, 21–40.
Stirrett, S., Rolfe, R., & Shewchuk, S. (2012). *The invisible hand's green thumb: Market-based instruments for environmental protection in Alberta*. Alberta: Canada West Foundation. Retrieved from http://cwf.ca/research/publications/
Strickert, G., Chuna, K. P., Bradford, L., Clarck, D., Goberd, P., Redd, M. G., & Payton, D. (2016). Unpacking viewpoints on water security: Lessons from the South Saskatchewan River Basin. *Water Policy*, *18*, 50–72.
Swainson, B., De Loë, R. C., & Kreutzwiser, R. D. (2011). Sharing water with nature: Insights on environmental water allocation from a case study of the Murrumbidgee catchment, Australia. *Water Alternatives*, *4*(1), 15–34. Retrieved from www.water-alternatives.org
United Nations. (2011). *UN-water policy brief: Water quality*. Author. Retrieved from http://www.unwater.org/publications/un-water-policy-brief-water-quality/
Valbuena, D., Bregt, A. K., McAlpine, C., Verburg, P. H., & Seabrook, L. (2010). An agent based approach to explore the effect of voluntary mechanisms on land use change: A case in rural Queensland, Australia. *Journal of Environmental Management*, *91*, 2615–2625.
Water Act, W-3 R.S.A. (2000).
Watts, S., & Stenner, P. (2012). *Doing Q methodological research*. London: Sage.
Zuo, A., Wheeler, S., Bjornlund, H., Edwards, J., & Xu, W. (2015). Exploring generational differences towards water resources and policy preferences of water re-allocation in Alberta, Canada. *Water Resources Management*, *29*(14), 5073–5089.

Minamata: how a policy maker addressed a very wicked water quality policy problem

Hikaru Kobayashi

I assumed the leading administrative role, that of Vice-Minister, in my final career assignment in the Ministry of the Environment from 2008 to 2010. During that period, the Japanese Diet enacted a new law directed at resolving problems that had arisen in settling the lingering issues of compensating the remaining victims of Minamata disease. This was a problem with a 60-year history. Working out how to implement this last attempt at a resolution was the job of the Ministry of the Environment. In other words, it landed on my desk. In fact, at the end of my term I was asked to continue for another year as a part-time senior councillor in charge of the negotiation with the victims.

More recently, I have had the opportunity to be a Fulbright scholar-in-residence in an American college, where I learned that Japan's struggle with Minamata disease is used in the curriculum as a worst-case study of the problems of rehabilitation from severe industrial pollution. I also have learned that many cities in the United States are facing similar problems of rehabilitation from the health-related industrial pollution legacies of the middle of the last century. Hence, I am offering my observations as a participant in the latest, probably overly optimistic, attempt to bring the Minamata legacy to closure.

Minamata disease is a neurological disorder caused by the ingestion of methyl mercury, usually via seafood from waters polluted by industrial discharges containing the chemical. It is named after the city in south-west Japan, where it was first identified over half a century ago. It cannot be cured, and can occur with symptoms of varying degrees of severity. The complex problems of victim compensation and liability have a long and tortuous history, and continue to pose serious questions of redress, responsibility and the role of government.

In this paper, I divide my argument into three sections, discussing: (1) the course of the disease and the manner of dealing with its long-term consequences; (2) the approach to implement the new law; and (3) an assessment of issues and accomplishments based on the introduction of initiatives linked to this approach.

How Minamata became a wicked problem

First, I discuss the background of Minamata disease and how the associated problems have grown in complexity and intensity over time. In the 1950s, after consuming large quantities of fish and shellfish, fishermen and other residents of the area near Minamata developed a disease which in many cases presented symptoms of impaired coordination, such as numbness and loss of sensation in the hands and feet, narrowing of the field of vision, and impaired walking and speech. The most severely afflicted, those who most commonly showed evidence of the disease during the period immediately following its discovery, suffered from seizures and an agonizing death.

The victims suffered from social discrimination as well. At first the disease was mistakenly thought to be contagious, or seen as an attempt to cause trouble in the company town. Certain parties claimed that revealing the source of the pollution would result in damage through misinformation. Moreover, the government's incremental and reactive approach itself aggravated tensions among different sectors of society and between the victims and the polluting company. These conflicts served to magnify the isolation, pain and suffering of the victims of Minamata disease.

Minamata disease was in a sense a product of Japan's period of rapid economic growth during the 1960s. At that time, scientific understanding of the environment, environmental measurement techniques, decontamination technology and anti-pollution administration were all still in the early stages of development. Economic growth came first. Therefore, it was not possible to diagnose the problem quickly, much less to find a solution.

Ever since the botched handling of the issue in its early stages, the consequences have posed a wide variety of recurring challenges in balancing redress for the victims, the economic viability of the major employer in the area, and the budgetary impacts on local governments. Yet the main parties concerned are making efforts to achieve reconciliation and recovery.

In the remainder of this section, I will conduct a chronological examination of the process leading up to the request to prepare the Act on Special Measures for Compensation of Minamata Disease Victims and Resolving the Issue of Minamata Disease.

A significant number of research papers and books (e.g. Harada, 1972; George, 2002) focusing on the problem of Minamata disease have identified problems during the period up to the Supreme Court ruling in the Kansai Lawsuit in 2004. However, almost all of these publications adopt a critical attitude towards the government's policy. There are very few contributions from individuals who were involved in it (exceptions being Hashimoto, 1988; Kojima, 1996). In this sense, this paper represents the discourse of a minority. In particular, evaluation of the sequentially expanding administrative system and the relationship between judicial and administrative support for the victims are explored here for the first time from the perspective of a governmental actor. From this perspective, the report written based on the victims' standpoint (No-more Minamata Lawsuit Archive Commission, 2012) is also informative.

Background of the pollution

The Chisso Corporation (formerly called Shin Nichitsu), the source of the methylmercury in Minamata, is a major chemical enterprise whose origin dates back to 1908. The methylmercury it produced was a by-product of acetaldehyde, a highly useful chemical compound that can be used to produce octanol, used in manufacturing plastics such as vinyl chloride, which gained in popularity in the post-war period. It was also a raw material for acetate, used in the production of acetate fibre. Although the company had manufactured acetaldehyde since before the Second World War, production expanded rapidly after the war, when the product generated a significant proportion of the company's profits. This enlarged scale of production probably led to a greater volume of wastewater discharged by the company, although no one measured such things back then.

The company used mercury sulphate as a catalyst to produce acetaldehyde. This is not the only method possible – while Chisso has stopped producing acetaldehyde, other firms do, using a different process – but at the time, chemists believed that inorganic mercury was stable, and indeed it is not highly toxic to humans in that form, even if ingested, unless its vapour is inhaled. Unfortunately, they did not realize that it could transform into toxic organic forms, such as methylmercury.

On 1 May 1956, Hajime Hosokawa, director of the Chisso company hospital, reported a patient suffering from neurological symptoms to the Minamata health office. The patient (a young girl) lived in a fishing village, where other people were already suffering from the same symptoms (Harada, 1972; Sakamoto, Murata, Tsuruta, Miyamoto, & Akagi, 2010). The affected population had a wide geographical distribution across an area of the Shiranui Sea roughly as wide as Tokyo Bay (approximately 30 km × 30 km).

Experts had been aware of the toxic nature of methyl and other forms of organic mercury long before the occurrence of Minamata. What they had not yet developed was an accurate means of measuring minute amounts of such compounds at the concentration of the methylmercury in the wastewater discharged by Nichitsu or to link it through bioaccumulation to concentrations in the food consumed by the victims. The linkages between organic and inorganic mercury were disputed by scientists, so experiments were carried out, notably by Nichitsu, which did not publicize the results they obtained. Hence it took a long time for the causal relationships to be established scientifically by outsiders (Harada, 1972).

For these reasons, no measures were taken to treat the methylmercury, and Nichitsu continued to discharge the wastewaters into Minamata Bay for more than 10 years after the discovery of the disease in the area, further polluting the sediment and various species of sea life.

Initially, victims of pollution were able to receive compensation only by accepting payoffs from the company, engaging in long precursory battles, or by being judged to deserve compensation due to individual circumstances and symptoms through legal trials. Victims of Minamata disease were subsequently indemnified under the Law Concerning Pollution-Related Health Damage Compensation and other Measures (enacted in 1969 and enforced in 1970), a system administered by the government to compensate victims of pollution regularly and relatively quickly without trial. Under this law, 2282 victims, including 376 still living, had received compensation as of March 2017 (Ministry of the Environment, 2017).[1]

In addition, a number of victims suffering from Minamata symptoms received financial aid through measures established in 1995 and through a special act for the provision of aid to Minamata sufferers enacted in 2010. As of 2017 it was estimated that 66,000 victims had received compensation under the above measures.

The damage at Minamata did not only affect the health of its victims. Local industries, such as fishing and tourism, suffered, and different sectors of the community became hostile and divided (Kobayashi, 1994; Study group, 1999).

Factors that hindered the resolution of the Minamata disease issues and the misery and gravity of pollution as a social issue

Recurring friction

After the discovery of Minamata disease in 1956, it was extremely difficult to ascertain the cause (Study group, 1999). In fact, there was pressure not to do so, but despite this, the local Kumamoto University School of Medicine continued to conduct research into the disease. Based on the results of this research, in November 1959 the Ministry of Health and Welfare's Food Sanitation Investigation Council Minamata Food Poisoning Subcommittee (represented by Kenshi Wanibuchi, former president of Kumamoto University) announced that a form of organic mercury caused Minamata disease. Although its report did not go as far as stating the source of this mercury, the committee was immediately dissolved, and the pro-business Economic Planning Agency took over investigating the cause of the disease. But the School of Medicine continued its research, and in 1963, it released a report concluding that the source of the disease was the volatilization of mercury released from the Chisso factory in Minamata. The Ministry of Health and Welfare confirmed this finding, and in 1968, Chisso stopped releasing hazardous wastewater. Finally, strict legal restrictions on the release of wastewater followed in 1970.

The problem of Minamata disease passed a number of significant milestones (Study group on social-scientific aspect of Minamata Issue (edition), 1999). The first was the decision by Chisso to offer 'sympathy money' in 1959 to end a campaign for compensation by approximately 80 victims who had developed serious Minamata disease. The families of deceased victims of fatal or acute Minamata disease received a one-off payment of 300,000 yen (about USD 850 at the time), while adult victims still alive received 100,000 yen (USD 280) per year. In return, the victims had to agree not to demand further compensation, even if it became clear in the future that Chisso wastewater had caused the disease. This agreement was later nullified as it violated standards of public order and morality. This led to a temporary easing of tensions.

However, in 1965, Minamata disease broke out again, this time in Niigata on the Japan Sea coast, and in 1967, a lawsuit was filed against the polluting company, Showa Denko. These events reignited tensions over the original Minamata disease incident. In 1969, a compensation trial was held at Kumamoto District Court. The decision of both district courts held the companies responsible, in 1971 for the Niigata case and in 1973 for the Kumamoto case. Further, in the early 1970s, the Japanese government recognized Minamata disease as a pollution-related disease, for which compensation was available through the mentioned Law Concerning Special Measures for the Relief of the Pollution-Related Health Damage (enacted in 1969 and enforced in 1970) and the Pollution-Related Health Damage Compensation Act (enacted in 1973) that expanded on this. Once again, an out-of-court compensation agreement (1973) was concluded through direct negotiations between organizations representing the patients, in particular one called Gojokai, and Chisso.

The compensation agreement and the government compensation system paved the way for the receipt of financial aid for Minamata disease victims who were suffering from comparatively obvious or typical forms of the disease. This formed the second milestone of the Minamata disease problem. This led to a second temporary easing of tensions.

However, as claims for compensation continued under these systems, people started to question whether the scope of the government system for determining who received compensation was too restrictive, and, as a result, new 'hidden' victims began to surface. In the 1980s, these hidden victims, and others not covered by the government system, filed a lawsuit that, in addition to confirming the liability of the Chisso Corporation, sought to hold national and prefectural governments accountable for failing to prevent the spread of damage.

The lawsuit movement expanded and intensified until, in 1995, a political settlement was reached between the plaintiffs, the patients' representatives, the polluting enterprise, and the Japanese government through an initiative involving the three major political parties at the time – the Japan Socialist Party (the ruling party, led by the Murayama cabinet), the Liberal Democratic Party and the New Party Sakigake.

Victims suffering from symptoms of Minamata disease and individuals who were thought to have consumed large amounts of polluted fish were awarded a one-off payment of 2.6 million yen (about USD 30,000) through a relatively quick process. Over 11,000 people received the one-off payment, the vast majority (10,353) from Kumamoto. In exchange, the plaintiffs then withdrew the lawsuit, an agonizing decision but deemed necessary considering the tremendous amount of time it would take to confirm the individual circumstances of each plaintiff if the trial were to continue. This, the 'political settlement', represents the third major milestone in the Minamata disease problem.

Later, in 2004, the Supreme Court delivered a verdict concerning the failure to act on the part of national and prefecture governments (Minamata disease Kansai Lawsuit) confirming that the government should have exercised its regulatory authority around the end of 1959. The negligence of the government resulted in continued exposure and a large number of unnecessary victims in the 1960s.

A great deal of research has sought to identify the reasons why the national government and the company responsible for the pollution both failed to take action until it was too late (Harada, 1972; Study group on social-scientific aspect of Minamata Issue (edition), 1999). This research reveals a number of important elements:

- Various experts in the field of chemistry opposed the theory that Chisso was to blame for the poisoning.
- Influenced by a tendency to prioritize industrial profit, the government postponed the implementation of measures to stop production processes due to a lack of definitive understanding of the scientific causal relationships involved.
- Chisso attempted to mitigate the spread of pollutants instead of taking radical measures, and this only widened the scope of damage.

- Although measures were taken to provide compensation, these measures tended to be limited to victims with acute or fatal Minamata disease.

While people often state the lack of scientific knowledge and consensus as one reason for opposing pollution-control measures, the experience of Minamata disease has taught us the importance of the precautionary principle, taking action to implement discriminating measures, even in the case of scientific uncertainty.

In addition, change occurred as people acquired a better understanding of the symptoms of the disease. Non-acute and non-fatal forms of the disease were discovered by doctors and by the victims themselves. These discoveries revealed that the disease was not only caused by the direct ingestion of fish, but could also be transferred through the mother's womb (foetal Minamata disease), and that there were also victims who, despite suffering severe damage, did not display deadly symptoms. In some cases, it became difficult to distinguish the symptoms from those of other diseases.

The range of initiatives to aid the recovery and compensation of victims expanded sequentially as more scientific knowledge became available. This expansion helped provide financial aid for victims as quickly as possible; however, since the initiatives were implemented sequentially, the qualification requirements for compensation and the content of the measures required many details that, instead of helping victims, the mechanisms for indemnification became difficult to understand.

In short, the measures adopted 50 to 60 years ago were by no means perfect. Actually, they left the door open to further conflict.

As new knowledge became available, the government gradually added more measures. This caused friction to recur periodically when new victims applied for compensation, and the company responsible for the pollution and the government fought the applications. It may seem ironic to say that the government fought new applications while adding more measures, so it may be useful to state how the above process looks from the viewpoint of a government administrator.

(1) A person who is convinced that s/he deserves compensation requests a certification as a victim from the government, but the government denies this request because his/her symptom differs from those recognized as applying at that moment. (2) S/he files a lawsuit, and the court rules that his/her condition is symptomatic of Minamata disease. (3) More people suffering from the same symptoms demand recognition from the government. (4) The government at first hesitates to expand the scope of the criteria related to Minamata victim designation. (5) Finally, backed by the court decision, the pressure from the sufferers wins the day, despite the reluctance of the government and Chisso to expand the scope of coverage. The resulting time lag between the judicial judgement and the expansion of scope for the official compensation itself tends to provoke various conflicts, including, ironically, that previously designated victims tend not to recognize the newly designated persons as proper victims of Minamata disease.

Initiation of relief movements for victims suffering wider symptoms following the decision of the Supreme Court

For a time, owing to the political settlement, there were very few claims for compensation under the Pollution-Related Health Damage Compensation Act. While most victims accepted the political settlement of 1995 and withdrew their lawsuits, a few dozen plaintiffs did not and continued litigating in a lawsuit (the 'Kansai lawsuit'), which went all the way to the Supreme Court. The court's decision, rendered in 2004, put Minamata disease back in the limelight, and on the government agenda.

During the lawsuit, the court tackled cases directly based on the individual circumstances of each plaintiff and considered the necessity for damages in each case. This process was conducted separately from the regular system for acknowledging victims of Minamata disease according to the Pollution-Related Health Damage Compensation Law (hereafter Compensation Law), which offered compensation without the need for a trial. A total of 22 years had passed since the Osaka District Court lawsuit in 1982. As a result of an examination of the individual conditions of the plaintiffs, the Supreme Court judgment confirmed that 51 out of the 58 cases before it should receive compensation for damage caused by methylmercury poisoning. This was hardly full redress, however, as the compensation approved by the court decision to cover damages in full was significantly smaller than the amount provided through the Compensation Law and the compensation agreements made with the Chisso Corporation.

Perhaps one reason for the smaller compensation was that not all of the 51 approved cases presented sensory disorders that caused numbness in the extremities – a symptom of Minamata disease, which was required to be proven to qualify for compensation under the previous government system. The court decision required inclusion of these cases as victims of methylmercury poisoning qualified to receive compensation. On the other hand, the court did not recognize plaintiffs whose symptoms had been discovered after a certain number of years had passed.

Since the Supreme Court decision of the Kansai Lawsuit recognized a wider range of symptoms than before, there was a sharp increase in the number of people in the Minamata area who began to wonder whether their own symptoms could also be attributed to the disease. Although the requirements of the 'political settlement' enabled victims with a wider range of symptoms to receive compensation, the settlements were restricted to victims suffering from symptoms that resembled typical Minamata disease. For example, victims who suffered severe nerve damage in three of their limbs, but had not lost all feeling in the fourth limb, fell outside the scope of financial aid. Many of these victims formed patients' support groups and continued to pursue financial aid through new channels.

Many victims applied for recognition through the Pollution-Related Health Damage Compensation Law because they anticipated a widening of the range of symptoms recognized by the law (Figure 1). Others went back to court. Since the Kansai Lawsuit decision affirmed the government's failure to act, and this placed it on the list of perpetrators, rather than seeking compensation through the government, a mammoth lawsuit (the No More Minamatas Lawsuit) was filed to make the perpetrators pay compensation according to the decisions of a neutral judiciary. Similar lawsuits were filed in Kumatoto, Tokyo, Osaka and Niigata, ultimately involving over 3000 plaintiffs.

The reason for this action is that the Japanese government typically establishes relief systems using fixed standards to enable comparatively swift decisions. Examples include the compensation systems for pollution-related health damage as well as the system governing industrial accidents. Cases not fitting the fixed standard because they require scrutiny of the individual conditions of victims are determined by the judiciary. It is

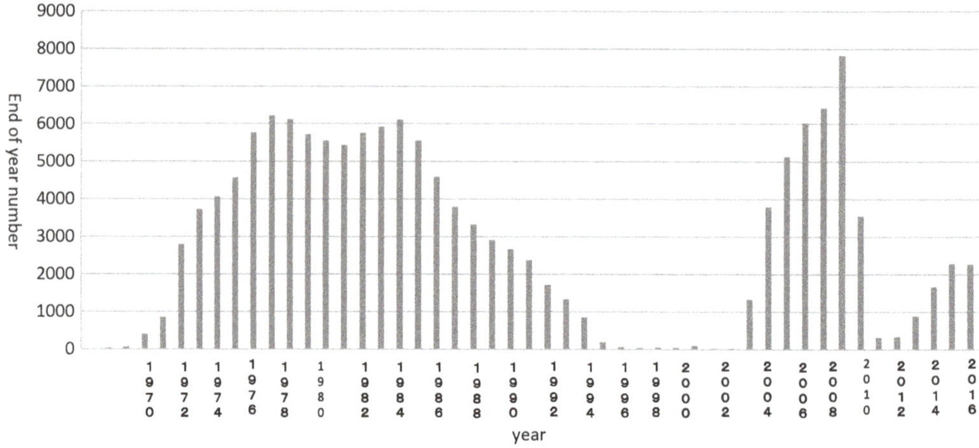

Figure 1. Number of pending applications for designation as a Minamata disease victim, by year. Source: Data from the Ministry of Environment, 2011-2016; Ministry of Environment (2011).

also possible to appeal, in cases where the fixed government relief system has not been applied appropriately. This does not mean that the government relief system is incorrect. Rather, the judiciary performs a role that only it can: the provision of relief on a case-by-case basis. Nevertheless, handling all 3000 cases individually would overwhelm the judicial process.

This represented a fourth stage of friction associated with the problem of Minamata disease.

In light of the conditions at the time, the ruling coalition (Liberal Democratic Party and Komeito) formed a task force to consider new measures to deal with the problem of Minamata disease. In 2007, taking victims who requested compensation following the political settlement of 1995 as a precedent, the task force proposed a relief package based on a one-off payment of 1.5 million yen (about USD 12,500) – even less than that provided in 1995.

Nevertheless, while a number of patients' groups unofficially indicated their intentions to accept the proposal, the plaintiff group involved in the large ongoing lawsuit believed that the proposal limited the responsibility of the perpetrators by enabling them to determine the scope of symptoms included in the compensation. Moreover, the amount of compensation offered fell well below the expectations of the group. In other words, it became clear that the content of the proposal would fall short of solving all of the disputes.

Meanwhile, as expected, Chisso Corporation was unable to accept the proposal since the financial burden of further one-off payments on top of the long-term 'hundred-year' debt it had already accrued was a burden the company could not afford. Although payments such as one-off payments and allowances for damages, fisheries compensation, and the costs of mercury dredging were mainly financed publicly, Chisso was expected to reimburse the government, saddling the company with a large long-term debt that it could ill afford to repay while remaining financially viable.

Prefecture (Kumamoto) issues bonds are purchased by the national government. The national government is not authorized to lend money to a private enterprise, but

Kumamoto Prefecture has done so probably taking into consideration its responsibility for failing to regulate the effluents of the Shin Nichitsu/Chisso Corporation.

The task force then made a further proposal, in June 2008, that involved dealing with the extraordinary loss and long-term debt that Chisso had incurred until then by dividing the company into two entities – one responsible for compensating victims and one financially secure unit. Then, the financially secure company would list its stocks and use the proceeds from sales to cover its extraordinary losses and secure funds for future compensation.

However, although the proposal may have satisfied the Chisso Corporation, it was met with strong opposition from a number of angles since it failed to guarantee that the perpetrator would continue its compensation payments and offered no prospects of relief for the 10,000 (at the time) sufferers who wondered if they too might be victims of Minamata disease as a result of the Supreme Court decision. The standoff continued for some time despite the Minister of the Environment's denunciation of Chisso for their refusal to negotiate.

The Act on Special Measures for Compensation of Minamata Disease Victims

The Act on Special Measures for Compensation of Minamata Disease Victims and Resolving the Issue of Minamata Disease (2009) (hereafter the Special Measures) was enacted with the hopeful intent to give satisfactory closure to the lingering compensation, political and social problems of Minamata disease. This section explains how this law was established, and its implementation.

Background to the formulation of the Special Measures

By 2008, three years had passed since the Supreme Court decision in the Kansai Lawsuit without any signs of a breakthrough; hence, the situation was becoming more and more chaotic. It was not possible to resolve the situation by one-off payments and other benefits since the situation had reached deadlock. The situation changed according to the following sequence of events.

To break the deadlock, the government team introduced a new policy by incorporating the perspectives of all parties involved and thereby gaining their respect. As a result, the new policy framework could tackle these issues collectively and thereby achieve real progress. This approach involved new legislation and then undertaking consistent multi-directional initiatives through administrative measures based on it. Specifically, the legislation:

- clearly recognized new patients seeking relief and receiving one-off payments (although complications may have existed in reality) and cemented policy for the swift provision of relief for these victims, which was limited to a fixed period;
- created a financially secure subsidiary (now called JNC) under a Chisso-affiliated holding company that was responsible for the payment of compensation for Minamata disease, and established a number of special legislative regulations for

corporate reconstruction that included measures for actively enabling Chisso to profit from investment;
- specified the process through which the Chisso Corporation could manage its liability for damages, which included new one-off payments in addition to ongoing payments of compensation; and
- established express provisions for encouraging stakeholders, including the national government, to make ongoing commitments to improve regional prosperity and regenerate the regional economy.

The aim of the act was to apply the unique weight of legislation to cement policy principles that, in terms of content, included every expectation and interest of its stakeholders (victims, other local citizens, and Chisso Corporation). The government developed a policy aiming at:

- understanding and addressing the needs of a diverse range of regional stakeholders that extended beyond the victims of the pollution;
- 'visualizing' the overall change in the regional society through the combined efforts of individual stakeholders by striking a balance between parallel measures to: compensate victims; enable the thorough execution of compensation payments by the polluting company; counterbalance the failure of national and prefectural governments to prevent the spread of damage; increase the prosperity of the region; and stimulate the regional economy, so that no individual measure is prioritized over another;
- establishing suitable mechanisms that enabled bottom-up activities and stakeholder initiatives to be incorporated and assigned special status; and
- gathering a wide range of opinions related to the intentions and needs of stakeholders.

Consultation between the ruling and opposition parties also began when the opposition party (the Democratic Party) proposed a bill with the same effect. At that time, due to the impending general election, all parties were aware that it would be politically unacceptable to neglect the issue during the run-up to the election and when forming the new cabinet. Coordination with opposition parties (the Democratic Party, which went on to win the election, and the People's New Party) concerning the text of the law also progressed rapidly. In particular, the law extensively incorporated the opinions of the Democratic Party, expected to be the next ruling party.

The Special Measures were finally adopted in July 2009.

Interaction between the Special Measures and the 'mammoth lawsuit'

The various lawsuits taking place in the district courts of Kumamoto, Tokyo, Osaka and Niigata had collectively become known as the 'mammoth lawsuit'.[2] The plaintiff group involved in the lawsuit, the Shiranui [Bay] Patients' Association, had expressed opposition to the Special Measures when the ruling and leading opposition parties were in the process of considering them. This opposition stemmed from the fear that, ultimately, the polluting company would escape liability following the division of the Chisso

Corporation into one financially secure company and one company to take responsibility for the payment of reparations, if the latter were allowed to dissolve.

However, the ruling and opposition parties also considered this issue during the adjustment process. The cash flow needed for the payments of one-off compensation to the newly designated victims can be secured by adjusting the yearly reimbursement from Chisso to Kumamoto Prefecture for underwriting the debt. Chisso is expected to complete the lump-sum compensations without selling the stock of their new affiliate company, JNC (for Japan New Chisso), but will thereafter liquidate the stock to reimburse their debt. Article 13 of the Special Measures provides that the approval of the minister of the environment is required well before the liquidation of the affiliate. In this way, the Special Measures were prepared so as to counter this concern and secure a funding mechanism to enable even victims who were not involved in the mammoth lawsuit to receive compensation swiftly.

The No More Minamatas Lawsuit sought to secure relief for all victims, including those who were not involved in the lawsuit itself; however, since the Special Measures enabled these victims to receive relief, from September 2009, both the plaintiffs and defendants expressed agreement to an out-of-court settlement.

In January 2010, the Kumamoto District Court recommended a settlement in light of the commitments of the plaintiffs and the defendants. Both plaintiffs and defendants then agreed to the settlement and cemented the details through discussions, while the court oversaw the progress.

The discussions served to establish conditions under which the agreement could be reached regarding methods used to determine requirements for mercury exposure, the scope of symptoms included in the one-off payments, and the methods used to determine these symptoms. For example, all parties reached agreement that the scope of symptoms required for one-off-payments – an aspect of the trial that had attracted a significant amount of attention – would include all victims with methylmercury poisoning as defined in the Supreme Court decision. However, the parties were unable to reach a compromise on the financial aspects of the settlement, such as the individual amounts received as one-off payments and the one-off payment to reimburse the plaintiff group for expenses it incurred through its activities, without help.

The entire negotiating process, including everything the two parties agreed on, was provided to the Kumamoto District Court in interim reports. Therefore, the court recommended a settlement policy based on the requests of the plaintiffs and defendants. On 15 March 2010, having received these requests, the Kumamoto District Court revealed its recommendations concerning the financial intricacies of the settlement in addition to the scope of the symptoms required for one-off-payments and the formalities for determining these symptoms. The recommendation included the payment of a one-off sum of 2.1 million yen (about USD 23,000) to each victim, presumed to be halfway between the payments insisted on by the opposing parties. Both the plaintiffs and the defendants accepted this recommendation. On 29 March 2010, the basic agreement of the settlement was established. This set a precedent for identical lawsuits active in other courts in Japan, which were settled quickly on the same terms.

In addition to the one-off payments, this settlement included promises related to the implementation of various administrative measures, difficult to achieve through a judicial decision. The effects of these measures also extended beyond the plaintiffs to include other potential victims.

For example, the measures were based on the cooperation and participation of regional stakeholders, including the plaintiffs involved in the lawsuit, and aimed to objectively reveal the health effects of methylmercury through research based on the latest medical knowledge and health examinations and guidance for those who felt concerned. These settlement terms were the result of a desire for health maintenance and improved welfare, which was not only expressed by the plaintiffs, but also shared by local residents.

The process of establishing policy for initiatives following the settlement of the mammoth lawsuit and differences from the past administrative style

Parallel to settling with the plaintiffs, the Japanese government examined methods for determining the recipients and the total amount for one-off payments, by holding local briefing sessions, engaging in direct dialogue with local residents, and conducting question-and-answer sessions, while hearing the opinions of other patient groups and autonomous bodies. While incorporating the local dialogue and investigation, the government decided to refer to the No More Minamatas settlement to determine the symptoms to consider in determining who would receive relief, and the size of the one-off payments.

On 16 April 2010, the cabinet approved the relief policy, which contained detailed information about the strategy to enhance regional welfare and achieve regional recovery (see Ministry of Environment 2011 for information about the basic policy). The above process can be summarized as following three stages:

- Legislation provided a general framework for initiatives involving all victims, including those who were involved in the lawsuit and those who were not.
- Settlement was reached following agreement with the plaintiffs based on detailed conditions, including principles governing the execution of regular administrative measures.
- The cabinet approved policy providing administrative measures to apply the content of the settlement and the results of discussions with local stakeholders to victims not involved in the lawsuit.

This procedure established a policy aimed at resolving the long-term deadlock. Previously, Minamata disease policy had been determined by the government and opposition parties. A distinctive characteristic of the new policy was that its design and approval progressed through opinion exchange and adjustment mechanisms that reached beyond Diet members to directly involve an extremely diverse range of stakeholders, including the team of lawyers and patient groups involved in the lawsuit, local governments, and local community and economic organizations.

Moreover, when compared to Minamata policy of the past, the implementation process of the Special Measures went beyond establishing measures for one-off relief payments as quickly as possible to addressing the benefits of welfare and regional economic recovery. It engaged a wide range of stakeholders involved in the

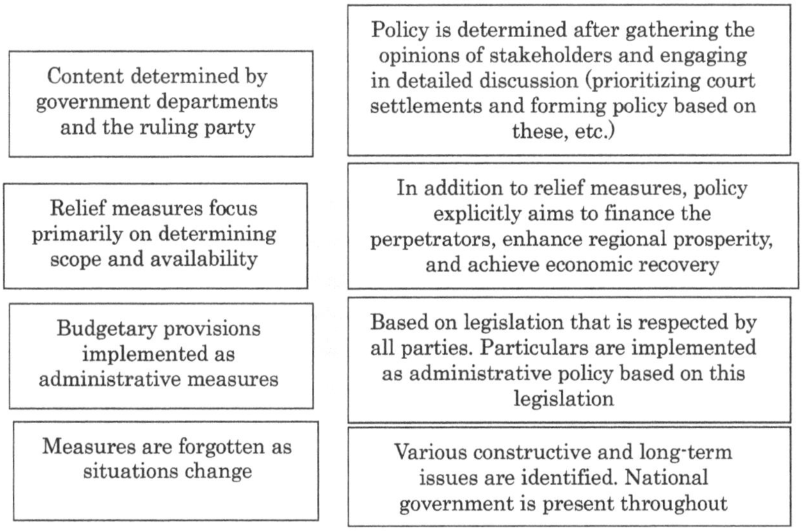

Figure 2. Differences between the approach of past initiatives and that of the Act on Special Measures for Compensation of Minamata Disease Victims and Resolving the Issue of Minamata Disease.

enhancement of these aspects. Here, it may be useful to refer to Figure 2, which offers a simplified comparison of the approach of the Special Measures and past approaches to the problem of Minamata disease.

The policy design evolved within the region itself, as a result of the long-term impoverishment and conflict that occurred in the region. The mayor of the local government adopted the approach to end the prolonged suffering caused by regional division and friction,[3] and it had already been established as a point of reference for the 'attitude' and 'behaviour' required when addressing the problems in the region.

The approach emphasized what came to be known as *moyai-naoshi* ('re-knotting' = mending relations), where all those involved in the regional effort show respect for each other, having built a firm understanding of the feelings of other local stakeholders. This included the victims, who, in addition to the anxiety caused by their own health problems, had experienced prolonged emotional turmoil, swayed by social rule changes regarding where to draw the line in terms of how much relief victims should receive and whether this should be determined on a case-by-case basis. The approach also emphasized the feelings of regular citizens, who expected that, while victims should be treated kindly, the perpetrator, Chisso Corporation, should strengthen its financial standing.

When attempting to counteract damage and achieve regional recovery in regions that have incurred significant damage due to pollution, it is beyond doubt that policy makers must implement excellent policy initiatives that foster regional economic revitalization and enhance regional welfare, in addition to providing suitable compensation. However, in order to connect these initiatives to tangible compromise and recovery in the region, members of the regional society, the polluting company, and national and local governments must think and behave constructively in a way that considers other parties, lend their support to concrete efforts, and continuously improve their behaviour.

The real key to success lies in understanding and accepting others. Although this may seem obvious, it is the lesson learned by those who have been involved in the tragic pollution saga of Minamata disease for over 50 years. I believe that this emotional understanding was the vital factor that enabled us to conduct the complex and time-consuming opinion exchange and coordination sessions mentioned above and forge connections to a dynamic problem-solving process.

Interim evaluation of the Special Measures

This section also offers a novel perspective. Here, I introduce movements that have occurred since the relief measures related to the Special Measures were approved by the cabinet and discuss whether it is possible to conclude that a self-sustained process of development towards the resolution of the Minamata problem is in progress.

Outline of the overall state of progress

Local initiatives began under the new framework in the spring of 2010 in both the Minamata and Niigata regions. It is important to note that not all of the lawsuits have been resolved, since legal proceedings seeking compensation from the national government, in addition to the polluting company, are still in progress in Kumamoto and Niigata district courts. Elsewhere, some parties have expressed the opinion that the relief measures, which offer small one-off payments for victims with a wide range of symptoms, should be given permanent, rather than provisional, status.

Since the introduction of the new framework, local measures that seek to counteract damage through relief and compensation have progressed steadily, as well as parallel measures that seek to hold the Chisso Corporation firmly responsible for the pollution.

Moreover, it is vitally important that elements explicitly written into the policy of the relief measures are established in addition to measures concerning damage recovery – in other words, measures that seek to secure welfare for ageing victims and victims of foetal Minamata disease and regional development measures that seek to revitalize and invigorate a region impoverished by Minamata disease.

How was the contentious issue of the application deadline for relief measures received?

As described in the preceding paragraphs, an extremely wide range of initiatives involving a diverse range of stakeholders are progressing steadily in accordance with cabinet-approved policy based on the Special Measures. Here I focus on two of these areas, the application deadline for relief measures and the strengthening of Chisso Corporation's financial standing, and consider the resilience of the process in which stakeholders coevolve through their joint efforts to solve the problem of Minamata disease. In a sense, even though these matters have been explained, from the perspective of the victims it appears as though the perpetrator stands to benefit while they do not.

Since the measures taken through the relief policy called for swift action, people were aware that the application period would soon be brought to a close. However, the policy did not establish a specific deadline; it opted to set the deadline at a later date, following consultation with patient groups and other stakeholders.

The Supreme Court decision of 2004 made the distinction between symptoms of methylmercury exposure that emerged during or immediately after the period of exposure, and other similar symptoms. Thus, although it is important to complete one-off compensation payments and payments for medical expenses swiftly, given that similar symptoms will increase with age in any population cohort, the act required that the availability of the publicly funded system have a time limit.[4] However, although all parties were aware of the temporary nature of the measures during the settlement talks, the issue of a specific time limit was left undecided.

Since the plaintiff group and the team of lawyers involved in the mammoth lawsuit intended to secure aid not only for the plaintiffs but also for other victims who were not involved in the lawsuit, considering the increasing difficulty associated with being identified as a victim, these parties stood to benefit from pushing this time limit out as far as possible. In addition, some experts believed that the government should add some of the distinguishing symptoms of methylmercury poisoning stipulated by the Supreme Court to the symptoms eligible for compensation according to the government compensation system – by incorporating these symptoms in the Compensation Law – to eliminate the differences between the court decision and the Special Measures. Moreover, these experts tended to argue that the 'Special Measures' should be permanent. As a result, questions concerning the timing of the deadline, and the process through which it should be established, became major outstanding issues for the national government, local stakeholders, and other parties involved in the debate.

Applications for relief measures began 1 May 2010, coinciding with the yearly memorial ceremony to mark the official discovery of Minamata disease. The prime minister attended the ceremony, where he offered an apology for the government's failure to prevent the spread of damage and affirmed the policy for providing swift relief, enhancing welfare, and achieving regional recovery based on the Special Measures. The prime minister also vowed to contribute to international mercury regulations. It was the first time that a Japanese prime minister had attended such an event. In this way, the application process began on a positive note in a kind of social ceremony that seemed to mark a turning point.

As stipulated in the Special Measures, 18 months later, at the end of 2011, the minister of the environment heard the opinions of a number of patient groups. Since the issue of the application deadline had been sufficiently disseminated throughout the groups, some parties argued for the immediate closure of the application period, whereas others called for the measures to be made available permanently. The victims' group that had been operating the longest and occupied a pivotal position agreed to the idea of establishing an application deadline. The minister considered the situation carefully, and in February 2012 announced that those who believed they were victims of the disease should apply by 31 July 2012.

A number of newspapers and other media sources argued that the decision was premature (Mainichi Shimbun, 2012), and several patient groups organized protests. At the same time, the decision encouraged patient groups to double their efforts within the given framework. Influential patient groups obtained the cooperation of doctors employed by private medical institutes in the area to ensure that potential victims were identified within the application period and provided group managed medical examinations and assistance with applications. The Ministry of the Environment and local governments in the region also made efforts to disseminate information regarding the deadline. As a result, 46,248 applications were made within the 26-month application period in Kumamoto and Kagoshima Prefectures, 14,087 of them in six months following the announcement of the deadline (Kumamoto Prefecture, 2012).[5]

Due to the large number of applications, the prefectures and cities in the region concluded that the objectives for relief provided by the Special Measures have largely been achieved[6].

In light of the process described above, it is possible to conclude that although a considerable amount of dissatisfaction arose, local stakeholders accepted and overcame the closure of the application period for relief measures relatively peacefully.

How was the strengthening of Chisso Corporation's financial standing and the decision to divide the company received?

Chisso Corporation's financial burden had soared due to the new relief measures. In March 2011, the company reached a settlement with the plaintiff group in the mammoth lawsuit and another settlement with patient groups not involved in the lawsuit. These settlements required Chisso Corporation to issue one-off payments of more than 3.6 billion yen (USD 45 million) to the plaintiff group, and 3.1 billion yen (USD 38 million) to patient groups who were not parties in the suit. Having completed the one-off group payments, Chisso Corporation is currently continuing to issue one-off payments to individual victims who do not belong to these groups. Assuming that one-off payments are made to 30,000 victims, the total amount paid would be approximately 60 billion yen (ca. USD 600 million). When the sum of the group payments is included in the total, the total financial burden will certainly run into tens of billions of yen, not a figure that can be sustained by a single company.

To make these payments, the Chisso Corporation took out long-term loans from a government fund. The minister of the environment has asked that banks cooperate by enabling the Chisso Corporation to fulfil its financial obligation to victims. However, Chisso must repay the money it has borrowed.

Excluding its extraordinary loss, Chisso's ordinary profit continued to rise, from around 10 billion yen (USD 94 million) in 2000 to 20 billion yen (USD 220 million) in 2010. However, as the company has carried forward an enormous extraordinary loss due to its research and development and capital investment, resulting in a capital deficit of over 100 billion yen (USD 1 billion), the company is not optimistic about its future.

There is also the danger that by incurring this new debt, Chisso will be unable to continue to pay the medical-care allowances that it promised to pay to victims in the future through the Compensation Law and other agreements. To prevent this, we decided to strengthen the financial standing of Chisso through the use of several special provisions in the Special Measures, in a manner which could only be achieved through legislative instruments, among them the previously discussed restructuring of the company into two components.

To reform the management structure of the company, the Special Measures followed examples in Japan's legislation for corporate restructuring, stipulating detailed provisions that established various special taxation measures. For example, typical enterprises cannot reduce taxable income by deducting losses incurred more than seven years earlier from profits. However, the law exempted companies who have incurred extraordinary losses through Minamata disease compensation payments from this restriction. We made this decision on the basis that allocating more income to victims was preferable to paying more tax to the national government.

The Special Measures secured the performance of Minamata disease obligations through mechanisms such as those discussed above. However, the measures were poorly received by patient groups and in various journalistic and academic circles, who sought to hold Chisso responsible as much as possible (Yokemoto, 2010).

Since a large number of loss-making corporations conduct corporate reconstruction through 'corporate separation' by spinning off profitable departments as new companies while closing down loss-making departments, critics feared that the management reform described above would enable the liquidation of the compensation-obligated Chisso, thus preventing the company from being brought to trial in the future. This fear stemmed from a proposal made by the Liberal Democratic Party's Minamata disease subcommittee, long before the Special Measures, that focused on the corporate separation of Chisso Corporation and the liquidation of its parent company. Therefore, the criticism was not directed at the corporate structure overhaul itself, but rather at possible events further down the line.

The final Special Measures, which were a compromise between the ruling and opposition parties, responded to this fear with the following stipulations. (1) The Chisso Corporation will be required to submit to the minister of the environment for approval a business restructuring plan which demonstrates how the company will achieve unhindered performance of its compensation obligations when transferring to the core company–holding company system. (2) Although the rights and profits of shareholders shall not be compromised in the transfer to the core–holding company system, by way of caution, the replacement permission of the court is required instead of that of the general meeting of stockholders. (3) The prior approval of the minister of the environment is required before the holding company is permitted to profit from the sale of core-company stock. (4) Profit from the sale of stock is primarily used to secure funds for compensating future claims brought by victims of Minamata disease. The law did not forbid the company from declaring bankruptcy, since this is legally impossible. Provision (3) also stipulated that the holding company could not simply list stock when stock market conditions were favourable, until compensation obligations had been completed. Thus, the minister of the environment would not grant this permission for the foreseeable future.

In addition to its existing obligations, Chisso incurred a substantial amount of new debt owing to the March 2010 settlement, through which it was made to pay one-off payments totalling almost 7 billion yen (ca. USD 70 million) to seven patient groups, and the commencement of applications for one-off payments to regular victims from May 2010. In light of this new debt, in June 2010, Chisso appealed to the minister of the environment for designation as a polluting company undergoing business restructuring. This designation was granted in July 2010.

In light of increased criticism from some patient groups and journalistic sources, the Ministry of the Environment invited representatives from Chisso to take part in public briefings to explain the details of the restructuring of business operations and its relation to future compensation payments as well as hearing the opinions of the company on this restructuring and the recovery of the local economy. Chisso even published a draft of its business restructuring plan, demonstrating that the company was willing to take on board opinions of other parties. The company made the necessary internal changes and submitted its plan in November 2011 for the approval of the minister of the environment.

Furthermore, the Ministry of the Environment accepted written opinions, and the minister deliberated the plan at great length while considering these written opinions, as well as opinions expressing criticism and concern, put forward at the local briefings. As a result, on 15 December 2010, the minister approved the restructuring plan.

Following the approval, in February 2011, replacement permission for the business transfer was granted by Osaka District Court, where Chisso is registered. Then in March 2011 the transfer was accepted by the board of directors, and the new 'core company' officially commenced operations (rebranded as JNC) on 1 April. During this process, a group of shareholders also applied for an injunction to prevent the restructuring on the grounds that it would result in a loss of profits for shareholders. However, Chisso won this lawsuit, which was the only legal countermeasure against the decision.

On the one hand, a great number of victims and residents felt calm and had high expectations for the healthy future of Chisso in terms of securing compensation and achieving regional recovery. However, some victims found it hard to accept that Chisso Corporation would be allowed to do business as usual while they continued to suffer (Nagamatsu, 2012). Despite the existence of complexly ambivalent feelings such as these, it is no exaggeration to say that local friction had been minimized by the considerate efforts of Chisso to engage in public exchanges of opinion, as well as the persistence of various patient groups in dealing with the situation rationally.

Conclusions

The resonances of Minamata disease have to end one day. Yet they must never be forgotten. We must learn from the problem of Minamata disease and bequeath the wisdom gained through our experiences to future generations, to ensure that a similar problem does not occur again, or if it does, we will know how to deal with it without making the problem worse or prolonging the agony. The problem of Minamata disease should be resolved in a way that transcends its negative legacy. This can be accomplished by connecting the lesson of the environmental destruction and the experience of the recovery effort to regional development based on shared ideals and on achievements that can be seen and felt in concrete terms.

When we envision such a future, it is undeniable that the region still has a long way to go. But with the implementation of further relief measures for several thousands of victims, those involved in the problem of Minamata disease have already overcome a number of great trials through a process of compromise and concession that, despite generating a certain degree of discord along the way, never reached the point of impasse (Nagamatsu, 2012). The first step of this process has also been achieved owing to the response of Chisso, which has been constructive and compromising, rather than aggressive. The victims also responded with a calm and constructive attitude toward solving the problem. As a result, an atmosphere of cooperation has been established in the local area. In terms of practical efforts, environmental community projects that could not have been implemented without the knowledge and participation of local citizens and entrepreneurs are already showing signs of success.

Then can matters be expected to progress smoothly in the future? It cannot be denied that the recent progress was partly due to the fact that justice was served, when tens of thousands of victims were simultaneously awarded swift relief. A major problem related to this aspect of justice concerns the possible emergence of a small number of individuals who were eligible for relief but were not able to receive it. In this case, would the strength cultivated through the processes to date enable those involved to treat these individuals warm-heartedly, without reverting to the insurmountable conflict of the past?

The aftermath of the tragedy of Minamata disease is also one of evolution towards understanding and social learning. As people, we must further develop our wisdom and behaviour to create a society that coexists with nature.

Notes

1. In Niigata, the other site of Minamata disease in Japan, 705 received compensation, 152 of whom were still alive, as of March 2017.
2. Kumamoto and Niigata were sites of poisoning. Some victims subsequently moved to Tokyo (Kanto) and Osaka (Kansai) and had standing in the courts there.
3. M. Yoshii, address at the third ceremony in memory of Minamata disease victims, May 1994.
4. I can only speculate on the reasons for this provision, but there would seem to be at least three. (1) The Supreme Court provided that recognized symptoms should be evidenced within 5 years of exposure to organic mercury. (2) It is difficult to distinguish symptoms associated with ageing from those associated with Minamata disease syndrome. (3) Compensation, by its nature, should be done as promptly as possible.
5. Figures refer to Kumamoto. Applications in Niigata linked to Minamata disease are not included.
6. Press conference, Governor Kabashima of Kumamoto, 25 July 2012.

Acknowledgments

This article is an adaptation in English of Chapter 5 in Kobayashi (2016), reproduced with the kind permission of the publisher, Shimizukoubundo-Shobou. The author thanks Dr James Nickum and his colleagues for their valuable feedback, comments and suggestions.

Disclosure statement

No potential conflict of interest was reported by the authors.

Funding

The original paper was partially supported by the Global Environmental System Leaders Program of Keio University.

References

George, T. S. (2002). *Minamata: Pollution and the struggle for democracy in postwar Japan.* Cambridge, MA: Harvard University Press.

Harada, M. (1972). *Minamata byou* [Minamata disease]. Tokyo: Iwanami Shoten.

Hashimoto, M. (1988). *Shishi kankyou gyosei* [History of my environmental administration]. Tokyo: Asahi Simbun Sha.

Kobayashi, H. (ed). (1994). *Kankyo kigyou ron josetsu* [Introduction of environmentally friendly enterprises]. Tokyo: Godo Shuppan.

Kobayashi, H. (2016). *Chikyu no yoi ichibu ni naru: kankyou kyousei keizai e no ikougaku* [Transitioning to an ecological economy as a positive part of thriving planet]. Tokyo: Shimizukoubundou.

Kojima, T. (1996). *Minamata byou no seiji kaiketsu* [Political settlement of the Minamata issue]. *Jurist, 1088*, 5–11.

Kumamoto Prefecture. (2012, August 30). *Actual number of petitions for special relief based on the new law* Kumamoto: Kumamoto Prefectural Government.

Mainichi Shimbun. (2012, July 17). Senzai kanja wo kirisuteruna [Don't cut out potential victims]. Editorial. *Mainichi Shimbun.*

Ministry of the Environment. (2011). *Minamata byou no kyoukun to nippon no suigin taisaku* [Lessons of Minamata disease and Japan's measures to cope with mercury]. Tokyo: Author.

Ministry of the Environment. (2017). *Kankyo no joukyo houkoku heisei 28* [Annual report on the state of environment in 2016]. Tokyo: Author.

Nagamatsu, T. (2012). *Kankyo higai no gabanansu- Minamata kara fukushima he* [Governance on Pollution Damages from Minamata to Fukushima]. Tokyo: Seibundo.

No-more Minamata Lawsuit Archive Commission. (2012). *No more Minamata soshou tatakai no kiseki* [Track of litigation of No-more Minamata case]. Tokyo: Nippon Hyouronsha.

Sakamoto, M., Murata, K., Tsuruta, K., Miyamoto, K., & Akagi, H. (2010). Retrospective study on temporal and regional variations of metylemercury concentrations in preserved umbilical cords collected from inhabitants of Minamata area, Japan. *Ecotoxiciology and Environmental Safety, 73*, 1144–1149.

Study group on social-scientific aspect of Minamata Issue (edition). (1999). *Minamata byou nikansuru shakaikagakuteki kenkyukai houkokusho* [Report of study group on social-scientific aspect of Minamata isuue]. Minamata: National Institute of Minamata Disease.

Yokemoto, S. (2010). Minamata byou hoshou kyusai no yukue [Anxious future of compensation and reliefs of Minamata disease]. *Kankyo to Kougai 40*(2), 59–63.

Mercury pollution in Colombia: challenges to reduce the use of mercury in artisanal and small-scale gold mining in the light of the Minamata Convention

Farith A. Diaz, Lynn E. Katz and Desmond F. Lawler

ABSTRACT

The Minamata Convention on Mercury, which entered into full force in August 2017, was signed by more than 100 countries, including several where artisanal and small-scale gold mining is practised, such as Brazil, China, Ecuador, Ghana, Peru and Colombia. Focused on Colombia, the country that likely faces the most difficult challenges to implement this convention, this paper discusses the extent of mercury pollution in the water sources of the country and the barriers Colombia must overcome to comply with the Minamata Convention. Many of these same barriers are also present in other countries impacted by artisanal and small-scale gold mining.

Introduction

The term 'artisanal miners' 'encompasses all small, medium, large, informal, legal and illegal miners who use rudimentary processes to extract gold from secondary and primary ores bodies' (Veiga et al., 1999, p. 1). According to the United Nations Environment Programme (UNEP), artisanal and small-scale gold mining, an activity extensively practised in developing countries, releases approximately 640–1350 tonnes of mercury per year to the environment (Telmer & Veiga, 2009).

Artisanal and small-scale gold mining contributes about 37% of the total annual anthropogenic mercury emissions to the atmosphere (Kirby et al., 2013). Artisanal and small-scale gold mining is intensively practised in countries in Africa, Asia and South America; and most of the mercury releases occur in South America (53%), East and Southeast Asia (36%), and Sub-Saharan Africa (8%) (UN Environment, 2019). Mercury used in artisanal and small-scale gold mining pollutes soil, freshwater sources, living organisms and communities around and downstream of gold mining sites (Kirby et al., 2013).

Despite being a hazardous and illegal practice in many countries, the use of mercury for gold amalgamation is still the most common technique used in artisanal and small-scale gold mining because mercury is traditionally used in many countries, is easy-to-use, is readily available and relatively inexpensive (Telmer & Stapper, 2012; Veiga & Baker,

Supplementary data for this article can be accessed at https://doi.org/10.1080/02508060.2020.1845936

2004). Nevertheless, the United Nations and other international organizations have made remarkable efforts to minimize the use of mercury in artisanal and small-scale gold mining in developing countries. One of these efforts was the Global Mercury Project (GMP), an initiative created in 2002 to 'help demonstrate ways of overcoming barriers to the adoption of best practices, waste minimization strategies, and pollution prevention measures that limit contamination of the international waters' (Veiga & Baker, 2004, p. i). The latest effort to control the use of mercury is the Minamata Convention on Mercury, which entered into full force in August 2017. Although it does not prohibit the use of mercury in artisanal and small-scale gold mining, it aims to reduce its use and where feasible to eliminate it.

Although metallic mercury has been used in artisanal and small-scale gold mining for around 450 years to separate gold particles during the amalgamation process (Veiga et al., 2006), environmental mercury pollution, as a consequence of artisanal and small-scale gold mining, only has been studied in the last decades. Communities located in regions affected by artisanal and small-scale gold mining use rivers and streams for multiple purposes, and even build water-intake structures to produce potable water. Despite the effort of multilateral organizations and local governments to reduce the use of mercury in artisanal and small-scale gold mining, new challenges are emerging that thwart the efforts to regulate this ancestral practice in developing countries such as Colombia, which is the focus of this paper.

After more than 50 years of an internal armed conflict and a recent peace agreement signed between the government and Revolutionary Armed Forces of Colombia (Farc) rebels (Presidencia, 2016), no other country is likely to face more difficult challenges to reduce or eliminate the use of mercury in artisanal and small-scale gold mining than Colombia. The involvement of organized crime and illegal armed groups (right-wing paramilitarists and leftist-guerrilla) in the gold mining industry (Redacción Negocios y Economía, 2016; Verité, 2016), the remote geographical regions where most of the informal and illegal miners operate (The Global Initiative, 2016; Verité, 2016), the illegal trade of metallic mercury (UN Environment, 2017; Unidad Investigativa, 2017), the already extensive damage caused by artisanal and small-scale gold mining throughout the country (Diaz Arriaga, 2014; IDEAM, 2015), and the number of illegal and informal miners for whom artisanal and small-scale gold mining is the primary source of income for their families (The Global Initiative, 2016) all pose serious challenges to a problem that has existed for centuries and will likely will continue for a long time.

Mercury pollution in Colombia as a result of artisanal and small-scale gold mining: a legacy that will last for decades

Until 1937, Colombia was the largest gold producer in the Americas (Burton, 2011). Today, Colombia is the fifth largest gold producer in Latin America, behind Peru, Mexico, Brazil and Argentina (Thomson Reuters, 2018). As Figure 1(a) shows, despite the drop in the past two years, the production of gold in Colombia during the last 11 years (2008–18) has been > 30 tonnes per year. Most of the gold produced in Colombia comes from six (out of 32) states in the country: Antioquia (46.1%), Chocó (28.2%), Bolivar (7.4%), Nariño (5.8%), Cauca (5.2%) and Caldas (3.4%) (Figure 1b), and it is mostly extracted from alluvial deposits, mainly in the states of Chocó, Antioquia and Nariño (PNUMA & MADS, 2012). As expected, many tonnes of mercury are also

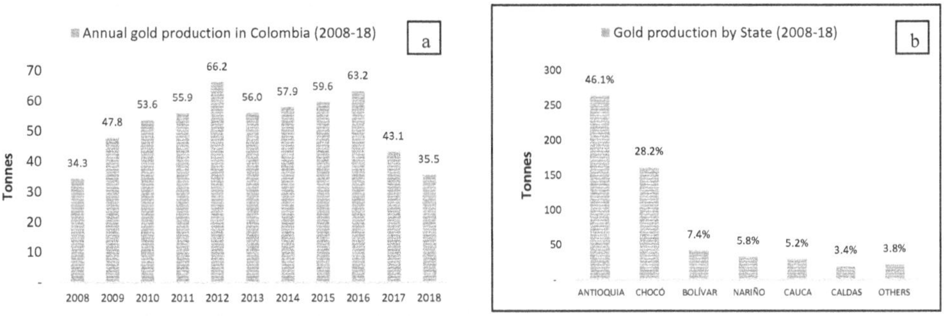

Figure 1. (a) Annual gold production in Colombia in the last 11 years; and (b) gold production by state during the same period. Source: Unidad de Planeación Minero Energética (UPME) (2018).

dumped in those states where gold mining is particularly intensive. Previous studies reported that the states with the highest use of mercury are Bolivar, Chocó and Antioquia, with 304, 195 and 170 tonnes per year, respectively (IDEAM, 2015; PNUMA & MADS, 2012).

In 2014, a governmental study determined that 193 tonnes of mercury per year were used in the gold mining industry in Colombia; from this, 105 tonnes were used in legal mining and 88 tonnes in illegal mining (MinEnergía, ANM, UPME & SGC, 2016; Universidad de Córdoba & MinEnergía, 2014). The negative impacts of both legal and illegal gold exploitation are evident in Colombian territory. Figure 2(a) shows the regions of the country where legal and illegal gold mining is taking place. Gold production in Colombia is concentrated along the Pacific Region and some areas of the Andean Region. Figure 2(b) shows that the same regions from where gold is extracted are also those where most of the metallic mercury is dumped. According to the Institute of Hydrology, Meteorology and Environmental Studies (IDEAM in Spanish), a Colombian government agency, 205 tonnes of mercury were discharged to the environment in 2012 in 179 municipalities (IDEAM, 2015). Out of the 205 tonnes of Hg discharged, 27.5% was dumped during the extraction of silver and 72.5% during the extraction of gold (IDEAM, 2015).

Colombia provides a clear example of the devastating and long-lasting effects of illegal and legal gold mining in natural environments. More than 200 rivers, 17 out of 32 states of the country and more than 300 municipalities are impacted by artisanal and small-scale gold mining (Cuevas Guarnizo, 2015; Diaz Arriaga, 2014; El Heraldo, 2013; Henao & Gonzalez Espinosa, 2016; Redacción El Tiempo, 2015). Figure 3 shows the destruction of natural ecosystems in the states of Chocó and Antioquia (Figure 3a–c), and a water intake structure in Ayapel Swamp (Córdoba) (Figure 3d) where atypical mercury concentrations have been detected (Redacción Vivir, 2013).

In Colombia, it was estimated that more than 200 rivers, out of a total of 1150 throughout the country, cross heavily mercury polluted ecosystems (Unidad Investigativa, 2017). As part of this investigation, the primary author collected water samples in June 2017 from five of those rivers, one swamp and one creek to determine the levels of mercury downstream of different active gold mining districts. The water bodies selected were the Magdalena River (Bolivar), the Nechí River (Antioquia), the Atrato River (Chocó), the Quito river (Chocó), the Tutunendo River (Chocó), the Ayapel Swamp (Córdoba) and the Villa Creek (Antioquia).

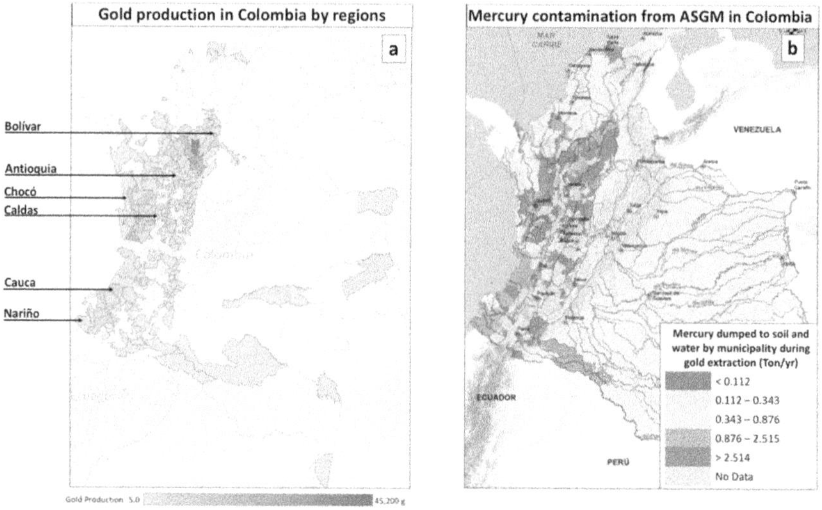

Figure 2. (a) Gold production in Colombia by regions; and (b) mercury contamination as a result of legal and illegal mining. Sources: (a): Subdirección de Minería-UPME (Agencia Nacional de Minería-ANM, September 2018) (UPME, 2018); and (b) Instituto de Hidrología, Meteorología y Estudios Ambientales (IDEAM) (map version: 1.0, 3 February 2018) (IDEAM, 2015).

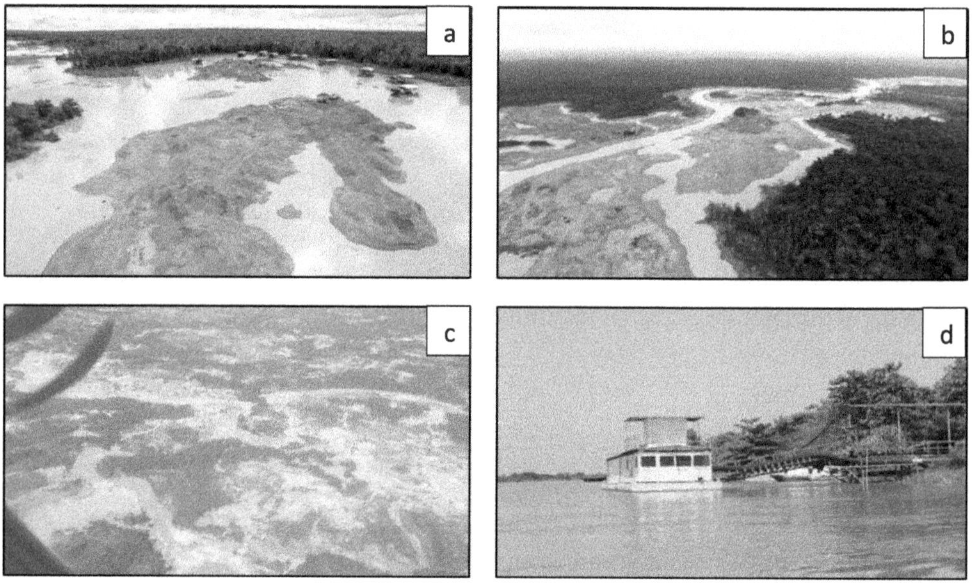

Figure 3. The destruction of natural ecosystems is one of the main consequences of illegal mining in Colombia: (a) illegal dredging operation in the Quito River (Chocó) spotted in June 2017; (b) illegal mining in the Quito River has changed the course of the river and affected transportation, fishing and agriculture activities; (c) similar consequences can be seen in many states of Colombia, as in the case of El Bagre (Antioquia); and (d) water intake for the water treatment plant in Ayapel (Córdoba). Mercury concentrations > 400 ng/L were detected in Ayapel Swamp during this study.

For a summary of the procedure, methodology and results of the sampling, see the supplemental data online.

Gold production and mercury used in gold mining around the world

In the Americas, mercury has been used for more than 450 years in gold mining to extract gold particles using an amalgamation process (Veiga et al., 2006). As part of the process, the gold–mercury amalgam is separated using retorts where metallic mercury is vaporized with the application of heat (Drace et al., 2012; Esdaile & Chalker, 2018; Gammons et al., 2006; Meech et al., 1998; Telmer et al., 2006). Part of the metallic mercury is lost into rivers and soils during the amalgamation process and tailing discharges, while another fraction is lost to the atmosphere during amalgam burning and gold purification *in situ* or in gold dealers' shops (Esdaile & Chalker, 2018). The amount of mercury in the amalgams varies between 40% and 80% by mass (Esdaile & Chalker, 2018; YTSMedia, 2013). The UNEP's report entitled *Mercury: Time to Act* (Kirby et al., 2013) includes some important data that helps visualize the global magnitude of artisanal and small-scale gold mining:

- Artisanal and small-scale gold mining employs millions of people; more than 10 million people work in artisanal and small-scale gold mining in Africa, Asia and South America.
- Mercury from artisanal and small-scale gold mining is transported long distances downstream in rivers and remains in the environment for decades, even after mining activities cease.
- New technologies that do not involve the use of mercury are available, but socio-economic barriers limit their implementation.

As indicated in Figure 4, Colombia is among the top 21 gold mining countries, along with China, Peru, Indonesia, Sudan, Ghana and Brazil, where mercury pollution seems to be particularly critical. Moreover, these countries, together with the Philippines, Democratic Republic of Congo, Tanzania and Bolivia, are usually listed as the main mercury polluters in the world (Artisanal Gold Council, 2009; Davis, 2018; UNEP, 2017). Previous publications indicate that China is the world's largest mercury polluter, with annual mercury emissions from artisanal and small-scale gold mining between 240 to 650 tonnes per year, followed by Indonesia with between 130 to 160 tonnes per year, and Colombia, in third place, with 50–100 tonnes per year (Artisanal Gold Council, 2009; Cordy et al., 2011).

More than 80% of the gold produced in Colombia comes from illegal mining (Burton, 2011; Redacción Negocios y Economía, 2016; Unidad Investigativa, 2017). A similar trend was observed in Venezuela, where the gold produced illegally was estimated at between 80% and 90% (The Global Initiative, 2016; Verité, 2016). Lower percentages were reported for other countries in Latin America: Ecuador (77%), Bolivia (31%), Peru (28%), Nicaragua (13%), Brazil (10%) and Mexico (9%) (The Global Initiative, 2016; Verité, 2016). As Figure 4 shows, it is clear that a good portion of the global gold production relies on the use of metallic mercury for this purpose. In fact, between 15% and 30% of the world gold's production is

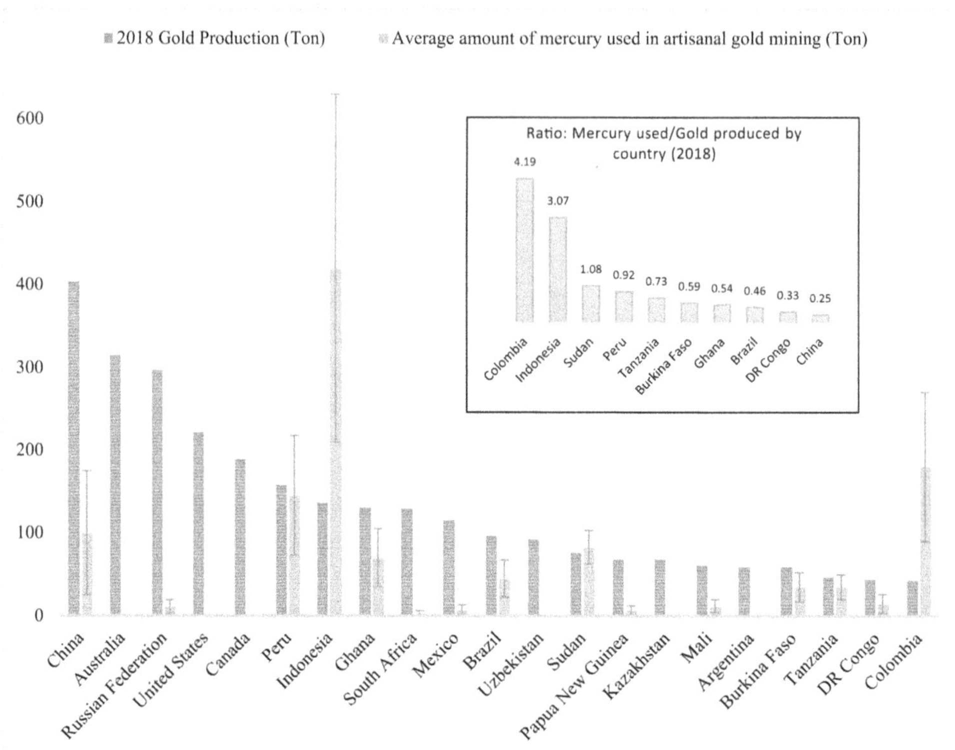

Figure 4. The year 2018 top 21 gold mining countries and average amount of mercury used in artisanal and small-scale gold mining. Sources: UN Environment (2017); and Thomson Reuters (2018).

provided by artisanal and small-scale gold mining (Cordy et al., 2013; Esdaile & Chalker, 2018; Swain et al., 2007).

The upper right insert of Figure 4 shows that in Colombia the average amount of mercury used in gold mining is approximately four times the annual gold production. In fact, Colombia has been considered the world's highest per capita mercury polluter (Cordy et al., 2011). In Indonesia, the average amount of mercury used in artisanal and small-scale gold mining is almost triple the amount of the gold produced by the country in 2018. An additional detail is that Indonesia is ranked seventh in global gold production, while Colombia is ranked 21st. The negative impact of artisanal and small-scale gold mining in the environment and human health are evident in these two countries. as has been reported previously (Cordy et al., 2011; Limbong et al., 2003; Marrugo-Negrete et al., 2008; Olivero-Verbel et al., 2002; Palapa & Maramis, 2015). For the remaining top 21 countries, the ratio mercury used/gold produced is ≤ 1.0.

Regardless, illegal gold mining places Colombia in the top 21 gold-producing countries in the world. The huge disproportion between the amount of gold produced in Colombia and the amount of metallic mercury used to produce it reflect the poor management practices associated with metallic mercury use in the gold industry by the miners and the lack of enforcement of the regulations that have been promulgated to control the use of this pollutant.

The main challenges to reduce the use of mercury in artisanal and small-scale gold mining regions: a study of the Colombian case

The singularities observed in the gold mining industry in Colombia make this country a suitable case scenario to analyse some of the barriers to reduce or eliminate the use of mercury in gold mining and to avoid the degradation of natural ecosystems as a consequence of mining activities. The long tradition of illegal and informal mining, the increasing number of miners whose livelihoods depend on this activity, the large number of municipalities where illegal mining is practised, the illegal trade of mercury, the remote location of many mining districts, the obstacles to legalize miners, and the involvement of illegal armed groups are some of the challenges faced by the Colombian government to control the use of metallic mercury in artisanal and small-scale gold mining. Some of these challenges are addressed in this section.

The presence of illegal armed groups and organized crime in artisanal and small-scale gold mining regions and its involvement in gold production

Despite the efforts made by former Colombian President Juan Manuel Santos to reduce the epidemic violence that has engulfed the country during the last 50 years, the involvement of mafia and illegal armed groups in the already illegal extraction of gold has hampered the attempts of the government to control illegal and informal mining activities (The Global Initiative, 2016). A major complication in the effort to control illegal activities in Colombia, such as gold mining or even coca crops, is that, for several decades, the country has been in the grip of a devastating internal armed conflict that exceeded the response capacity of the state (LeGrand, 2003; Presidencia, 2016). During this time, the state was not able to protect its own citizens, particularly in rural areas dominated by illegal armed groups, let alone control Colombian territory. Besides fighting against leftist rebels, the Colombian armed conflict grew as more actors entered the scene, including right-wing paramilitary groups, drug cartels, and corrupt governmental and military officials (LeGrand, 2003). Years of conflicts left extensive areas of Colombian territory unattended by the state. Even today, many areas are under the control of mafias and illegal armed groups, which are also expanding towards territories once occupied by the Farc rebels (Otis, 2018).

Nevertheless, in the last years, the Colombian army and national police have carried out important military operations against illegal gold miners in remote areas of the country. The Colombian air force also supports the fight against illegal mining activities either by using drones or though military overflights in conflictive artisanal and small-scale gold mining regions. To date, dozens of dredges have been destroyed by the Colombian army, and mining equipment and machinery have been confiscated along the country. These dredges are used for dredging sediments and to perform other mining activities around riverbeds.

The military operations carried out by Colombia's military forces have also exposed the presence of foreigners involved in illegal activities in the country. Citizens from Peru, Venezuela and Brazil have been seized in illegal gold mining districts by the Colombian authorities. The presence of foreigners in gold mining activities was evident during the trip to collect water samples to analyse mercury in several rivers and streams around the country. The porous border between Colombia and its neighbouring countries makes it easy for illegal migrants to cross the border, and for mafias to smuggle arms, machinery, illicit drugs and metallic mercury. Nevertheless, small-scale miners can also be victims of human trafficking and often are seen as criminals rather than as the victims they are (Verité, 2016).

In 2018, Colombia's ranking fell three positions (from 18th to 21st) in the top 21 World Gold Mining Rank. Colombian gold output fell 15.0% in 2018 (Thomson Reuters, 2018); based on data from the Unidad de Planeación Minero Energética (UPME) (Figure 1), the country's gold production fell from 43.1 tonnes (2017) to 35.5 tonnes (2018), which represents a 17.6% decrease (UPME, 2018). This drop could be explained by two factors. First, coca crops in Colombia, particularly in conflict-dominated areas and/or artisanal and small-scale gold mining regions, increased dramatically in the previous year, reaching a historic record of 171,000 ha, a 17% increase compared with 2016 (SIMCI, 2018). Second, the continuous efforts by the government to reduce illegal gold mining activities and the destruction of machinery and seizure of illegal mines was also a factor that played a role in the reduction of gold production (Thomson Reuters, 2018).

The increase in coca crops suggests that inhabitants from rural areas switched from traditional productive crops and illegal gold mining to coca crops despite the more than US $10 billion spent in the last two decades to fight illegal drugs in the country (Otis, 2018). This increase in coca production in turn has been attributed to multiple factors (Llorente & Garzón Vergara, 2018): the re-establishment of drug mafias in areas formerly controlled by Farc rebels, the low levels of eradication of illegal crops, the increasing seizure of cocaine by the Colombian authorities, and the failures in programmes intended to replace illegal crops. In the last years, the Colombian government has been promoting, without success, the aerial spraying of glyphosate to increase the eradication of coca crops.

The illegal trade of metallic mercury in the light of the Minamata Convention

The Minamata Convention on Mercury, a legally binding treaty aiming to protect human health and the environment, became effective on 16 August 2017. So far, 128 of the 193 United Nations member countries have signed this initiative, including several where artisanal and small-scale gold mining is practised, such as Brazil, China, Ecuador, Ghana, Peru, Bolivia and Colombia. This agreement requires that participating countries take several measures to reduce or eliminate the use and release of mercury from artisanal and small-scale gold mining, to control air mercury emissions in industrial activities, to reduce the use of elemental mercury in mass consumption goods such as batteries, cosmetics and lights as well in manufacturing processes such as chlor-alkali production, to address the supply and trade of mercury, and to develop strategies to remediate mercury-contaminated sites (UNEP, 2017).

Today, many mercury-free techniques for gold extraction, which are safer for both miners and the environment, are available. For example, gravimetric concentration methods such as panning, sluicing, shaking tables, spiral concentrators, vortex concentrators and centrifuges have been shown as an alternative for miners in artisanal and small-scale gold mining regions (USEPA, 2018a). Nevertheless, the socioeconomic conditions, the need for easy money, the easy accessibility of metallic mercury and the lower efficiencies compared with mercury-based processes seems to be the main barrier for miners to adopt a more environment-friendly practice that is also non-hazardous to human health. (Telmer & Stapper, 2012).

Some of the countries impacted by artisanal and small-scale gold mining created new regulations to control the use of metallic mercury in industrial activities. Such was the case of Colombia, where Law 1658/2013 was issued to control the use and merchandizing of mercury in industrial and mining activities. This law was fully implemented by 2018 for the mining sector, but it has not really been enforced. In Brazil, Decree 97.507/1989 prohibits the use of mercury and cyanide in artisanal and small-scale gold mining, except when its use is authorized by a competent environmental organization (Sousa et al., 2011). In Ecuador, the mining laws (Ley de Minería) prohibit the use of mercury in artisanal and small-scale gold mining and restrict the use and distribution of metallic mercury in the country (Environmental Regulation No. 003, Official Registry No. 909, 21 March 2013). And in Peru, a multisectoral action plan was approved to implement the Minamata Convention on Mercury (Decree No. 010-2016-MINAM). Among these countries, Brazil seems to be the one with older, stringent and more comprehensive regulations to control artisanal and small-scale gold mining and the use of mercury in this activity (Sousa et al., 2011). At the same time, Brazil is a good example of the difficulties to enforce these environmental regulations considering the vastness of their territory, the complexity of the existing laws, the lack of training on legal issues to miners, and the absence of the government where artisanal and small-scale gold mining is practised (Sousa et al., 2011). Colombia faces similar challenges as Brazil, but additionally the fight against illegal armed groups.

Despite these governmental efforts to reduce the use of mercury in artisanal and small-scale gold mining in Latin America, the historical use of mercury in illegal mining is expected to continue due to the creation of illegal markets for mercury trade that will supply the mercury needed in illegal mining, at least for a few decades until local governments take full control of the supply chain of the metal. Bypassing the mercury controls has proven to be easy in Latin America. In Colombia, the smuggling of mercury is expected to increase, as well as its cost. Approximately 272 kg of mercury, valued at more than US$70,000, were seized by the local police in Antioquia in May 2018 (Restrepo, 2018). Some of the mercury smuggled to the country enters via the Colombia–Ecuador border. The amount of mercury smuggled may surpass 50 tonnes per year (Unidad Investigativa, 2017). In Ecuador, artisanal miners still use mercury despite its prohibition. Most use this metal in their own houses to evade the control of the authorities. For instance, local miners use a secret network to buy mercury from smugglers (Diario Expreso, 2016). In Brazil, it was reported that 99.3% of artisanal miners use mercury and cyanide without legal authorization (Veiga, 2017). Illegal mercury trade has also been reported in West Africa, the Philippines, Malaysia and Indonesia (UN Environment, 2017).

The main strategies for the illegal and undocumented mercury trade seem to involve hiding mercury among other goods, mislabelling the containers that transport mercury and shipping high-quality mercury as cheap product, among others (UN Environment, 2017). This illegal trade in mercury is one of the main challenges faced by developing countries affected by illegal mining activities in the implementation of the Minamata Convention (UN Environment, 2017).

The arduous task to formalize and legalize illegal/informal miners

Although the definition of artisanal and small-scale gold mining includes both informal and illegal miners, a legal differentiation is necessary to avoid the criminalization of informal miners (Redacción Nacional, 2014). Informal or artisanal gold mining has been a sustainable activity practised by many communities worldwide. In those communities, villagers use ancestral traditions to extract gold from rivers, streams and soils manually, using pans, trays or non-mechanical equipment. An informal miner is one who works outside the law but has the option to become legal through appropriate mechanisms to formalize his/her practice (Redacción Nacional, 2014). In contrast, an illegal miner is one whose mining practice is impossible to legalize because either it is associated with criminal groups or operates in prohibited areas (Redacción Nacional, 2014). Consequently, informal and illegal miners operate without mining rights and environmental licences. Under this scenario, miners lack social protection and occupational health and safety training (PNUMA & MADS, 2012). In contrast, formal mining, besides meeting mining, environmental and labour regulations, provides royalty revenues and involves taxation (PNUMA & MADS, 2012).

A report from 2016 estimated that approximately 50,000 people work in illegal mining in Colombia; this figure is only surpassed by Peru (about 100,000 miners) and Brazil (about 75,000 in the Brazilian Amazon basin) in Latin America (The Global Initiative, 2016). Fewer people are involved in illegal mining in countries such as Bolivia (about 45,000), Ecuador (about 10,000) and Venezuela (about 4000 on the border between Colombia and Brazil) (The Global Initiative, 2016).

In Colombia, the term 'mining production unit' (UPM in Spanish) refers to 'a set of tasks, with facilities, operations and/or equipment, that make up an economic and administrative infrastructure, aimed at the exploitation of minerals' (MinEnergía, 2016b, p. 60). Of the 14,357 mining production units identified by the government during the 2010–11 census, only 37% involve ownership of mining rights (or mining title) (MinEnergía, 2012). Of the 4133 mining production units dedicated to gold exploitation in 16 states of the country, only 550 (13.3%) own a mining title and 899 (21.8%) pay royalty revenues (MinEnergía, 2012).

The Colombian government has tried multiple times, without success, to legalize informal miners. Some of the laws created to accomplish these goals are Law 141/1994 (Decree 2636), Law 685/2001, Law 1382/2010, Law 1685/2013 and Law 1753/2015. Such laws provide mechanisms to apply for a mining title, obtain a concession rights contract, delineate areas for mining operation and request the legalization of mining production units (UPMs), among others. The framework for the legalization process for the worst-case scenario, a UPM without mining rights, is described in Figure 5. The process starts with a dialogue between the government and informal miners and ends with either the legalization of UPMs, miners moving to a different industrial activity or with the closure of the UPMs.

According to the Colombian National Miners Confederation (Conmineros in Spanish), which brought together approximately 250,000 miners from different activities and 18 states, the main barriers faced by miners to formalize their practice are the excessive bureaucracy, lack of information, absence of sufficient economic, technical and legal support provided by the government, and preferential treatment to multinationals and large-mining companies, which compete for the mining lands and benefit from legislation that allows them to obtain mining rights in a short period of time.

Among the benefits associated with regulating informal miners are the reduction of the environmental impacts on natural ecosystems; the protection of water sources used for drinking water, recreation, fishing or agricultural purposes; the reduction of

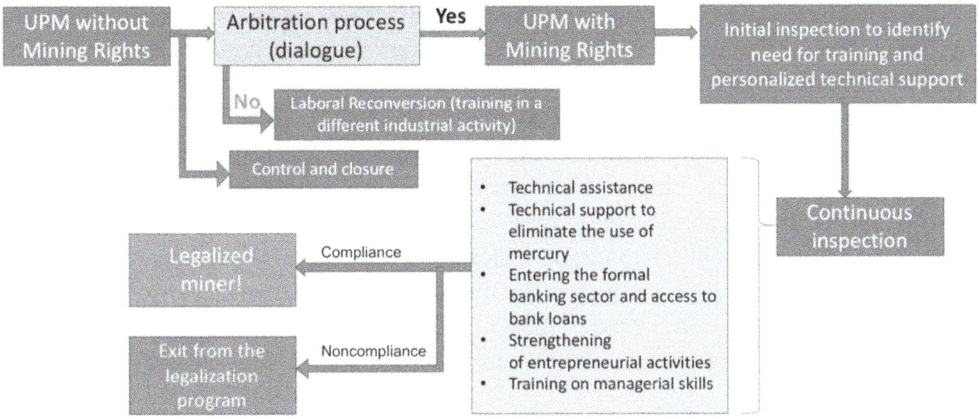

Figure 5. Framework for the legalization process of a mining production unit in Colombia that lacks mining rights. Source: Ministerio de Minas y Energía de Colombia (MinEnergía) (2016a).

deforested areas; the protection of biodiversity; the minimization of the mercury used; and the control of the criminal activities surrounding this practice.

The lack of funding for environmental protection in the face of more urgent needs

The protection of human health, wildlife and natural environments demands a large amount of financial, technological and human resources that are only affordable to developed countries. In the United States, the Comprehensive Environmental Response, Compensation, and Liability Act (CERCLA), commonly known as Superfund, was created for 'cleaning up abandoned or uncontrolled contaminated sites that affect human health and the environment' (USEPA, 2018b). Nevertheless, in the last decades, the annual appropriations to this programme, managed by the US Environmental Protection Agency (USEPA), have been reduced from about US$2 billion to US$1.1 billion (GAO, 2015). In Canada, a Federal Contaminated Sites Action Plan (FCSAP) was created to help provinces manage contaminated sites under their jurisdiction. This action plan started in 2005 and will run until 2020. It was estimated that, as of March 2013, the remaining environmental liability associated with federal contaminated sites was approximately US$10.6 billion (Story & Yalkin, 2014). Limitations in budget allocation to remediate contaminated sites seem to be critical even in developed countries such as Canada and the United States. These countries, with stronger economies and larger human and technological resources compared with countries such as Colombia, also face difficulties when allocating budgets for environmental protection. Both the United States and Canada apply the 'polluter-pays' principle to force corporations, private companies and individuals to provide financial compensation for environmental damage. This polluter-pays principle is more difficult to apply in artisanal and small-scale gold mining regions. Polluters are usually poor families or communities that use illegal or informal gold mining as their primary source of income. In other cases, the polluters are criminal organizations whose leaders are not easy to identify and are much less interested in paying compensation for destroying the environment.

In Colombia, the lack of funding to protect the environment and to enforce regulations on the use of mercury in artisanal and small-scale gold mining is critical. After the peace agreement between the government and Farc, Colombia needed approximately US$31 billion to finance post-conflict programmes over a 10-year period (EFE, 2016). Such programmes are designed to finance projects related to rural development, agrarian jurisdiction, local governance and rural property registration. No major emphasis was placed on the restoration of the environmental damage caused by the over 50 years of armed conflict which included terrorist attacks that caused oil spills, deforestation, illegal mining, fauna trafficking, among others. Consequently, the appropriation of enough funds to protect the environment, to enforce regulations aimed to prevent the use of metallic mercury in illegal mining and to remediate ecosystems affected by artisanal and small-scale gold mining seems unlikely in the short term.

Conclusions

Despite the actions taken for the implementation of the Minamata Convention and the fact that the use of metallic mercury in artisanal and small-scale gold mining is already prohibited in many countries, a significant decrease in the use of mercury seems unlikely in the short term, particularly in developing countries such as Colombia. The main barriers for a successful reduction of metallic mercury in Colombia are the remote locations where illegal gold mining is practised, the illegal trade of metallic mercury, the involvement of mafias and illegal armed groups in the gold industry, the difficulties in the legalization of illegal or informal miners, and the lack of funding to enforce environmental regulations.

The Minamata Convention does not ban the use of mercury in artisanal and small-scale gold mining, but aims to reduce or eliminate its use. Consequently, participant governments are those called to create new legislations, to take actions to prohibit mercury in mining and industrial activities, and to enforce the law. For instance, the Colombian government published an ambitious plan in 2014 (Plan Nacional de Mercurio) aiming to reduce the use of mercury in the mining sector by 2018 and in other industrial activities by 2023. This plan includes four programmes to address the issue of mercury from different perspectives: (1) the programme for institutional strengthening, (2) the programme for social, technological and environmental management, (3) the educational and communication programme and (4) a research programme on environmental, health and technological innovations. All the same, any successful plan needs to count on the commitment of miners to use mercury-free alternatives and eliminate the use of mercury from their mining practice. For this, a government-funded programme is needed that offers financial incentives to persuade miners to change their poor mining practices. The effects of mercury pollution in the environment are expected to last for decades. The Colombian government may need decades to create and consolidate action plans aiming to reduce the use of mercury from mining activities, but also to remediate the thousands of hectares already affected by artisanal and small-scale gold mining.

It is obvious, however, that the main goal should be to eliminate the use of mercury in mining activities. Achieving this goal will require nations to implement the strategies developed by the UNEP as well as other tailored strategies based on the individual needs of each country. Educating future generations in artisanal and small-scale gold mining regions, developing affordable and easy-to-use mercury-free technologies, and providing communities with alternatives to make a living can help communities reduce the use of mercury in gold mining and prevent the continuous pollution of their territories. Most of these polluted sites are located in communities that are at a noticeable disadvantage due to poverty and illiteracy. To protect these communities, a three-part programme should include (1) a mechanism to identify and list impaired ecosystems, (2) an environmental clean-up programme for soils, sediments and water remediation, and (3) a community health programme to identify and treat individuals affected by mercury toxicity. Any action plan taken is needed immediately.

Acknowledgments

Thanks to Professor Nazly Rubio (Universidad Tecnológica del Chocó), Bismark Quinto and Harry Escobar (Coodechocó), Juan Carlos Valdelamar, Yuber Torres and Dr Jesús Oliveros (Universidad de Cartagena), David Navas and Dr José Marrugo (Universidad de Córdoba), and Harry Gutierrez and Dr Gladis Morales (Universidad de Medellín) for their help during the sampling stage of this project. Special thanks to the International Office, The University of Texas at Austin, for supporting this project.

Disclosure statement

The authors declare that they have no known competing financial interests or personal relationships that could have appeared to influence the work reported in this paper.

Funding

As part of this research, water samples were collected from five rivers in Colombia to determine the levels of mercury downstream different gold mining districts. This water sampling was supported by the International Office, The University of Texas at Austin.

ORCID

Farith A. Diaz http://orcid.org/0000-0003-3350-8065
Desmond F. Lawler http://orcid.org/0000-0002-4219-7994

References

Artisanal Gold Council. (2009). *Mercurywatch: Mercury released from ASGM.*

Burton, P. (2011). The Colombian gold mining industry. *World Gold Analyst, 87.* GFMS World Gold Ltd.

Cordy, P., Veiga, M., Crawford, B., Garcia, O., Gonzalez, V., Moraga, D., ... Wip, D. (2013). Characterization, mapping, and mitigation of mercury vapour emissions from artisanal mining gold shops. *Environmental Research, 125*, 82–91. https://doi.org/10.1016/j.envres.2012.10.015

Cordy, P., Veiga, M. M., Salih, I., Al-Saadi, S., Console, S., Garcia, O., ... Roeser, M. (2011). Mercury contamination from artisanal gold mining in Antioquia, Colombia: The world's highest per capita mercury pollution. *Science of the Total Environment, 410–411*, 154–160. https://doi.org/10.1016/j.scitotenv.2011.09.006

Cuevas Guarnizo, A. M. (2015, August 12). Bañados en mercurio! *El Espectador.* https://bit.ly/3ble15C

Davis, K. (2018). *Data visualiation on global mercury emissions.* UNEP. https://bit.ly/3e2Yg4j

Diario Expreso. (2016, April 5). *Mercurio, prohibido pero en venta en Ecuador.*

Diaz Arriaga, F. A. (2014). Mercurio en la minería del oro: Impacto en las fuentes hídricas destinadas para consumo humano. *Revista de Salud Pública, 16*(6), 947–957. https://doi.org/http://dx.doi.org/10.15446/rsap.v16n6.45406

Drace, K., Kiefer, A. M., Veiga, M. M., Williams, M. K., Ascari, B., Knapper, K. A., ... Cizdziel, J. V. (2012). Mercury-free, small-scale artisanal gold mining in Mozambique: Utilization of magnets to isolate gold at clean tech mine. *Journal of Cleaner Production, 32*, 88–95. https://doi.org/10.1016/j.jclepro.2012.03.022

EFE. (2016, September 22). *¿Cuánto le costará el posconflicto a Colombia?* Revista Dinero. https://bit.ly/3apSlDa

Esdaile, L. J., & Chalker, J. M. (2018). The mercury problem in artisanal and small-scale gold mining. *Chemistry – A European Journal, 24*(27), 6905–6916. https://doi.org/10.1002/chem.201704840

Gammons, C. H., Slotton, D. G., Gerbrandt, B., Weight, W., Young, C. A., McNearny, R. L., ... Tapia, H. (2006). Mercury concentrations of fish, river water, and sediment in the Río Ramis–Lake Titicaca watershed, Peru. *The Science of the Total Environment, 368*(2–3), 637–648. https://doi.org/10.1016/j.scitotenv.2005.09.076

GAO. (2015). *SUPERFUND trends in federal funding and cleanup of EPA's nonfederal national priorities list sites.* United States Government Accountability Office. https://www.gao.gov/assets/680/672734.pdf

The Global Initiative. (2016). *Organized crime and illegally mined gold in Latin America.* The Global Initiative against Transnational Organized Crime. https://bit.ly/3eE74xS

Henao, J. C., & Gonzalez Espinosa, A. C. (Eds.). (2016). *Minería y Desarrollo. Tomo 2.* U. Externado de Colombia.

El Heraldo (2013, August 23). Habitantes de Ayapel estarían condenados a morir por el mercurio. *El Heraldo.* https://bit.ly/3atsHxn

IDEAM. (2015). *Estudio Nacional del Agua 2014.* https://bit.ly/34Whtmb

Kirby, A., Rucevska, I., Yemelin, V., Cooke, C., Simonett, O., Novikov, V., & Hughes, G. (2013). Time to act. *UNEP, 1*(8), 483. https://doi.org/10.1038/ngeo275

LeGrand, C. C. (2003). The Colombian crisis in historical perspective. *Canadian Journal of Latin American and Caribbean Studies, 28*, 55–56. https://doi.org/10.1080/08263663.2003.10816840

Limbong, D., Kumampung, J., Rimper, J., Arai, T., & Miyazaki, N. (2003). Emissions and environmental implications of mercury from artisanal gold mining in north Sulawesi, Indonesia. *The Science of the Total Environment, 302*(1–3), 227–236. https://doi.org/10.1016/S0048-9697(02)00397-2

Llorente, M. V., & Garzón Vergara, J. C. (2018). *Por qué siguen aumentando los cultivos de coca en Colombia?* Fundación Ideas para la Paz. http://www.ideaspaz.org/publications/posts/1686

Marrugo-Negrete, J., Benitez, L. N., & Olivero-Verbel, J. (2008). Distribution of mercury in several environmental compartments in an aquatic ecosystem impacted by gold mining in northern Colombia. *Archives of Environmental Contamination and Toxicology, 55*(2), 305–316. https://doi.org/10.1007/s00244-007-9129-7

Meech, J. A., Veiga, M. M., & Tromans, D. (1998). Reactivity of mercury from gold mining activities in darkwater ecosystems. *Ambio, 27*(2), 92–98.

MinEnergía. (2012). *Censo Minero Departamental 2010–2011.* Ministerio de Minas y Energía. https://bit.ly/2XVPwav

MinEnergía. (2016a). *Plan de Acción de Formalización.* Ministerio de Minas y Energía. https://bit.ly/2VpRtuj

MinEnergía. (2016b). *Política Minera De Colombia – Bases para la minería del futuro.* Ministerio de Minas y Energía. https://bit.ly/3lHdijT

MinEnergía, ANM, UPME, & SGC. (2016). *Plan estratégico sectorial para la eliminación el uso del mercurio.* https://bit.ly/2V0VL9O

Olivero-Verbel, J., Johnson, B., & Arguello, E. (2002). Human exposure to mercury in San Jorge river basin, Colombia (South America). *Science of the Total Environment, 289*(1–3), 41–47. https://doi.org/10.1016/S0048-9697(01)01018-X

Otis, J. (2018). *Colombia is growing record amounts of coca, the key ingredient in cocaine.* NPR. November 7, 2018, from https://n.pr/34S5Hag

Palapa, T. M., & Maramis, A. A. (2015). Heavy metals in water of stream near an amalgamation tailing ponds in Talawaan–Tatelu gold mining, North Sulawesi, Indonesia. *Procedia Chemistry, 14*, 428–436. https://doi.org/10.1016/j.proche.2015.03.058

PNUMA, & MADS. (2012). *Sinopsis Nacional de la Mineria Aurífera Artesanal y de Pequeña Escala.* Ministerio de Ambiente y Desarrollo Sostenible. https://bit.ly/3anXsUz

Presidencia. (2016). *Final agreement to end the armed conflict and build a stable and lasting peace.* Presidencia de la República de Colombia. https://bit.ly/2QJGHvH

Redacción El Tiempo. (2015, August 4). Mercurio en los ríos de Colombia. *El Tiempo.* https://bit.ly/2VmIwlq

Redacción Nacional. (2014, October 28). Es necesario diferenciar la minería ilegal de la informal. *El Espectador.* https://bit.ly/2VM7byG

Redacción Negocios y Economía. (2016, December 12). El 80% de la producción de oro colombiano sigue en manos ilegales. *El Espectador.* https://bit.ly/3brl3Fb

Redacción Vivir. (2013, August 26). Mercurio en el agua de Ayapel. *El Espectador.* https://bit.ly/2VHGPhr

Restrepo, V. (2018, July 3). Antioquia aún no logra erradicar el mercurio. *El Colombiano.* https://bit.ly/2RTSoRD

SIMCI. (2018). *Informe de Monitoreo de Territorios Afectados por Cultivos Ilícitos 2017.* Oficina de las Naciones Unidas contra la Droga y el Delito (UNODC). https://bit.ly/2Vmj5QR

Sousa, R., Veiga, M., Van Zyl, D., Telmer, K., Spiegel, S., & Selder, J. (2011). Policies and regulations for Brazil's artisanal gold mining sector: Analysis and recommendations. *Journal of Cleaner Production, 19*(6–7), 742–750. https://doi.org/10.1016/j.jclepro.2010.12.001

Story, R., & Yalkin, T. (2014). *Federal contaminated sites cost.* Office of the Parliamentary Budget Officer. www.pbo-dpb.gc.ca

Swain, E. B., Jakus, P. M., Rice, G., Lupi, F., Maxson, P. A., Pacyna, J. M., . . . Veiga, M. M. (2007). Socioeconomic consequences of mercury use and pollution. *Ambio, 36*(1), 45–61. https://doi.org/10.1579/0044-7447(2007)36[45:SCOMUA]2.0.CO;2

Telmer, K. H., Costa, M., Rômulo Simões, A., Araujo, E. S., & Maurice, Y. (2006). The source and fate of sediment and mercury in the Tapajós River, Pará, Brazilian Amazon: Ground- and space-based evidence. *Journal of Environmental Management, 81*(2), 101–113. https://doi.org/10.1016/j.jenvman.2005.09.027

Telmer, K. H., & Stapper, D. (2012). *A practical guide: Reducing mercury use in artisanal and small-scale gold mining.* UNEP. https://bit.ly/2VN7dWX

Telmer, K. H., & Veiga, M. M. (2009). World emissions of mercury from artisanal and small scale gold mining. In R. Mason & N. Pirrone (Eds.), *Mercury fate and transport in the global atmosphere.* Springer. https://doi.org/10.1007/978-0-387-93958-2_6

Thomson Reuters. (2018). *GFMS gold survey 2018.* https://bit.ly/2RgYwCR

UN Environment. (2017). *Global mercury supply, trade and demand.* United Nations Environment Programme, Chemicals and Health Branch. https://bit.ly/334VerN

UN Environment. (2019). *Global mercury assessment 2018.* United Nations Environment Programme, Chemicals and Health Branch. https://wedocs.unep.org/handle/20.500.11822/27579

UNEP. (2017). *Minamata Convention on mercury: Text and Annexes.* www.mercuryconvention.org

Unidad Investigativa. (2017, April 9). Así entra al país el mercurio que envenena pueblos y ríos. *El Tiempo.* https://bit.ly/3aqMf5M

Universidad de Córdoba, & MinEnergía. (2014). *Estudio de la Cadena del Mercurio con énfasis en la actividad minera aurífera Colombiana.* Ministerio de Minas y Energia. https://bit.ly/3cAu7ba

UPME. (2018). *Unidad de Planeación Minero Energética*. Unidad de Planeación Minero Energética. Retrieved November 7, 2018, from http://www1.upme.gov.co/Paginas/default.aspx

USEPA. (2018a). *Artisanal and small-scale gold mining without mercury*. Retrieved November 7, 2018, from https://bit.ly/3eENTnJ

USEPA. (2018b). *Superfund: CERCLA overview*. Retrieved November 6, 2018, from https://www.epa.gov/superfund/superfund-cercla-overview

Veiga, M. M. (2017). *Mercurio en la Minería Artesanal de Oro: Problemas Ambientales y de Salud*. Webinar hosted by Lumex Instruments.

Veiga, M. M., & Baker, R. F. (2004). *Protocols for environmental and health assessment of mercury released by artisanal and small-scale gold miners*. UN Publications. https://bit.ly/34e91ML

Veiga, M. M., Hinton, J., & Lilly, C. (1999). Mercury in the Amazon: A comprehensive review with special emphasis on bioaccumulation and bioindicators. *NIMD forum'99* (pp. 19–40). Minamata, Japan.

Veiga, M. M., Maxson, P. A., & Hylander, L. D. (2006). Origin and consumption of mercury in small-scale gold mining. *Journal of Cleaner Production*, *14*(3–4), 436–447. https://doi.org/10.1016/j.jclepro.2004.08.010

Verité. (2016). *The nexus of illegal gold mining and human trafficking in global supply chains: Lessons from Latin America*. https://bit.ly/2VpkvdB

YTSMedia. (2013). *Mercury: The burning issue*. https://www.youtube.com/watch?v=x9u9ngtoP6Q

Water laws of Georgia, Moldova and Ukraine: current problems and integration with EU legislation

Yuliya Vystavna, Maryna Cherkashyna and Michael R. van der Valk

ABSTRACT

The water laws of Georgia, Moldova and Ukraine are reviewed and compared with the Water Framework Directive of the European Union with a focus on water quality, water pollution and water management. Theoretical aspects and the implementation of the laws are discussed in terms of integration with European water legislation. Discrepancies are identified that should be addressed in future national legislation.

Introduction

In countries with a so-called transitional economy, such as post-Soviet non–European Union members, water-intensive economic activity is often combined with weak water management and pollution control, leading to a significant anthropogenic impacts on water resources (Cherkashina & Vystavna, 2009; Hagemann, Klauer, Moynihan, Leidel, & Scheifhacken, 2014; Leidel, Niemann, & Hagemann, 2012; Organization for Economic Cooperation and Development [OECD], 2011; Opopol, 2006). In recent decades, the water quality in Georgia, Moldova and Ukraine has deteriorated (OECD, 2011; World Bank, 2003), while an observed reduction in river discharges has compounded shortages in drinking water supplies (Vystavna & Diadin, 2015). In Georgia, Moldova and Ukraine, more than 70% of drinking water is derived from surface water, mostly from altered river basins where water quality has a 'polluted' or 'highly polluted' status. Water quality improvement and better access to drinking water can be attained through appropriate water monitoring, water quality and quantity assessments, identification of pollution sources, and pollution prevention. However, the implementation of these measures requires the updating of environmental legislation, and water laws in particular (Hagemann et al., 2014; OECD, 2007, 2011; United Nations Development Programme, 2013).

A driving force of water legislation development in Georgia, Moldova and Ukraine is adaptation to the water legislation of the European Union (EU) according to signed association agreements (European Commission, 2014a, 2014b, 2014c). This article reviews and compares the water laws of these three countries with the Water Framework Directive (WFD) 2000 of the EU to outline current problems associated with practical implementation of these laws and their integration with EU legislation.

General overview of water legislation in Georgia, Moldova and Ukraine

During the Soviet period, water legislation in Georgia, Moldova and Ukraine was based on the general law Fundamentals of Water Legislation of the USSR and Union Republics, introduced in 1971 (OECD, 2007, 2011). In that period, national water laws in the form of water acts were focused on the interests of principal water users, mainly economic sectors (Leidel et al., 2012; Opopol, 2006). After the collapse of the Soviet Union, each independent republic developed and initiated water laws taking into account specificities of local water resources (mainly regional water balance) and water use (mainly regional economic structure) (Law of Georgia on Water, 1997; Law of the Republic of Moldova on Water, 2011; Water Code of Ukraine, 1995). However, these newly developed water laws contained only minor changes from Soviet water law.

Current water laws in Georgia, Moldova and Ukraine include only general legislative provisions, without direct measures and implementation. For their execution, numerous regulations and instructions were developed, including governmental regulations, standards, building codes, norms, rules of water use and other regulatory documents. Payment (fees and charges) for water extraction and wastewater discharge was applied in the form of taxes (e.g., Tax Code of Ukraine (2014).

Water governance

According to the water laws presently in place, water governance in Georgia, Moldova and Ukraine is provided by state authorities, environmental ministries and agencies (Law of Georgia on Water, 1997; Law of the Republic of Moldova, 1993; Law of Ukraine, 1991; Resolution of the Cabinet of Ministers of Ukraine, 1996a, 1998a, 2014a, 2015; Resolution of the Government of Moldova, 2009) (Table 1). Some of these institutions have overlapping responsibilities. Studies (Hagemann et al., 2014; Nazarov, Cook, & Woodgate, 2000) suggest that a system which distributes functions over multiple institutions reduces the efficiency of water management and regulation in the country.

Also, in line with their respective environmental laws (Law of the Republic of Moldova, 1993; Law of Ukraine, 1991), Ukraine and Moldova have special committees that focus on the development of regional water protection programmes, in particular transboundary water basins (Resolution of the Cabinet of Ministers of Ukraine, 2014a; Resolution of the Government of Moldova, 2009). Activities of these committees are regulated by state environmental agencies, which are a part of national environmental ministries (Table 1). State environmental agencies are also responsible for permissions on water withdrawal and wastewater discharges.

At the moment, Georgia is reforming its national environmental legislation and water protection sector. However, responsibilities are still highly fragmented between water authorities (Table 1). For instance, the Ministry of Environment Protection and Natural Resources carries out water management and provides general environmental control, together with the development of regular environmental monitoring. This

Table 1. Water governance institutions in Georgia, Moldova and Ukraine.

Country	Water monitoring	Water quality control	Water quantity control	Water management
Georgia	Ministry of Environment Protection and Natural Resources and its agencies	Ministry of Environment Protection and Natural Resources; Ministry of Labour, Health Protection and Social Security; Ministry of Agriculture; Ministry of Regional Development and Infrastructure	Ecological inspection by the Ministry of Energy and Natural Resources; Ministry of Environment Protection and Natural Resources	Ministry of Environment Protection and Natural Resources; Ministry of Labour, Health Protection and Social Security; Ministry of Agriculture; Ministry of Energy and Natural Resources
Republic of Moldova	Ministry of the Environment; Ministry of Health Protection; Agency of Geology and Mineral Resources	Ministry of the Environment; Ministry of Health Protection; State Ecological Inspection	State Ecological Inspection; Agency of Geology and Mineral Resources	Ministry of the Environment; Agency of Geology and Mineral Resources; Ministry of Health Protection; State Ecological Inspection; Agency 'Apele Moldovei'; regional basin committees
Ukraine	Ministry of Ecology and Natural Resources; Public Service of Geology and Mineral resources; State Agency of Water Resources; State Sanitary and Epidemic Service, Ministry of Regional Development, Building and Municipal Economy; State Emergency Service.	Ministry of Ecology and Natural Resources; State Agency of Water Resources; State Sanitary and Epidemic Service; Ministry of Regional Development, Building and Municipal Economy	State Ecological Inspection; Public Service of Geology and Mineral Resources.	Cabinet of Ministries of Ukraine; Ministry of Ecology and Natural Resources; State Agency of Water Resources; Public Service of Geology and Mineral Resources; State Ecological Inspection; State Sanitary and Epidemic Service; Ministry of Regional Development, Building and Municipal Economy; State Emergency Service

ministry regulates water policy and controls, sets thresholds for contaminants in wastewater, reforms environmental law and develops the water registry that includes the main hydrological parameters of national water resources. The National Environmental Agency, which is a part of the Ministry of Environmental Protection and Natural Resources, is responsible for surface water monitoring. The Ministry of Energy and Natural Resources is responsible for permissions on groundwater extraction and development of water supply and sewage works, together with the Ministry of Regional Development and Infrastructure. Drinking water quality is regulated by the Ministry of Labour, Health Protection and Social Security. Similar problems of fragmentation of responsibilities are observed in Moldova and Ukraine.

Consequently, there is no comprehensive database or understanding of the linkages between existing multilevel legal institutions in the three countries. Water governance in Georgia, Moldova and Ukraine is fragmented and needs significant structural changes.

Water use and water users

In the water laws of Georgia, Moldova and Ukraine (Law of Georgia on Water, 1997; Law of the Republic of Moldova on Water, 2011; Water Code of Ukraine, 1995), two major water-use categories are identified: 'general' and 'special'. The 'general' category includes all uses which satisfy basic needs of the population, including bathing, swimming, water withdrawal (without pumping or additional equipment), recreational boating and fishing, and water for livestock. In Moldova and Ukraine, water for general use can be supplied from surface water and shallow aquifers (e.g., wells, springs). In Georgia, water for general use can be supplied from surface water and shallow and deep aquifers, if it is accessed without advanced facilities (i.e., pumps), avoiding effects on groundwater level and contamination. Otherwise, the water use falls into the 'special' category. In these three countries, general water use is free of charge, without the need to become a 'legal water user' or obtain special permissions for natural resource use. Local environmental authorities (Table 1) regulate and control general water use, and can limit it in some cases, e.g., disaster events (floods, pollution, etc.).

The 'special' water-use category represents the use of water resources through advanced water supply and discharge facilities. In the water laws of the three countries, wastewater discharge is always in the special category. Special water use requires permission from environmental authorities and entails payments. Permits for water use are given for periods from 3 to 25 years. Short-term permits, of 3–5 years, are mainly for wastewater discharge, while longer periods are for household supply.

In Georgia, Moldova and Ukraine, water supply from hydro-systems (such as channels and water reservoirs), water supply to water-scarce regions, and protection from water disaster (mainly flooding) are not considered special use. Also, in the three countries, waters with health benefits (mineral waters) can only be used for curative bathing purposes, with wastewater discharges being prohibited into these waters. Some particular differences are also observed in the regulation of water use between the three countries. For example, the Georgian water law considers water 'special' if its use influences depletion and/or pollution of water resources (§33). In the Moldavian water law, water supply for drinking is set as a priority among other types of water use (§24). Types of surface water use have the following priority order: agriculture; industrial, including mining; fishery; hydropower; sport and recreation. The priority for groundwater use is drinking water supply; food and beverage production; agriculture.

Therefore, in the water laws reviewed here, water-use categories and types are described in detail and have a certain hierarchy in terms of their priority, emphasizing the preservation of water quality and quantity for drinking purposes. However, inconsistencies in the description of water-use categories lead to the limitation of the 'general' category, mainly for bathing and swimming. Therefore, people have restricted access to 'general' use that contradicts the principal human right to use natural resources, as described in the Constitutions of Georgia (1995), Moldova (1994) and Ukraine (1996).

Property right to water resources

In the water laws of Georgia, Moldava and Ukraine, water property rights are key queries that identify directions of water management. There are minor differences between these countries in the regulation of property rights on water bodies. For instance, according to Ukrainian water law (Constitution of Ukraine, 1996; Water Code of Ukraine, 1995), all water bodies are the property of the Ukrainian people and can only be rented. Yet, according to the land law (§59, Land Code of Ukraine, 2001), small confined reservoirs (less than 3 ha) can be owned, per appropriate decision of executive authorities or the local government. According to the Moldavian water law (Law of the Republic of Moldova on Water, 2011), water is a public good, but individuals or legal entities have the right to use water under a legislative framework. For example, the land under a water reservoir can be subject to both public and private ownership. An individual or legal entity can build an artificial lake on a plot of land they own under current legislation. The Georgian water law (Law of Georgia on Water, 1997) give exceptional public powers over water bodies in the territory of the country. Moreover, this act stipulates that property rights in land do not give rights to use surface water and groundwater on that land. These waters can be used only by the public (§23).

Therefore, the property rights on water are contradictory across different legislative acts in Georgia, Moldova and Ukraine. This presents a substantial problem for the practical implementation of water legislation and quite often leads to disputes between users of natural resources.

Water quality and quantity standards

In Georgia, Moldova and Ukraine, water standards are important parts of environmental and water laws (Resolution of the Government of Moldova, 2009; Law of the Republic of Moldova, 1993; Law of the Republic of Moldova on Water, 2011; Law of Georgia on Water, 1997; Water Code of Ukraine, 1995). These standards were mainly developed for a certain type of water use. Environmental safety standards, water quality standards, permissible standards for water pollutant discharges, branch standards for water contaminants for economic sectors, and technical standards on water use are set out in the environmental laws of Georgia, Moldova and Ukraine. These standards were set for more than 1000 parameters in the 1980s and are generally based on maximum allowable concentrations (MAC) of contaminants established for drinking water supplies, household use, fisheries and other needs of the population. This number of parameters is significantly higher than defined in the WFD or the Environmental Quality Standards Directive (European Union, 2008). Practically, water quality monitoring is limited to about 80 parameters, with only about a third of them being similar to those in the WFD. A general problem is a lack of environmental monitoring tools (Nazarov et al., 2000; Hagemann et al., 2014; Vystavna & Diadin, 2015). Also, Georgian, Moldavian and Ukrainian water laws do not classify water bodies according to water quality or quantity. The water quality assessment is based on the assumption that if one of the parameters exceeds the MAC, the water body cannot be used for certain purposes (e.g., swimming, bathing and fishing). The discharge of water contaminants that have no MAC values is not allowed. However, temporary exceptions can be made with permission of control institutions and are valid for the period required to develop and approve new MAC values. All expenses for the development of new MAC values must be covered by the legal water user.

Reflecting the focus of the water policies in the three countries on the achievement of 'better water quality', a maximum allowable discharge (MAD) is set for wastewaters entering natural waters. The MAD is the mass load of a pollutant that may be discharged during a certain time period, measured as grammes per second or tonnes per year. The MAD is developed to achieve water quality standards. MACs of some hazardous compounds are different from the WFD (Table 2). For most water users, water quality standards are too strict, and high capital investments in sewage facilities are required to follow them. In most cases, water quality standards cannot be achieved, and water users pay higher taxes for wastewater discharge and water pollution. Generally, taxes provide less than the investment needed. Also, these water quality standards still refer to outdated economic activities and to chemical substances that are no longer in use in industries or households (Alan et al., 2006; Vystavna et al., 2012). Moreover, technical water quality standards were established for specific components that occur in particular technological processes and that can be discharged in water bodies or public sewerage. These standards are being developed by appropriate institutions with agreement of the Ministry of Environmental Protection and Natural Resources in Georgia, the Ministry of the Environment in Moldova or the Ministry of Ecology and Natural Resources in Ukraine. In these technical standards, discharge of wastewater which could be reused is prohibited.

In the water laws of Georgia, Moldova and Ukraine, control of wastewater discharge into water bodies is based on limits of wastewater discharge. These are calculated taking into account the wastewater volume and the concentration of contaminants measured by MAD, the discharge conditions (flow rate, type of equipment from which wastewater is discharged, etc.) and the water quality of the receiving water body. In the water laws of Georgia, Moldova and Ukraine, discharges of wastewater to shallow and deep aquifers, and on the soil surface, are prohibited except by special permission of the environmental authority. All industrial wastewater must be discharged into public sewerage after pre-treatment at an industrial facility to within established limits. Wastewater can be used for irrigation only with special permission from sanitary and veterinary authorities.

The water laws of the studied post-Soviet countries are mainly concerned with point sources of water pollution. In many cases, however, diffuse pollution has become a larger problem. This is observed in nutrient loading, where point sources contribute only 8–10% of the total nitrate contamination, in mainly rural areas that are not connected to sewerage and where wastewater leaks from septic tanks (Vystavna et al., 2017; Yakovlev, Vystavna, Diadin, & Vergeles, 2015). In Moldova, for example, about

Table 2. Comparison of maximum allowable concentrations (in mg/L) in the Water Code of Ukraine (1995) and the EU Water Framework Directive.

Compound	Ukraine	EU
Benzene	0.5	0.05
Benzo(a)pyrene	0.000005	0.0001
Cadmium and its compounds	0.0013	0.00045–0.0015
Hexachloro-benzene	0.05	0.00005
Hexachloro-butadiene	0.01	0.0006
Hexachloro-cyclohexane	0.004	0.00004
Mercury and its compounds	0.00053	0.00007
Pentochloro-phenol	0.01	0.001

70% of households are not connected to sewerage or wastewater treatment facilities, and therefore constitute a major source of organic and microbial pollution.

Current water quality standards need to be reformed and updated to comply with the EU standards. The legislative basis for water quality monitoring has no clearly defined parameters, no quality targets, no applied measurement and analysis methods, and no provision for harmonization between different water agencies. It also lacks a mechanism of regular and standardized internal quality management (Hagemann et al., 2014).

Adaptation to EU water legislation

According to the signed EU integration acts (European Commission, 2014a, 2014b, 2014c; Law of Ukraine, 2010; Decision, 2013; Resolution of the Parliament of Ukraine, 2014b), Georgia, Moldova and Ukraine should develop environmental legislation that more closely aligns with European legislation.

Moldova is actively taking measures to adapt its national legislation to the EU legislation. The evidence can be found in the Law of the Republic of Moldova on Water (2011), which is partially harmonized with EU legislation on urban wastewater treatment, the protection of water against nitrate pollution from agricultural sources, bathing waters, assessment and management of flood risks, and environmental quality standards in the field of water policy. Adaptation of the Moldavian water legislation to EU requirements resulted in better sewage control, protection of water resources from pollution, greater responsibilities of wastewater treatment plants, development of a basin approach and introduction of environmental risk assessment and management.

To adapt to EU water legislation, Georgia is also changing the Soviet standards and norms, developing research on conservation and sustainable use of water, and safe water use, as well as control and supervision of practical implementation of the legislation.

The goals and objectives of environmental policy in Ukraine and its transition towards the environmental standards of the EU are also stipulated in legislation. They point out the need to develop appropriate legal acts and to adapt to European environmental norms and standards.

The main principles of EU water management are participation of water users, central and local public authorities, civil society and other stakeholders in planning and decision making regarding use and protection of water resources (WFD, 2000; Leb, 2015; Voulvoulis, Arpon, & Giakoumis, 2017). The following principles are partly adopted in the legislation of Georgia, Moldova and Ukraine: the polluter-pays principle, according to which the costs of pollution prevention or elimination of the consequences of water pollution should be covered by the polluter; the precautionary principle, which is based on prevention measures; the principle of sustainable water use, which addresses the needs of present and future generations to use and protect water resources; and the principle of the economic value of water, which recognizes the economic value of water resources.

Taking into account that Ukraine and Moldova share transboundary river basins (the Danube and Dniester), and that Ukraine and Georgia share marine zones (Black Sea) with the EU, some of the principles of the WFD have been already included in the international conventions on water protection (e.g., Convention on the Territorial Sea and the Contiguous Zone, 1958; Convention on the Protection of the Black Sea against

Pollution, 1994; Law of Ukraine, 1995; Resolution of the Cabinet Ministers of Ukraine, 1996b).

In spite of current and past actions towards integration with EU legislation, important strides still need to be made to harmonize the legislation of Georgia, Moldova and Ukraine with EU law. These are mainly related to water resources management, which should be transformed to match the legislation.

Comparative analysis of the WFD and the water law of Ukraine

Another substantial obstacle to adaptation of the EU legislation in the three countries is discrepancies in the terminology that is used in laws, norms and standards. For example, 'pollution', 'pollutant' and 'water quality' are key legislative categories in both the WFD and the Water Code of Ukraine. But these categories are interpreted differently (Table 3) in a way that significantly complicates practical application of international conventions on transboundary water use and adaptation to the EU legislation. For instance, in the Ukraine water code, water is primarily considered as a natural resource for different types of human uses, such as swimming, fishing, or drinking. Therefore, preservation of water quality and quantity is based on the satisfaction of the needs of the population and economy. In

Table 3. Comparison of the terminology in the Water Code of Ukraine (1995) and the EU Water Framework Directive.

Category	Water Code of Ukraine	Water Framework Directive
Pollution	'Pollution' is an entry of water pollutants into a receiving water body (§1).	'Pollution' means the direct or indirect introduction, as a result of human activity, of substances or heat into the air, water or land which may be harmful to human health or the quality of aquatic ecosystems or terrestrial ecosystems directly depending on aquatic ecosystems, which result in damage to material property, or which impair or interfere with amenities and other legitimate uses of the environment (§33).
Pollutant	'Pollutant' is a substance that is introduced into a water body as a result of human activities (§1).	'Pollutant' means any substance liable to cause pollution, in particular those listed in the annexes of the WFD. 'Hazardous substances' means substances or groups of substances which are toxic, persistent and able to bio-accumulate, and other substances or groups of substances which give rise to an equivalent level of concern (§29). 'Priority substances' means substances identified by the competent authorities and listed in the WFD. Among these substances there are 'priority hazardous substances' which means substances identified by the competent authority and for which special prevention measures have to be taken (§31).
Water quality	'Water quality' is a characteristic of water composition and properties that defines its suitability for particular purposes of water use (§1). 'Category of water quality' is a measure of water pollution status that is set according to composition and properties of water, and is mandatory for a certain period.	'Environmental quality standard' means the concentration of a particular pollutant or group of pollutants in water, sediment or biota that should not be exceeded to protect human health and the environment (§35).

contrast, the WFD views water resources as a heritage (§1). At the same time, both legislations aim at reducing pollution and meeting appropriate water quality standards. The Water Code of Ukraine considers water sanitation and protection of public health as a priority. The WFD defines environmental performance and ecosystem preservation as the main targets of water protection.

In the Water Code of Ukraine, pollution is understood as the introduction of anthropogenic substances, 'pollutants', into surface water and groundwater. These pollutants make the natural waters unsuitable for a certain type of water use (drinking, swimming, bathing, fishing, etc.). The WFD is not primarily focused on water users, but considers water resources as aquatic and riparian ecosystems that should be protected and conserved (§33). The WFD considers water pollution more broadly than the Water Code of Ukraine, looking at both direct and indirect impacts on natural waters (Howarth, 2006; Voulvoulis et al., 2017). The indirect impact is associated with natural and human factors and takes into account the initial state of environmental media (air, soil and water). The river basin is treated as a holistic ecosystem.

As mentioned above, water protection in Ukraine is associated with human and economy needs only; the ecosystem approach, as a link between water bodies and other environmental components (soil, air, biota), is completely missing. In the WFD, water pollution is considered in relation to the natural state of the ecosystem.

According to both the WFD and the Water Code of Ukraine, the main causes of water pollution are wastewater discharge and economic activity. In Ukraine there are many cases where enterprises discharge wastewater into surface water without permission or beyond permitted limits (limits of wastewater discharge and MAD). These cases are considered unauthorized water use, constitute a violation of water legislation, and are subject to juridical responsibilities (e.g., Resolution of the Cabinet of Ministers of Ukraine, 1998a, 1999, 2002).

The differences in terminology between EU legislation and the legislation of Ukraine, Moldova and Georgia is an important barrier to the integration of the EU legislation and the practical implementation of water laws. Essential steps must be taken to shape the core terminology of water laws according to the EU legislation and national features.

Summary

Water legislation in Georgia, Moldova and Ukraine is under transformation from Soviet to EU environmental standards. This shift requires significant changes in the terminology, structure and content of water laws. At present, the main shortcomings of the water legislation of the three countries and its practical implementation are: (1) fragmentation of water legislation, which translates into poorly defined responsibilities; (2) significant discrepancies between different laws on property rights and the responsibilities of natural resource users; and (3) outdated water quality standards. General problems for integration with EU legislation are significant discrepancies not only in terminology but also in the overall concept of water resource protection. Future legislation in Georgia, Moldova and Ukraine should reflect new international practices and approaches in water protection. Effective mechanisms of practical implementation must also be introduced into various water sectors.

Disclosure statement

No potential conflict of interest was reported by the authors.

ORCID

Yuliya Vystavna http://orcid.org/0000-0002-1366-0767
Maryna Cherkashyna http://orcid.org/0000-0002-8892-5440

References

Alan, I. J., Mills, G. A., Vrana, B., Knutsson, J., Holmberg, A., Guigeus, N., … Greenwood, R. (2006). Strategic monitoring for the European water framework directive. *Trends in Analytical Chemistry*, *25*(7), 704–714. doi:10.1016/j.trac.2006.05.009

Cherkashina, M., & Vystavna, Y. (2009). Jurydichni problem skydu stochnikh vod v poverkhnostnie vonie objekty [The legislative problems of wastewater discharges in surface water]. *Naukovyi Visnyk Chernivetskogo Universytetu*. *35*, 85–91. Retrieved from http://lawreview.chnu.edu.ua/visnuku/st/2009_518visnuk.pdf

Constitution of Georgia (1995). Retrieved from http://www.parliament.ge/files/68_1944_951190_CONSTIT_27_12.06.pdf

Constitution of the Republic of Moldova (1994). Retrieved from http://www.presedinte.md/eng/constitution

Constitution of Ukraine (1996). Retrieved from https://www.kmu.gov.ua/document/.../Constitution_eng.doc

Convention on the Protection of the Black Sea against Pollution, 3939-XII (1994). Retrieved from http://zakon5.rada.gov.ua/laws/show/z0478-95

Convention on the Territorial Sea and the Contiguous Zone (1958). Retrieved from http://zakon5.rada.gov.ua/laws/show/995_178

Decision of the National Security and Defense Council of Ukraine On urgent measures for the European integration of Ukraine (2013).

EU Directive of the European Parliament and the Council. (2008). Concerning the environmental quality standards in the field of water policy, amending and subsequently repealing council directives 82/176/EEC, 83/513/EEC, 84/156/EEC, 84/491/EEC, 86/280/EEC and amending directive 2000/60/EC of the European parliament and of the council. *Official Journal of the European Union L 348*, 84–97.

European Commission. (2014a, May 29). Association agreement between the European Union, the European Atomic Energy Community and their Member States, of the one part, and Ukraine, of the other part. *Official Journal of the European Union L, 161*(3), 0001–2132. Retrieved from http://www.europarl.europa.eu/RegData/docs_autres_institutions/commission_europeenne/com/2013/0290/COM_COM(2013)0290(PAR2)_EN.pdf

European Commission. (2014b, August 30). Association agreement between the European Union, the European Atomic Energy Community and their Member States, of the one part, and the Republic of Moldova, of the other part *Official Journal of the European Union L, 260*, 001–740. Retrieved from http://eur-lex.europa.eu/legal-content/en/TXT/PDF/?uri=CELEX:22014A0830(01)

European Commission. (2014c, August 30). Association agreement between the European Union, the European Atomic Energy Community and their Member States, of the one part, and Georgia, of the other part *Official Journal of the European Union L, 260*, 001–740. http://eur-lex.europa.eu/legal-content/en/TXT/PDF/?uri=CELEX:22014A0830(02)

Hagemann, N., Klauer, B., Moynihan, R. M., Leidel, M., & Scheifhacken, N. (2014). The role of institutional and legal constraints on river water quality monitoring in Ukraine. *Environmental Earth Sciences*, *72*(12), 4745–4756. doi:10.1007/s12665-014-3307-5

Howarth, W. (2006). The progression towards ecological quality standard. *Journal of Environmental Law*, *18*(1), 3–35. doi:10.1093/jel/eqi049
Land Code of Ukraine. (2001). No. 2768-III Retrieved from https://www.wto.org/english/thewto_e/acc_e/ukr_e/WTACCUKR88_LEG_1.pdf
Law of Georgia on Water (1997). No. 936 Retrieved from https://matsne.gov.ge/ru/document/download/33448/19/en/pdf
Law of the Republic of Moldova about Environmental Control (1993). Retrieved from http://www.law-moldova.com/eng/legislation_republic_moldova.html
Law of the Republic of Moldova on Water. (2011). No. 272. Retrieved from http://www.lawyer-moldova.com/2012/10/water-law.html
Law of Ukraine on Environmental Protection (1991). No. 1264-XII Retrieved from https://www.ecolex.org/details/legislation/law-no-1264-xii-on-environmental-protection-lex-faoc045101/
Law of Ukraine on the Exclusive (Marine) Economic Zone (1995). No. 162/95-BP.
Law of Ukraine on the Main Principles (Strategy) of the National Environmental Policy of the Ukraine for the Period until the Year 2020 (2010). No. 2818-VI. Retrieved from https://www.ecolex.org/details/legislation/law-of-no-16295-bp-on-the-exclusive-maritime-economic-zone-of-ukraine-lex-faoc021661/
Leb, C. (2015). One step at a time: International law and the duty to cooperate in the management of shared water resources. *Water International*, *40*(1), 21–32. doi:10.1080/02508060.2014.978972
Leidel, M., Niemann, S., & Hagemann, N. (2012). Capacity development as a key factor for integrated water resources management (IWRM): Improving water management in the Western Bug River Basin, Ukraine. *Environmental Earth Sciences*, *65*(5), 1415–1426. doi:10.1007/s12665-011-1223-5
Nazarov, N., Cook, H. F., & Woodgate, G. (2000). Water pollution control issues in an independent Ukraine. *Water and Environment Journal*, *14*(2), 117–123. doi:10.1111/j.1747-6593.2000tb00237.x
OECD. (2007). *Surface water quality regulation in Moldova: Policy aspects of the reform*. Paris: OECD. Retrieved from http://www.oecd.org/env/outreach/41833059.pdf
OECD. (2011). *Development in Eastern Europe and the South Caucasus: Armenia, Azerbaijan, Georgia, Republic of Moldova and Ukraine*. Paris: OECD. doi:10.1787/9789264113039-en
Opopol, N. (2006). Water management in Republic of Moldova. In *Principal threats NATO security through science series C: Environmental security* (pp. 167–176). NATO Public Diplomacy Division.
Resolution of the Cabinet of Ministers of Ukraine (1996a). On approval of the state water monitoring, №815.
Resolution of the Cabinet of Ministers of Ukraine (1996b). On approval of the rules of internal sea waters and territorial sea from pollution and contamination, № 269.
Resolution of the Cabinet of Ministers of Ukraine (1998a). On approval of the state system of environmental monitoring, №391
Resolution of the Cabinet of Ministers of Ukraine (1999) On approval of the rules of surface water from contamination feedback waters, №465 .
Resolution of the Cabinet of Ministers of Ukraine (2002) On approval of the issuance of permits for special water use and changes to the resolution of the cabinet of ministers of Ukraine of 10 August 1992 № 459, №321.
Resolution of the Cabinet of Ministers of Ukraine (2014a). On approval of the state water resources agency of Ukraine, №393
Resolution of the Cabinet of Ministers of Ukraine (2015). On approval of the ministry of ecology and natural resources of Ukraine, №32.
Resolution of the Government of Moldova. (2009). Approval of the regulations on the organization and functioning of the ministry of environment, structure and number of authorized posts of its central office, №847.
Resolution of the Parliament of Ukraine (2014b).On confirmation of Ukraine's course toward integration into the European Union and immediate measures in this direction, No. 874-VII.

Tax Code of Ukraine (2014).

United Nations Development Programme. (2013). POST – 2015. Ukraine. The future we want. National Report. Retrieved from http://www.ua.undp.org/content/ukraine/en/home/library/mdg/Post_2015_Ukraine/

Voulvoulis, N., Arpon, K. D., & Giakoumis, T. (2017). The EU water framework directive: From great expectations to problems with implementation. *Science of the Total Environment*, *575*, 358–366. doi:10.1016/j.scitotenv.2016.09.228

Vystavna, Y., & Diadin, D. (2015). Water scarcity and contamination in eastern Ukraine. *IAHS-AISH Proceedings and Reports*, *366*, 149–150. doi:10.5194/piahs-366-149-2015

Vystavna, Y., Diadin, D., Grynenko, V., Yakovlev, V., Vergeles, Y., Huneau, F., & Knöller, K. (2017). Determination of dominant sources of nitrate contamination in transboundary (Russia/Ukraine) catchment with heterogeneous land use. *Environmental Monitoring and Assessment*, *189*, 509. doi:10.1007/s10661-017-6227-5

Vystavna, Y., Huneau, F., Grynenko, V., Vergeles, Y., Celle–Jeanton, H., Tapie, N., & Le Coustumer, P. (2012). Pharmaceuticals in rivers of two regions with contrasted socio-economic conditions: Occurrence, accumulation and comparison for Ukraine and France. *Water Air and Soil Pollution*, *223*(5), 2111–2124. doi:10.1007/s11270-011-1008-1

Water Code of Ukraine (1995). №213-95

Water Framework Directive (WFD). (2000). Directive 2000/60/EC of the European parliament and of the council of 23 October 2000 establishing a framework for Community action in the field of water policy. *Official Journal of the European Union L*, *327*(22), 275–346. 12/2000,0001–0073.

World Bank. (2003). *Water resources management in South Eastern Europe. Volume II.* Country water notes and water fact sheet. Author. Retrieved from http://siteresources.worldbank.org/INTALBANIA/Resources/Water_Resources_Management_in_South_Eastern_Europe-Volume_Two.pdf

Yakovlev, V., Vystavna, Y., Diadin, D., & Vergeles, Y. (2015). Nitrates in springs and rivers of East Ukraine: Distribution, contamination and fluxes. *Applied Geochemistry*, *53*, 71–78. doi:10.1016/j.apgeochem.2014.12.009

OPEN ACCESS

The evolution of markets for water pollution allowances in China: a case study of Jiaxing

Hao Wang, Sander Meijerink, Erwin van der Krabben and Huaguo Yang

ABSTRACT

The central question this paper asks is: How can we understand the evolution of markets for water pollution allowances? To answer, we have conducted an in-depth study of the development of water pollution allowance markets in Jiaxing, China, which was one of the first cities to experiment with these markets. We have completed a longitudinal analysis and conclude that market design is a continual learning process, with adjustments to solve issues and to respond to external changes.

Introduction

Water pollution has degraded ecosystems, causing water scarcity in parts of China. Hence, the Chinese government's goal is to improve water quality. In addition to regulatory policies, markets for water pollution allowances have been created to control water pollution. A pollution allowance is a government-issued permit to emit a certain amount of a pollutant; the holder of this permit may use it to pollute legally or sell it for a profit ("Pollution allowance", 2020). The rationale for having such markets is that polluting entities with relatively low abatement costs can reduce their pollution discharges below permitted maximum levels, enabling them to sell their pollution allowances to entities with higher abatement costs (Selman et al., 2009). In principle, these markets could lead to an improved allocation of scarce pollution allowances. In addition, with the possibility of selling unused allowances, entities have incentives to reduce more discharge if they can identify benefits in allowance trading. In pilot areas (hereafter, 'pilots') across China, markets for pollution allowances have been operating for several decades.

Markets for water pollution allowances in these pilots differ in their design and performance. After Minxing district in Shanghai began trading water pollution allowances in 1991 (Wu, 2011), several other pilots were initiated by the Chinese national government. Various rules for trading pollution allowances have been developed for these pilots. Some pilots are examples of how such markets have contributed positively to the environment; others have encountered problems, such as low market participation and unclear market demarcation. These differences in design and performance have inspired this paper.

Jiaxing has had a relatively long market history among the pilots claiming to have good market performance. Since its first experience in 2002 in Xiuzhou district, the institutional design has changed several times, from the official start in 2007 to its current status reflected by well-organized transactions. These changes are a response to both external developments, such as new legislation and internal reflections on market performance. Furthermore, the markets for water pollution allowances in Jiaxing operate as a polycentric system, in which Jiaxing city sets the general rules for its seven counties (also called districts/county-level cities[1]). These counties have the capacity to design and adjust operational rules for their specific markets, as long as they fit within the general framework. This polycentric system can be conceived as an experimental setting in which parties gain experience using market designs, and through which they learn about the effectiveness and efficiency of these market designs.

The markets in Jiaxing encountered underperformance problems during development. At an early stage, there was very little trade, contrary to policymaker expectations. According to Shen (2004), this could be explained by the fact that the government controlled both the quantity and cost of allowances, hindering the function of the market. He argues that when the total amount of pollution allowances and property rights are clear, the market should determine the price. This is because price is meant to indicate the scarcity of pollution allowances and control pollution. However, the overall pollution was not reduced after this market mechanism was introduced. There are many reasons for this. The non-cooperation between upstream and downstream actors complicated downstream pollution control, since upstream pollution was not sufficiently monitored, and non-point source pollution from agricultural lands and residential areas increasingly contributed to the overall pollution (Yu, 2007). With the development of markets, a shortage of allowances was recognized. Suggestions to increase allowance supply include providing funding for some manufacturing enterprises (hereafter, 'enterprises') to reduce pollution through technological improvement; establishing the mechanism of short-term lease; involving the large-scale cattle-raisers in the market, since there is the potential to reduce their significant contribution to pollution; and systematically using the allowance storage achieved by pollution reduction (Yu, 2008).

Shen et al. (2013) divided the history of markets for pollution allowances in Zhejiang province into the following three phases, by presenting related documents and statistics: autonomous exploration, deep practice and expansion, at the district, city and province levels, respectively. While this literature is helpful in understanding the market development of water pollution allowances in China, it does not present a theoretically informed analysis of how the institutional design and performance of these markets have developed.

In this study, we conducted a longitudinal analysis of markets for water pollution allowances in Jiaxing from 2002 to 2018. We evaluated how the city has changed its institutional design in response to internal reflections on observed market failures and external developments, such as changes in national or provincial legal and policy frameworks. Moreover, we identified how this type of market in Jiaxing has developed into a well-functioning market.

The remainder of the paper is structured as follows. In the next section we expand our theoretical framework based on institutional–economic theories, the literature on institutional change and the literature on water pollution allowance trading. We then briefly discuss the research methodology. This is followed by the results of the case analysis, which discusses changes in institutional design and addresses how these changes have occurred in response to external developments or the internal reflections on underperformance. In the last section we present the discussion and conclusions.

Theoretical framework

In reality, institutional design may change over time. There are different approaches to analysing institutional design changes. Buitelaar et al. (2007) concluded that the existing position of the institutional arrangement could be affected and made malleable by external and internal developments. Regarding external developments, they refer to societal developments that strain present institutional arrangements; considering internal developments, they refer to the stream of reflection, alternative ideas (solutions) and actions of the institutional bricoleurs who challenge the present stable situation.

Considering the situation of markets for water pollution allowances in a city, externally, institutional design changes can be driven by legislation or other policies of the state or provincial authorities, and shock events in the environment or society. Internally, the changes can be driven by the designers' reflections on market failures (performance that does not meet expectations). Institutional design or change requires careful implementation and learning from experience, as argued by Huitema et al. (2009).

The effectiveness of changing the design based on the observed performance relies on the theory that the institutional design of the markets might influence their performance. Lai and Hung (2008) argued, after Coase (1960, p. 209), that 'where transaction cost is not zero or property rights are unclear or poorly defined, the assignment of rights and liabilities would affect resource allocation' and that 'certain resource allocation would increase (decrease) efficiency by reducing (increasing) transaction costs'. A market is an approach to allocate resources (Jaeger, 2011), and institutional design affects its performance.

To learn and cope with the institutional design change, polycentric systems are expected to substantially promote capacity (Huitema et al., 2009). The markets for water pollution allowances could be implemented and influenced by governments of the state, province, city and county. For instance, different institutional designs are possible in different regions, such as counties, even though the market mechanism can be initiated by the same government at a higher level. The development of diverse markets in this polycentric system (Skelcher, 2005) is similar to an experimental setup for different institutional design testing, offering opportunities for learning across cities and regions. The polycentric nature of the markets for water pollution allowances in Jiaxing is depicted in Figure 1. The city, together with its counties, is considered in the case study. Trading is organized at the county level but supervised at the city level. The city sets a general framework for markets. The design at the city level can be influenced by external developments, such as changes in legislation at the national or provincial level. Within the city, counties work independently within the general framework but with the possibility of adjusting their market according to local situations and their own market performance. If, in some cases, the counties' rules and regulations have proven effective

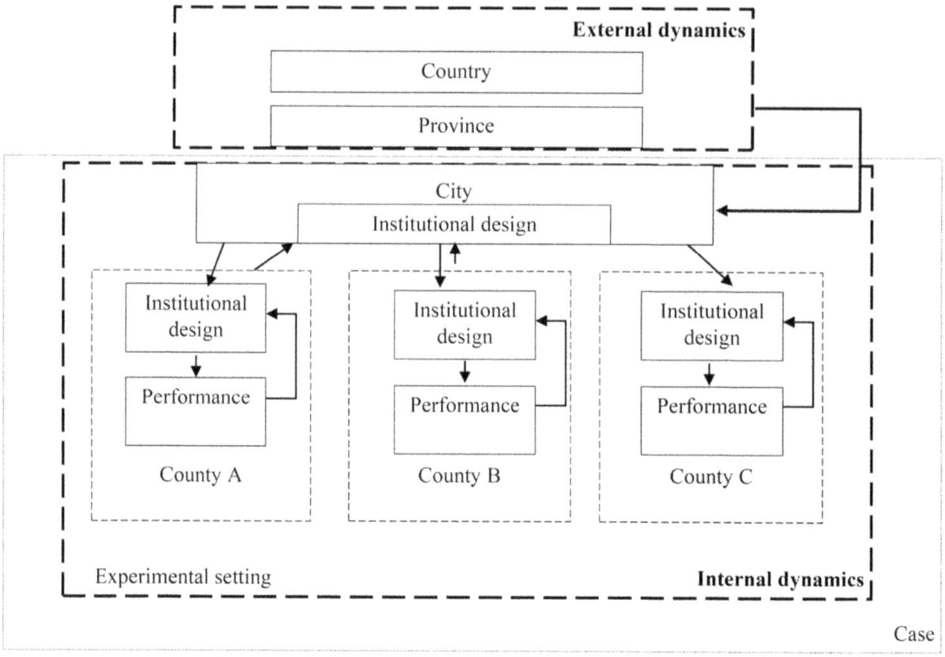

Figure 1. Theoretical framework.

in improving performance, they could subsequently be incorporated into the city rules. This incorporation could also affect the design and performance of markets in other counties.

To gain a concrete idea of institutional design, we referred to Lai and Hung (2008). They distinguish the dimensions of law, governance, means of coordination and contractual arrangements, and subsequently translate them into concrete planning terms. Although these dimensions have been developed in the field of planning and real estate research, we consider them equally relevant to the study of markets for water pollution allowances. By 'law', we refer to the official documents that define and structure the rules of markets for water pollution allowances. By 'governance', we refer to the role of the government or non-governmental organizations which operate the market, and their financial management of the markets. By 'means of coordination', we refer to the interaction between organizations, including allowance distribution and conflict management. By 'contractual arrangements', we refer to the details, including pollution indicators to trade, buyers and sellers, prices, duration of the allowances, and trading methods.

Concerning the dimensions of performance, we started the evaluation from the perspective of broad markets for watershed ecosystem services, based on the fact that 'water purification and waste treatment' is a form of ecosystem service (e.g., MEA, 2005). Building on a review of the literature, we distinguished four dimensions of performance of markets for watershed ecosystem services – namely, the preservation of ecosystem (services) (e.g., Brouwer et al., 2011; Wunder et al., 2008), distributional impacts (e.g., Shilling & Osha, 2003; Wunder, 2005), cost-effectiveness (e.g., Vatn, 2010) and participation (e.g., Bremer et al., 2014).[2] We formulated the details for each dimension to assess the performance of markets for water pollution allowances. By 'preservation of ecosystem services', we refer to the influence on water quality or quantity; by 'distributional impacts', we refer to the influence on the fiscal revenue used to promote environmental protection; by 'cost-effectiveness', we refer to whether the market is obtaining the best possible profit compared with the cost, or if it is more cost-effective than other policy instruments; by 'participation', we refer to the participation rate or number of participants.

Experiences have been gained with markets for water pollution allowances in the United States, Europe, Australia, Canada and other areas (Selman et al., 2009; Tietenberg, 2006). The research on these markets indicates several dimensions of institutional design that are relevant to understanding market performance, especially with regard to 'law'. An effective administrative and legal system is a prerequisite (Kraemer et al., 2004), because the market scheme operates within the context of an existing water pollution control system (Zhang et al., 2012). The assurance and acceptance of pollution rights and the accompanying trading rules are essential. Moreover, it is advisable to avoid revisions which might jeopardize the processed investments or challenge the value of permit systems (Kraemer et al., 2004).

Considering 'coordination', the definition and enforcement of pollution budgets have proven to be the most important regulatory impetus for water pollution trading by outstanding cases. Hence, waste control flexibility, which allows the pollution source to decide how reductions will be made, is necessary to create sufficient trading volume (Fisher-vanden & Olmstead, 2013). Considering 'contractual arrangements', the involvement of potential participants with significant abatement cost differences and the introduction of market advantages to them work well in stimulating the trading. Additionally, clearly defined units of trade and measurability of the units have proven important for the trading of pollution allowances (Fisher-vanden & Olmstead, 2013; Kraemer et al., 2004).

In addition to the market design, monitoring and reliable data help enforce mechanisms, resolve conflicts, facilitate trade and evaluate programmes (Kraemer et al., 2004; Tietenberg, 2006). A solid scientific understanding of the pollution factors and the behaviour of the catchment have evidently facilitated the most successful schemes in water pollution trade. It has also been noted that a thorough pilot phase for testing the rules, conditions, and opportunities for consultation with potential participants can be vital for the acceptability and success of a scheme (Kraemer et al., 2004).

Methodology

To investigate changes in the design and performance of the markets for water pollution allowances in Jiaxing city and to investigate the key factors which may explain these changes, firstly, we conducted a literature review using the literature database of China National Knowledge Infrastructure and obtained an overview on these markets. Secondly, we analysed legal and policy documents and media articles and carried out semi-structured interviews with Chinese scholars, government officials and business entrepreneurs who traded allowances.

We examined documents from the official websites of the state, Zhejiang province, Jiaxing city and its counties, for document analysis. For documents that were unavailable online, we applied individually to the counties' administration through fax or email messages. For the interviews, we contacted relevant Chinese scholars by email, based on the literature concerning pollution allowance trading in Jiaxing city or Zhejiang province. We were able to contact local government officials and business entrepreneurs. As a result, we conducted 15 semi-structured interviews, including 11 Chinese scholars who have conducted research on this type of market, four government officials who have been involved in the markets at city and county levels, and a chief financial officer who has been representing their company in the market from the printing and dyeing industry.

We used ATLAS.ti 8.0 to code the articles, documents and interview transcriptions. The codes relate to the development stages, counties, institutional design dimensions, performance dimensions, external factors and internal reflections. Finally, we analysed how changes in design were linked to both external and internal factors.

Case study

Brief introduction

Jiaxing city is situated in the north-east of Zhejiang province. Located in the Yangtze River Delta, this city is less than 100 km from the cities of Shanghai, Hangzhou, Suzhou and Huzhou. It is flanked by the East China Sea, Qiantang River and Tai Lake. It covers land and sea areas of 3915 and 4650 km^2, respectively. Jiaxing city comprises two districts (Nanhu and Xiuzhou), two counties (Jiashan and Haiyan), and three county-level cities (Pinghu, Haining and Tongxiang).[3] The registered population on 30 November 2018 was 3.6044 million, and gross domestic product (GDP) per capita was 103,858 CNY (US $15,695) (Jiaxing Statistic Bureau, 2019).

Water pollution in Jiaxing has become a constraining factor for sustainable economic and social development. It is caused by industrial pollution, agricultural non-point pollution, upstream sewage and domestic wastewater (Jiaxing Water Resources Bureau, 2018). In a strategic attempt to control water pollution, Jiaxing established markets for water pollution allowances. To analyse how external developments and internal reflections on performance have affected institutional design change, we reconstructed the historical evolution of these markets into four distinct stages.

Paid use of water pollution allowances in Xiuzhou district, 2002–07

External pressure from the national level mainly originates from the total emission control[4] system for major pollutants. The system states that 'total control is required for the emission of pollutants from river basins, regions, cities, districts and factories' (see Appendix A, A2).[5] The system was formally established in 1996 through the plan on total emission control, at the start of the 'Ninth 5-Year Plan (1996–2000)' (National Bureau of Statistics of China, 2019). Because Grades IV and V water amounted to 69.5% and 0.5% by 2000,[6] respectively (Government Office A, 2018), the entire Jiaxing city was under pressure to reduce pollution.

As a district of Jiaxing, Xiuzhou was leading the economic development of Jiaxing. Further economic development required greater emission capacity. In response to the demand for greater emission capacity and existing severe water pollution, Xiuzhou developed an approach called 'Environmental Capacity Replacement', which was meant to fund domestic wastewater treatment facilities through the paid use of water pollution allowances by polluting industries. Consequently, the emission capacity achieved by this treatment could be used for industrial pollution (Guo et al., 2002).

The Environmental Protection Bureau (hereafter, 'the Bureau') of Xiuzhou initiated the development of this approach. The Bureau spent one year liaising with the departments of finance, price[7] and accounting to analyse the possibilities of controlling the total discharge of pollutants through paid use of water pollution allowances (Chai et al., 2005). The official document, which set down the rules for the control of total water pollutants and the paid use of water pollution allowances, was promulgated in April 2002.

The implementation of the paid use of water pollution allowances initially encountered considerable resistance. Enterprises in the textile, printing and dyeing industries were concerned that this extra expenditure would reduce their competitiveness compared to enterprises in other counties of Jiaxing, which were not asked to buy water pollution allowances (Xinhua News Agency, 2005). Relying on these labour-intensive industries made it less possible for Xiuzhou to upgrade its industrial structure[8] in the short run. The Bureau hosted meetings and organized visits to convince entrepreneurs of the necessity of implementing a pollution quota. It also depicted a promising future for those possessing pollution allowances so that they could be seen as assets. During the initial meetings, some entrepreneurs put their fists down on the table in defiance of the officials' decision to implement the system (Xinhua News Agency, 2005).

The trading system was arranged such that the government of Xiuzhou sold the allowances of chemical oxygen demand (COD) at a fixed and reduced price if payment was made as a lump sum. The volume of allowances was based on the pollutant discharges specified in the enterprise environmental impact assessment approval and the wastewater treatment capacity of these enterprises when the Tai Lake Basin met the governance standard in 2000 (Jiang et al., 2005). The duration of the allowance was set at 50 years (Shen et al., 2014). Within these 50 years, enterprises were allowed to trade their allowances within Xiuzhou at a negotiated price (Pan, 2007).

Gradually, enterprise participation increased. In October 2002, 11 enterprises bought pollution allowances at the opening ceremony (Shen et al., 2014). From 2002 to 2003, the number of participating enterprises increased to 50 (Research Institute A, 2014). By July 2005, all 54 enterprises had paid, and the turnover was approximately 7 million CNY (Xinhua News Agency, 2005). Up to that point, no direct trade of water pollution allowances between enterprises had taken place; however, possessing such allowances was considered a valuable asset when enterprises were sold (Jiang et al., 2005).

As the value of water pollution allowances was increasingly recognized, more enterprises consulted the government on acquiring additional pollution allowances. Thus, it is possible for enterprises to make a profit by reducing pollution. If we consider a papermaking enterprise as an example, after the purchase of 2000 tons of COD allowances, the enterprise installed a new system to recycle wastewater, based on the advice received from the Bureau; this installation reduced COD emissions to 50 tons. The remaining 1950 tons of allowances were in high demand from other enterprises (Chai et al., 2005). Despite the high demand for pollution allowances, most enterprises did not want to sell their allowances (Xinhua News Agency, 2005).

The paid use of pollution allowances in Xiuzhou was deemed essential for controlling domestic wastewater and meeting pollution reduction targets set by the government at comparatively low costs (Shen & Xie, 2010). From 2002 to 2007, the discharge of industrial wastewater in the entire district was lower than the pollution quota given by the city. Through this scheme, a new way to protect the environment was established. In addition, the scheme contributed to upgrading the industrial structure[9] by establishing 'economic barriers' to polluting enterprises (Shen & Xie, 2010). However, despite the reduced discharge from industries, the general water quality in Xiuzhou deteriorated in 2005. This can be traced to the polluted water which continued to be discharged into Xiuzhou from upstream regions (Chai et al., 2005). Xiuzhou had created a precedent for the paid use of water pollution allowances in the city and the whole province.

Water pollution allowance trading for new projects, 2007–10

Various state and province documents clarified the importance of pollution reduction and quantified targets for COD reduction. Jiaxing was faced with the request of the province to reduce COD by 15.1% during the 'Eleventh 5-Year Plan (2006–2010)' to meet the national plan. Meanwhile, the Bureau of Zhejiang province clearly stipulated that the pollution allowances for new projects, or expansion of existing projects, should only be obtained under total emission control (Research Institute A, 2014). This indicates that the new demand should only be fulfilled by the existing allowances. Because of these new state and province policies, pollution allowances have become an even more scarce resource (Lin, 2007).

Enterprises with unused pollution allowances were not obliged to abandon these allowances. Along with stricter emission policies by the state and province, no pollution allowances could be granted to new projects (Li, 2007). Therefore, the city of Jiaxing set up a market to utilize all existing allowances, expecting a reduction in emissions and upgrading industrial structure (Zhejiang Environmental Protection Department, 2007). In contrast to other pilots, which discussed how to fairly assign the allowances among the existing projects, Jiaxing started with new projects.

The basic trading system at this stage was one in which enterprises with plans for new projects (including new construction, reconstruction and expansion of existing projects after 1 November 2007) were able to buy allowances from the government; enterprises which possessed unused pollution allowances, for example, due to pollution reduction, were able to sell these allowances to the government.

In 2007, two official documents promulgated the official start of the Jiaxing pollution allowance trading system (see Appendix A, A3 and A4). The market was based on research conducted by the Yangtze Delta Region Research Institute of Tsinghua University and delegated by the Jiaxing government (Zhejiang Environmental Protection Department, 2007). In 2008, a further document specified additional details on prices, trading procedures and the organization that operates the market (see

Appendix A, A5). Following the general rules set by the city, the seven counties published specific rules on allowance trading between 2007 and 2010 (see Appendix A, A6–A13).

The organization Jiaxing Municipal Pollution Allowance Reserve and Trading Centre (hereafter, 'the Centre') was set up to regulate the trading of the allowances under the authority of the Bureau. The Centre had a special bank account to deal with trade-related payments, and the money was used by the Centre to purchase the allowances from enterprises (hereafter, 'repurchase') and to execute environmental management-related projects, such as environmental quality improvement, ecological protection and environmental infrastructure construction. The Centre also set up subcentres in the counties. All pollution allowances had to be obtained under the jurisdiction of the subcentres. The Centre was responsible for the transactions in some counties before they had local subcentres. All transactions took place within local subcentres by 2009.

The coordination between the Centre and subcentres was also clarified. For instance, transactions must be periodically reported to the Centre by subcentres. Initially, the Centre partly controlled the use of subcentres' finances. Later, the subcentres gained full control of their finances. Conflicts that arose during the trading process were mediated by the Bureau or by a lawsuit in the People's Court.

The Bureau determined the basic price of COD allowances, with related departments being charged 80,000 CNY/ton for 20 years, according to the pollution abatement costs in various industries. It was also indicated that public auctions could be conducted. Enterprises with approved project proposals were allowed to join the auctions to buy the available allowances announced by the subcentre. In addition, to meet the emission reduction targets, such enterprises had to buy at least 1.2 times of the allowances they proposed; the multiple had to be 1.5 for heavily polluting industries such as chemical, pharmaceutical, tanning, printing and dyeing, and papermaking. Even though the allowances were sold for a period of 20 years, the Centre was allowed to repurchase the allowances. If the allowances had not been in use for two (the allowances achieved by the pollution reduction of the enterprise) or five years (the allowances achieved by trading), the Centre could obtain them for free. If enterprises were shut down, went bankrupt or moved out of Jiaxing, the Centre received their allowances for free or for a certain price, depending on whether the enterprise had gained the allowances at a cost or not.

Jiaxing then set up a general framework for trading pollution allowances in the city. The counties' rules were made in accordance with these city rules, and some of the counties' rules stated explicitly that 'if some issues are not covered by the local rules, please refer to the city rules' (see Appendix A, A14). Although the arrangements in the counties were usually quite similar to those in the general framework, there were still some notable differences. For example, Nanhu started the paid use of pollution allowances for existing projects simultaneously with those for new projects, for which the existing and new projects were distinguished by 1 January 2008 (see Appendix A, A9). Moreover, Nanhu started an allowance auction in 2008, which was the first auction on pollution allowances in China (Government Office A, 2018). In addition, the counties had different subsidy rules when the subcentre repurchased the allowances. This polycentric system provides counties the flexibility to develop rules which fit their specific local context.

From 2007 to 2010, the number of transactions in the counties increased gradually (Research Institute A, 2014) and the water quality improved. In 2008, water that was found to be worse than Grade V decreased by 25% compared with 2007, and ammonia nitrogen (NH_3-N) and phosphorus were found to be decreasing (Zhe Li Jiaxing, 2007). Due to the strict national and provincial targets to control emissions, along with the enterprises' reluctance to sell allowances, the market was in supply shortage (Wang, 2010a).

Paid use of allowances for existing projects, 2010–13

Together with the ongoing national and provincial pollution reduction requests, Zhejiang province issued official documents regarding a trading platform and trading methods in 2009, 2010 and 2011 (see Appendix A, A15–A17).

In the city of Jiaxing, with the ongoing pollution allowance trading system for new projects, the owners of existing projects (before 1 November 2007) wished to participate in the market after they had observed the increasing value of the allowances. Simultaneously, instigators of new projects argued that it was unfair to leave existing projects uncharged for pollution. Jiaxing authorities conceded that trading restricted to new projects limited the liquidity of pollution allowances considering that the capacity occupied by existing projects could not be fully utilized. As a result, Jiaxing started to charge pollution allowances in existing projects.

The basic trading system entailed enterprises with existing projects for which allowances are paid based on the volume confirmed by the Bureau and entrepreneurs. The official document was promulgated in 2010 (see Appendix A, A18 and A19). After the city set down the general rules, the seven counties published trial rules themselves (see Appendix A, A20–A25), except for Nanhu, which had already started allowance trading for both new and existing projects.

The Bureau of Jiaxing held a seminar on the paid use and trading of pollution allowances, and it determined that a practical way to assign allowances among existing projects was to check and admit the discharge volume using the Enterprise Environmental Impact Assessment Approval. Since approval is a legally binding document, paid use based on this document was comparatively easily accepted by enterprises.

To coordinate the pollution allowances paid between 2002 and 2003, the Bureau of Xiuzhou organized three meetings in 2010 in Xiuzhou with the related entrepreneurs and people responsible for environmental protection in the towns, to discuss the duration of allowances (Jiaxing Municipal Pollution Allowance Reserve and Trading Centre, 2010).

The basic price of COD allowances per ton was 20,000 CNY for five years or 80,000 CNY for 20 years.[10] Different discounts were offered according to the payment time required to encourage purchases. Specifically, the main differences among the counties' contractual arrangements were the dates when the trading system was implemented and discounts were offered. In addition, the Bank of Jiaxing cooperated with the Bureau to allow enterprises to mortgage their allowances. The loan could be used for manufacturing or infrastructure (Jiaxing Environmental Protection Bureau, 2008; Shen, 2015).

Between 1 July and 29 October 2010 the paid use of pollution allowances for 77% of all enterprises in Jiaxing had been completed (Wang, 2010b). The capital collected by the paid use of allowances contributed to governmental revenue, which had been designed to be used for repurchasing, environmental quality improvement, ecological protection and environmental infrastructure construction (see Appendix A, A4 and A26) By applying the paid use of pollution allowances to existing projects, Jiaxing had an updated overview of the capacity in use, and the capacity that could be released for potential new projects. This also contributed to a dataset with more systematic information, which further influenced market design.

Impact of the environmental resource marketization reform, 2013–18

Based on the experiences gained in the pilots, which included Jiaxing, the Chinese state government continued to develop its policies on market governance instruments to achieve its environmental goals. In 2014, the Office of the State Council issued a document on the paid use and trading of pollution allowances (see Appendix A, A27). The establishment of a provincial platform (see Appendix A, A28) for allowance trading was aimed at achieving consistency and standardization. In addition to consistent reduction requests for COD, the reduction request for NH_3-N was introduced by the state (Xinhua News Agency, 2011) in 2011 and by Zhejiang province in 2013 (see Appendix A, A29).

In 2014 and 2015, several official documents were introduced by the city of Jiaxing concerning the market-oriented allocation of environmental resources, the evaluation of the pollution allowance use, the repurchasing of pollution allowances, and the trading of NH_3-N (see Appendix A, A30–34). The counties translated these rules in the following years (see Appendix A, A35–A48).

At this stage, Jiaxing aimed to change the role of the government in pollution markets by separating government management from the market and leaving the trading of allowances to the market. Compared with the previous stage, the roles of the involved departments and organizations did not change significantly but were clearly defined. The Bureau monitors the markets and sets trading rules, and the finance department monitors the funds and sets budgets for purchasing allowances from enterprises.

To coordinate with the market-oriented allocation reform of environmental resources, all counties were advised to establish a joint conference system consisting of the bureaus of development and reform, finance, and environmental protection. There were methods created to coordinate the conflicts related to the markets, such as applying for mediation or filing a lawsuit, depending on the type of stakeholders involved in the conflicts. Short-term leases were introduced to coordinate the temporary shortage of allowances of some enterprises. This allows some enterprises to rent allowances from another enterprise for a few months to meet their temporary demand for allowances.

To have a complete and documented control of the pollution capacity and coordinate the allowances, Jiaxing introduced the annual budget management of allowances. The primary goal was to facilitate the allowance distribution among counties, industries, time and methods with annual plans. In principle, the budget should ensure that all allowances fall under the total emission control; the allowances in the budget may be derived from repurchased or stored allowances, and then to new projects. The budget is organized on the city and county levels. The city integrates the allowance capacity assigned by the province, the annual reduction targets and the estimated reduction by pollution reduction projects, before distributing these among the counties. The county plans the budget based on local economic development conditions, the pollution reduction plan, allowance storage and new project demand, before sending the plan to the city.

Since 2007, trading methods, including bidding, auctioning, listing and transfer by agreement, have been developed. New projects are required to acquire allowances on the official trading platform by bidding, auctioning and listing, in which the prices are determined by the market, and certain major projects (of public interest), such as those on environmental infrastructure and affecting people's livelihoods, have the possibility of obtaining allowances from the stored allowances of the Centre or subcentres, at prices determined by the government. At this stage, the allowances are sold for a maximum of five years or until the end of the current five-year plan. If the allowances are not in use for three years, the Centre or subcentres will repurchase them at the average price listed in the previous year or the price determined by the government. The repurchase rules were further detailed on the conditions and prices of repurchases.

To encourage enterprises to utilize their allowances and to award preference to those enterprises that have made use of advanced technology and have generated large profits, Jiaxing introduced a system for evaluating allowance use. Based on the results obtained from the evaluation, officials differentiated the prices and methods of trading. This strategy, literally translated from Chinese as 'Three–three Reforms', has been applied to heavy polluting industries, such as printing and dyeing, paper, leather, chemical, electroplating and thermal power. The evaluation system is based on the tax paid to the Chinese government divided by the amount of allowances that the enterprises used in the past year. The allowances for calculation vary with industries. The evaluation will then produce a ranking list of enterprises that can be applied to each industry. As of 2015, the evaluation system was operational. Besides the Three–three Reforms differentiation, Jiaxing also comprehensively evaluates all industrial enterprises, which refers to land, labour, pollution and technology. Within this polycentric system, counties have the freedom to adapt to specific indicators and combine these two differentiation systems.

In addition to the different design of differentiation management, the counties have various experience in the application of various trading methods. For instance, according to the data up to 2018, Nanhu has operated an allowance auction for approximately 10 years and an online auction for approximately six years. Therefore, it had a more developed practice of auctions compared with other counties.

Reflecting on market performance, the counties have improved their designs. For instance, Nanhu changed the minimum participant number of online auctions from two to one, because there was a situation in which one project had waited for several months until the second enterprise joined the auction, which unnecessarily delayed the project (see Appendix A, A49). Besides reflecting on their own market performance, the counties have also learned from each other, which confirms the advantage of this polycentric system. For instance, Jiashan learned from Pinghu, Haining and Haiyan on differentiation policies related to low-ranking enterprises (see Appendix A, A50).

The systematic allocation of environmental resources, including pollution allowances among enterprises, has effectively promoted the flow of environmental resources to low-pollution, low-energy and high value-added industries. The level of technological equipment and pollution control in Jiaxing enterprises has also generally improved (Zhe Li Jiaxing, 2007).

Jiaxing also applied other instruments to support market operations. For instance, by September 2013, Jiaxing was able to control the wastewater discharge of most enterprises by using top-up cards, for which the enterprises needed to charge with the allowance periodically at the Bureau (Wang, 2014). Because of all these efforts, including markets for water pollution allowances, during the 'Twelfth 5-Year Plan (2011–2015)' period, the emissions of COD and NH_3-N in Jiaxing were reduced by 18.49% and 15.74%, respectively, which amounted to more than the original pollution reduction targets (Zhe Li Jiaxing, 2007).

Discussion and conclusions

In this study we conducted a longitudinal analysis of the evolution of the markets for water pollution allowances in the city of Jiaxing, China. Using this analysis we investigated how the institutional design has changed corresponding to external developments and internal developments in this polycentric system. We based our theoretical framework on the literature on institutional change and markets for water pollution allowances. The framework introduced the dimensions we applied to analyse institutional design and performance, and external and internal factors, which may explain the institutional design changes.

We analysed the market development through four stages (Figure 2): (1) paid use of water pollution allowances in Xiuzhou district (2002–07); (2) water pollution allowance trading for new projects (2007–10); (3) paid use of allowances for existing projects (2010–13); and (4) impact of the environmental resource marketization reform (2013–18).

At stage 1, the district of Xiuzhou facilitated the paid use of water pollution allowances for COD under external pressure from the total emission control by the state, and internal demand for increased emission capacity for local economic development. During those years, the discharge of industrial wastewater in the entire district was lower than that of the quota. Xiuzhou demonstrated a positive precedent for this type of market.

At stage 2, total emission control remained in operation with detailed goals on COD reduction, in line with which the province decided to allow new enterprises only if they were able to get a share of the existing quota. To further utilize the existing quota, the city of Jiaxing started COD allowance trading for new projects. To conduct trading, Jiaxing published guiding rules and established the Centre. In the guiding framework set by the city, a fixed price was decided for the transfer of allowances, and the repurchase of unused allowances by the Centre was designed. The seven counties then published their own rules, and local subcentres were established to conduct transactions. Before the setup of a county's subcentre, the Centre was responsible for its transactions. Even after the whole system was put into practice, the Centre was still playing a monitoring role. Nanhu started paid use of allowances among existing projects simultaneously and introduced an allowance auction for new projects instead of using a fixed price. An increase in transactions and an improvement in water quality were recorded at this stage.

At stage 3, the total emission control remained in operation. In addition, the province published guiding rules for this type of market, which externally confirmed the positive attitude towards the trading of allowances. Together with the market performance at the previous stage and increasing demand for allowance liquidity, Jiaxing was motivated to start the paid use of pollution allowances for existing projects. Fixed price and fixed allowance duration were defined in the city-guiding rules. With the following localized

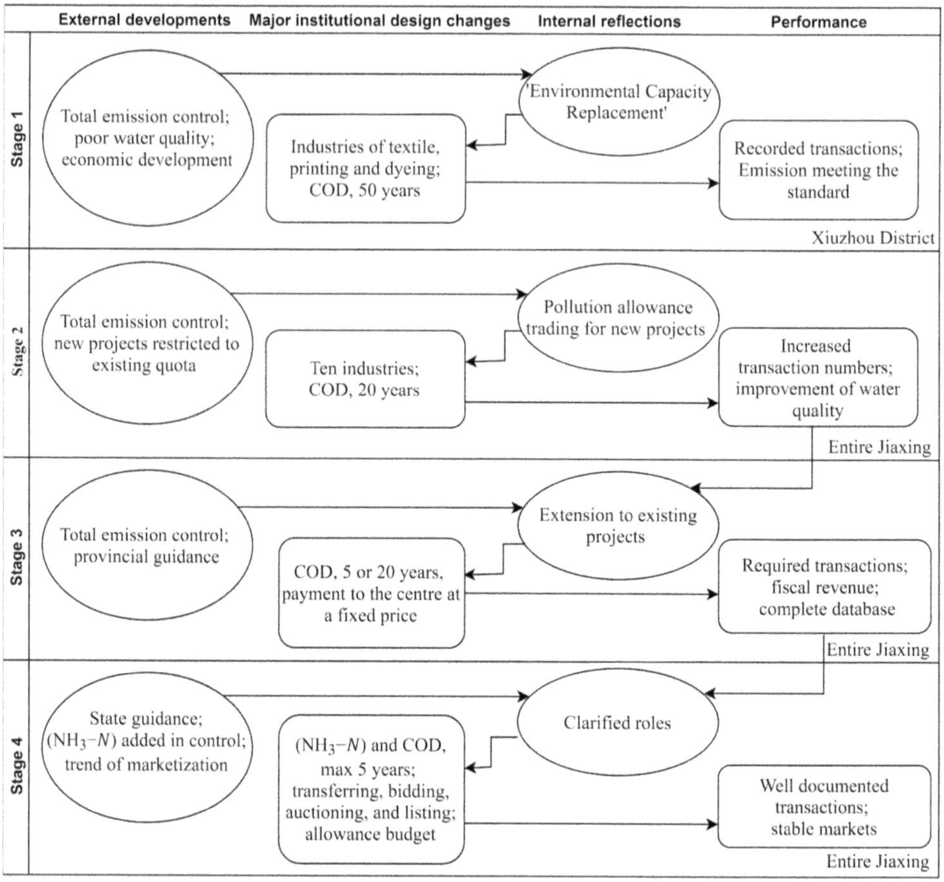

Figure 2. Main institutional design changes in response to the external developments and internal reflections on performance.

rules, the counties (expect for Nanhu) were able to introduce existing projects' allowances into the market. Different discounts and allowance mortgages were designed to motivate enterprises to complete their paid use.

Stage 4 is the stage at which transactions are standardized and well-documented. To some extent, a provincial online platform enforced the standardization of trading, on which subcentres (in Jiaxing's case) could conduct their local transactions. Besides COD, Jiaxing extended the allowance trading to NH_3-N as it was included in total emission control by the state and province. In line with the reform of market-oriented environmental resource allocation, Jiaxing tried to separate government management from the market by clarifying the monitoring role of government and restricting the trading of allowances for new projects to bidding, listing and auctioning. To increase the liquidity of existing allowances or make allowances more affordable, ranking for allowance use evaluation among enterprises, shorter allowance duration, short-term lease and auctions were introduced or extended. Similar to previous stages, the counties differed in detail regarding these innovative aspects, and they had the flexibility to learn from each other.

An annual budget for allowances was established at this stage to control all allowances in this city–county system. This further links the markets for water pollution allowances of the city and counties.

We were able to determine some links between institutional design and performance, some of which also match the presumptions drawn in the literature outlined in the theoretical framework.

In the law and legislation system, the introduction of total emissions control has directly contributed to the improvement of water quality. This control of the overall pollution is also vital to the proper functioning of the market by creating scarcity, which ensures the market participation of enterprises to some extent. In addition, the up-to-date steering documents published by Jiaxing and its counties have contributed to the implementation of institutional design changes. These documents cover the trial or formal rules on trading procedures, initial allowance allocation, allowance reserve and repurchase, allowance mortgage and allowance use evaluation.

Regarding governance, Jiaxing's case demonstrated the positive role that a government can play in the development of these markets. In Jiaxing, the government provided an effective administrative and legal system and ensured the consistency and improvement of the markets. In contrast to some views that government control limits the development of water pollution allowances or causes inactive trading (Li & Fang, 2012), the Jiaxing government has activated trade participation by adjusting some rules. Illuminating examples include the introduction of short-term leases, mortgages and one-participant auctions. Furthermore, funds from the paid use of allowances for existing projects implemented by the government has led to increased governmental revenue for solving environmental issues. Contrarily, it would be difficult to maintain overall control or consistent market development without the government, and the time required to achieve similar results may increase. It is difficult to compare the current costs borne by the government and the potential costs borne by the environment and society in future. In addition, compared with pollution fee regulation, this type of market motivates the enterprises to further utilize the pollution allowances by making allowances tradable. However, the important role played by the local government also indicates that the performance of the market might vary if it is extended to other areas, because its influence on water quality, to a large extent, depends on the support of other water policies.

The coordination between the city and counties is noteworthy in Jiaxing's case. City and county budgets have been set up to systematically control pollution allowances. Moreover, the fact that Jiaxing allows its seven counties to run independent markets allows for new rule testing and learning. For instance, the allowance auction began in Nanhu, and then it was published as one of the trading methods in the city document, after which it was gradually adopted by other counties. The case also shows that when transactions are conducted within the counties, instead of a larger territory, market development does not have to be limited. When it comes to expanding one market to a larger territory, one concern is that the influence of one pollution unit might differ by location. This leads to the result that the larger the area covered by the same market, the higher the uncertainty there can be. In Jiaxing, even though the pipe system for waste water can average the influence, there might be an unequal distribution of pollution if the trading is allowed between the counties. Furthermore, the market at the county level fits the current administrative system, because the enterprises are connected to various governmental offices at the county level. However, because of this administrative system, some counties might not take the initiative to learn about the innovations of other counties until the official document was published by the city. This might delay market development to some extent.

The updating of contractual arrangements has avoided the distortion of the previous development, which reduced the reluctance of enterprises to participate in the market. Whenever new rules such as change of allowance duration, were introduced, there were specific lines to ensure that the enterprises' exiting rights defined by the old rules would still be guaranteed. By introducing and updating the rules of repurchase and price differentiation, Jiaxing has motivated enterprises to improve technology and make full use of the allowances, which further contributes to the overall improvement of water quality.

Market design is a continuous learning process. It took more than a decade to develop well-functioning markets in Jiaxing, and the design of the markets has been adjusted continuously to resolve issues such as low market participation and unclear market demarcation. In addition, Jiaxing has continuously responded to changes in national and provincial legislation, such as enriching pollution indices. As some counties have been more successful in improving market performance than others, this has provided room for learning. Currently, the markets for water pollution allowances in Jiaxing are only for industrial point pollution. Of the documents we consulted, only Jiashan indicated that some large-scale farms which were approved by the Bureau could be the suppliers of allowances (see Appendix A, A8), but we did not find any cases in practice. In many watersheds, the majority of the pollution is created by non-point sources (Cooley et al., 2015; Fisher-vanden & Olmstead, 2013), and some practice and research have already been conducted on transactions involving non-point pollution. The high costs of measuring actual effluent for non-point polluters (Breetz et al., 2005), and the resulting reluctance of farmers to generate allowances (Selman et al., 2009), have challenged the incorporation of non-point polluters.

In summary, the following lessons stand out from this case study, which can be used as references for other locations: (1) control of overall pollution capacity is necessary to create scarcity in this market; (2) developing a well-functioning market is a learning process, the design of which needs to be altered in response to external developments and internal reflections; (3) in this process, changes could be tested in a subarea to gain experience, but changes that may distort the previous development should be avoided; and (4) different levels of governments can contribute to market development by ensuring legal stability and consistent rules.

Notes

1. In China, there are cities with subordinate districts, for instance, Jiaxing; cites at the same level of county, for instance, Tongxiang; and cities that are municipalities directly under the central government, for instance, Shanghai.
2. They are not exclusive from each other.
3. Hereafter, only the names of the districts, counties or city-level counties will be referred to, for instance, 'Nanhu' instead of 'Nanhu district'.

4. Based on the emissions of specific pollutants reported by provinces, the national office compiles a national plan that balances and decomposes the next-stage emission targets among the provinces. The provinces then decompose the plan to the lower levels of governments (see Appendix A, A1). However, up until now, environmental capacity has not yet been employed as the reference for the plan (Song, 2017). (By 'provinces', we also include the other administrative regions that are at the same level with provinces.)
5. To keep a relatively short reference list, the policy documents cited in this paper are provided in Appendix A.
6. According to 'Environmental quality standards for surface water (GB3838-2002)', the water quality in China is rated from highest to lowest as, III, IV and V, based on a group of indicators including temperature, pH, dissolved oxygen, chemiluminescence detection of permanganate index, chemical oxygen demand (COD), five-day biochemical oxygen demand, ammonia nitrogen, total phosphorus, total nitrogen, copper, zinc, fluoride, selenium, arsenic, mercury, cadmium, chromium (hexavalent), lead, cyanide, volatile phenol, petroleum, anionic surfactants, sulphide and *Enterococcus faecalis*.
7. The District Development and Reform Commission has taken responsibility for the former Department of Price.
8. Optimizing and upgrading the industrial structure indicates the process of production factor flow from the low value-added production sectors with poor efficiency and high consumption to high value-added sectors with high efficiency and low consumption. The production factors include capital, labour, land and technology (Shi & Wang, 2014).
9. See note 8.
10. The price of COD is similar to that during the period 2007–10, according to the document; the price was decided by the Bureau and other departments according to the pollution abatement costs in industries (see Appendix A, A4).

Acknowledgments

The fieldwork was supported by Radboud University. Hao Wang was supported by the China Scholarship Council (CSC).

Disclosure statement

No potential conflict of interest was reported by the authors.

ORCID

Hao Wang http://orcid.org/0000-0003-3392-1562
Sander Meijerink http://orcid.org/0000-0002-4249-6230
Erwin van der Krabben http://orcid.org/0000-0002-2566-890X

References

Breetz, H. L., Fisher-vanden, K., Jacobs, H., & Schary, C. (2005). Trust and Communication: Mechanisms for increasing farmers' participation in water quality trading. *Land Economics*, *81* (2), 170–190. https://doi.org/10.3368/le.81.2.170

Bremer, L. L., Farley, K. A., & Lopez-Carr, D. (2014). What factors influence participation in payment for ecosystem services programs? An evaluation of Ecuador's SocioPáramo program. *Land Use Policy*, *36*, 122–133. https://doi.org/10.1016/j.landusepol.2013.08.002

Brouwer, R., Tesfaye, A., & Pauw, P. (2011). Meta-analysis of institutional–economic factors explaining the environmental performance of payments for watershed services. *Environmental Conservation, 38*(4), 380–392. https://doi.org/10.1017/S0376892911000543

Buitelaar, E., Lagendijk, A., & Jacobs, W. (2007). A theory of institutional change: Illustrated by Dutch city-provinces and Dutch land policy. *Environment and Planning A, 39*(4), 891–908. https://doi.org/10.1068/a38191

Chai, C., Fu, P., & Sun, B. (2005). *Taking environmental capacity as a commodity, Jiangsu and Zhejiang try pollution allowance trading [in Chinese]*. E20 Environment Platform-Chndaqi. http://www.chndaqi.com/news/39463.html

Coase, R. H. (1960). The problem of social cost. *The Journal of Law and Economics*, 3, 1–44. https://doi.org/10.1086/466560

Cooley, H., Rippman, H., Heberger, M., & Cohen, M. (2015). *Incentive-based instruments for water Management*. The Rockefeller Foundation. https://www.rockefellerfoundation.org/report/incentive-based-instruments-for-water-management/

Fisher-vanden, K., & Olmstead, S. (2013). Moving pollution trading from air to water: Potential, problems, and prognosis. *Journal of Economic Perspectives, 27*(1), 147–172. https://doi.org/10.1257/jep.27.1.147

Government Office A. (2018). *Implementation and Development of Pollution Allowance Trading [in Chinese]*. Unpublished confidential document.

Guo, R., Xu, H., & Zhang, J. (2002). Discussion on the paid use mechanism of emission allowances Chinese. *China Environment Management*.

Huitema, D., Mostert, E., Egas, W., Moellenkamp, S., Pahl-Wostl, C., & Yalcin, R. (2009). Adaptive water governance: Assessing the institutional prescriptions of adaptive (co-)management from a governance perspective and defining a research Agenda. *Ecology and Society, 14*(1), 26. https://doi.org/10.5751/ES-02827-140126

Jaeger, W. K. (2011). *Ecosystem services and the potential role for markets*. EM 9033. Oregon State University Extension Service, 1–29. https://catalog.extension.oregonstate.edu/em9033

Jiang, F., Chai, J., & Xu, S. (2005). *Pollution Quota Can Be Traded; Yangtze River Delta Starts The Pollution Allowance Trading in China [in Chinese]*. Website of the People's Government of Zhejiang Province. http://zfxxgk.zj.gov.cn/xxgk/jcms_files/jcms1/web37/site/art/2012/12/28/art_2491_572000.html

Jiaxing Environmental Protection Bureau. (2008). *Jiaxing City Is the First to Implement Mortgage on Pollution Allowances in China [in Chinese]*. http://www.jepb.gov.cn/News/NewsDetail.aspx?ID=3490

Jiaxing Municipal Pollution Allowance Reserve and Trading Centre. (2010). *Jiaxing environmental protection bureau held the meeting with enterprises that bought pollution allowances between 2002 and 2003 [in Chinese]*. Previous website of Jiaxing Environmental Protection Bureau, not accessible anymore

Jiaxing Statistic Bureau. (2019). *Statistical communique of Jiaxing city on the economic and social development in 2018 [in Chinese]*. http://tjj.jiaxing.gov.cn/art/2019/4/7/art_1512321_31375488.html

Jiaxing Water Resources Bureau. (2018). *Water Resources Overview [in Chinese]*. http://slj.jiaxing.gov.cn/col/col1506253/index.html

Kraemer, R. A., Kampa, E., & Interwies, E. (2004). *The Role of Tradable Permits in Water Pollution Control*. Working paper. https://www.ecologic.eu/13004

Lai, L. W. C., & Hung, C. W. Y. (2008). The inner logic of the Coase Theorem and a Coasian planning research agenda. *Environment and Planning B: Planning and Design, 35*(2), 207–226. https://doi.org/10.1068/b33057

Li, F. (2007, September 5). Paid use of pollution allowances in Shaoxing started [in Chinese]. *Shaoxing Daily*.

Li, Z., & Fang, Y. (2012). Conception of establishing emissions trading centre in Zhejiang province [in Chinese]. *Economic Research Guide, 9*, 180–181. http://www.cnki.com.cn/Article/CJFDTotal-JJYD201209077.htm

Lin, J. (2007). *First pollution allowances reserve and trading centre opened [in Chinese]*. E20 Environment Platform-h2o-China. https://www.h2o-china.com/news/64023.html

MEA. (2005). Ecosystems and human well-being. In *Millennium Ecosystem Assessment* (p. 2). https://doi.org/10.1196/annals.1439.003

National Bureau of Statistics of China. (2019). *The effect of environmental protection continues to show; ecological civilisation is building up day by day–the fifth report in the series of reports on the achievements of economic and social development [in Chinese]*. http://www.gov.cn/xinwen/2019-07/18/content_5410785.htm

Pan, Z. (2007). What pollution allowance trading will bring to environmental protection [in Chinese]. *Sina News*. http://news.sina.com.cn/c/2007-12-03/115014439565.shtml

Pollution allowance. (2020). *Columbia Electronic Encyclopedia*, 6th edn, 1.

Research Institute A. (2014). *Research on the Policies of Paid Use and Trading of Emission Allowances in Jiaxing [in Chinese]*. Unpublished confidential document.

Selman, M., Greenhalgh, S., Branosky, E., Jones, C. Y., & Guiling, J. (2009). ISSUE BRIEF: Water quality trading programs: An international overview. *World Resources Institute*. https://wriorg.s3.amazonaws.com/s3fs-public/pdf/water_trading_quality_programs_international_overview.pdf

Shen, M. (2004). The studies of the institutions of water rights transactions–the case studies in China [in Chinese]. Doctoral dissertation, Zhejiang University. China National Knowledge Infrastructure.

Shen, M. (2015). *Two enterprises in Tongxiang got tens of millions of mortgages on pollution allowances [in Chinese]*. Website of Jiaxing Environmental Protection Bureau. http://old3w.cnjxol.com/xwzx/jxxw/qxxw/tx/content/2015-01/27/content_3262088.htm

Shen, M., & Xie, H. (2010). Eco-economisation: Positive and normative analysis – A case study on compensated use of emission rights in Jiaxing [in Chinese]. *Journal of China University of Geosciences (Social Sciences Edition)*, *10*(6), 27–34. http://www.cnki.com.cn/Article/CJFDTOTAL-DDXS201006005.htm

Shen, M., Xie, H., & Zhou, N. (2013). Zhejiang model for the emission system reform [in Chinese]. *Journal of Zhejiang Provincial Party School*, *6*, 29–36. http://www.cnki.com.cn/Article/CJFDTOTAL-ZSWD201306006.htm

Shen, M., Zhou, S., Xie, H., Wei, J., Xu, S., Zhou, N., Xu, M., Shen, W., Qiu, N., Yu, D., Zhang, B., Huang, W., Chen, Q., & Shen, L. (2014). *The studies of supervision mechanism for pollution rights*. China Environmental Science Press. [in Chinese]

Shi, Y., & Wang, X. (2014). *Optimizing and Upgrading Industrial Structure. Development Research Center of the State Council of the People's Republic of China*. http://www.chinadaily.com.cn/m/drc/2014-05/23/content_17537042.htm

Shilling, J. D., & Osha, J. (2003). *Paying for environmental stewardship*. Paying for Environmental Stewardship. WWF Macroeconomics for Sustainable Development Program Office.

Skelcher, C. (2005). Jurisdictional integrity, polycentrism, and the design of democratic governance. *Governance*, *18*(1), 89–110. https://doi.org/10.1111/j.1468-0491.2004.00267.x

Song, F. (2017). *The origin of total emission control system and the practice in China [in Chinese]*. Doctoral dissertation, Chinese University of Political Science and Law Press. China National Knowledge Infrastructure.

Tietenberg, T. H. (2006). *Emissions Trading, Principles and practice* (2nd ed.). Resources for the Future.

Vatn, A. (2010). An institutional analysis of payments for environmental services. *Ecological Economics*, *69*(6), 1245–1252. https://doi.org/10.1016/j.ecolecon.2009.11.018

Wang, E. (2010a). *Deputy director of Jiaxing environmental protection bureau: The difficulty of pollution allowance trading is the legal recognition [in Chinese]*. Sina. http://green.sina.com.cn/2010-09-07/134321055375.shtml

Wang, E. (2010b). *Further improvement in Jiaxing's pollution allowance trading: 2,600 enterprises will pay for the initial pollution allowances [in Chinese]*. Jiaxing Zai Xian. http://www.cnjxol.com/xwzx/jxxw/wmkjx/content/2010-07/27/content_1441488.htm

Wang, J. (2014, June 10). Zhejiang Sample of Pollution Allowance Trading (Part One). *China Financial and Economic News*. https://xuewen.cnki.net/CCND-CCJB201406100011.html [in Chinese].

Wu, T. (2011). *Pollution allowances practices and improvement: A case study of Shaoxing, Doctoral dissertation, Zhejiang Gongshang University [in Chinese]*. China National Knowledge Infrastructure.

Wunder, S. (2005). *Payments for environmental services: Some nuts and bolts. CIFOR Occasional Paper No. 42.* Center for International Forestry Research: Bogor, Indonesia. https://www.cifor.org/publications/pdf_files/OccPapers/OP-42.pdf

Wunder, S., Engel, S., & Pagiola, S. (2008). Taking stock: A comparative analysis of payments for environmental services programs in developed and developing countries. *Ecological Economics*, *65*(4), 834–852. https://doi.org/10.1016/j.ecolecon.2008.03.010

Xinhua News Agency. (2005). *Xiuzhou, Jiaxing, Zhejiang, started pollution allowance research [in Chinese]*. http://www.h2o-China.com/news/40040.html

Xinhua News Agency. (2011). *Twelfth Five-Year Plan for national economic and social development of the People's Republic of China [in Chinese]*. http://www.gov.cn/2011lh/content_1825838_2.htm

Yu, L. (2007). *Research on construction of water pollutant emission right trading system in Zhejiang province, Doctoral dissertation, Zhejiang Gongshang University [in Chinese]*. China National Knowledge Infrastructure.

Yu, X. (2008). Improving the pollution allowance trading mechanism in the context of emission reduction: Taking Jiaxing, Zhejiang, the First Pilot City in China as an Example [in Chinese]. *Proceedings of the 4th China (International) Forum on Environment and Development*, 436–440. Modern Education Press.

Zhe Li Jiaxing. (2007). *Reform and Opening in Jiaxing: Motivate Green Development–First Implementation of Pollution Rights Trading in China [in Chinese]*. Website of Jiaxingren. http://www.jiaxingren.com/folder22/folder286/2018-12-06/430627.html

Zhang, Y., Zhang, B., & Bi, J. (2012). Policy conflict and the feasibility of water pollution trading programs in the Tai Lake Basin, China. *Environment and Planning C: Politics and Space*, *30*(3), 416–428. https://doi.org/10.1068/c11124

Zhejiang Environmental Protection Department. (2007). *All the new projects in Jiaxing should pay for the pollution allowances, ending the history of free occupation of the natural resources [in Chinese]*. http://zfxxgk.zj.gov.cn/xxgk/jcms_files/jcms1/web37/site/art/2012/12/28/art_2491_572895.html

Appendix A

A1. State Environmental Protection Administration of China. (2008). *Assessing the Total Emission Reduction of Major Pollutants [in Chinese]*. http://www.mofcom.gov.cn/aarticle/bh/200802/20080205391184.html

A2. Environmental Protection Commission of the State Council of the People's Republic of China. (1986). *Provisions on Technical Policies for Prevention and Control of Water Pollution [in Chinese]*. http://cyfd.cnki.com.cn/Article/N2015110032002634.htm

A3. Jiaxing People's Government. (2007). *Pollution Allowance Trading of Major Pollutants in Jiaxing (Trail Rules) [in Chinese]*.

A4. Jiaxing People's Government. (2007). *Details on the Pollution Allowance Trading of Major Pollutants in Jiaxing (Trail Rules) [in Chinese]*.

A5. Jiaxing Environmental Protection Bureau (2008). *Further Rules on Normalizing Pollution Allowance Trading [in Chinese]*.

A6. Haining People's Government. (2008). *Pollution Allowance Trading of Major Pollutants in Haining City (Trial Rules) [in Chinese]*.

A7. Haiyan People's Government. (2008). *Pollution Allowance Trading of Major Pollutants in Haiyan County (Trial Rules) [in Chinese]*.

A8. Jiashan People's Government. (2007). *Pollution Allowance Trading of Major Pollutants in Jiashan County (Trial Rules) [in Chinese].*
A9. Nanhu People's Government. (2007). *Pollution Allowance Trading of Major Pollutants in Nanhu District (Trial Rules) [in Chinese].*
A10. Nanhu People's Government. (2008). *Supplementary Rules on Pollution Allowance Trading of Major Pollutants in Nanhu District (Trail Rules) [in Chinese].*
A11. Pinghu People's Government. (2008). *Pollution Allowance Trading of Major Pollutants in Pinghu City (Trial Rules) [in Chinese].*
A12. Tongxiang People's Government. (2007). *Pollution Allowance Trading of Major Pollutants in Tongxiang City (Trial Rules) [in Chinese].*
A13. Tongxiang People's Government. (2008). *Details on Pollution Allowance Trading of Major Pollutants Tongxiang City (Trial Rules) [in Chinese].*
A14. Xiuzhou People's Government. (2010). *Pollution Allowance Trading of Major Pollutants in Xiuzhou District, Jiaxing City (Trail Rules) [in Chinese].*
A15. Zhejiang People's Government. (2009). *Opinions by Zhejiang People's Government on the Pilots of Paid Use and Trading of Pollution allowances [in Chinese].*
A16. Zhejiang People's Government. (2010). *Temporary Rules on the Implementation of the Pilots on the Paid Use and Trading of Pollution Allowances in Zhejiang Province [in Chinese].*
A17. Zhejiang Environmental Protection Department. (2011). *Details on the Temporary Rules on the Implementation of the Pilots on the Paid Use and Trading of Pollution Allowances in Zhejiang Province [in Chinese].*
A18. Jiaxing People's Government. (2010). *Paid Use of the Initial Pollution Allowances of Major Pollutants in Jiaxing City (Trial Rules) [in Chinese].*
A19. Jiaxing Environmental Protection Bureau. (2010). *Details on Paid Use of the Initial Pollution Allowances of Major Pollutants in Jiaxing City (Trial Rules) [in Chinese].*
A20. Haining People's Government. (2011). *Paid Use and Trading of Pollution Allowances of Major Pollutants in Haining (Trail Rules) [in Chinese].*
A21. Haiyan People's Government. (2010). *Paid Use of the Initial Pollution Allowances of Major Pollutants in Haiyan (Trial Rules) [in Chinese].*
A22. Jiashan People's Government. (2012). *Pollution Allowance Trading of Major Pollutants in Jiashan County (Trial Rules) [in Chinese].*
A23. Pinghu People's Government. (2010). *Paid Use of the Initial Pollution Allowances of Major Pollutants in Pinghu District (Trial Rules) [in Chinese].*
A24. Tongxiang People's Government. (2011). *Temporary Rules on the Paid Use of the Initial Pollution Allowances in Tongxiang City [in Chinese].*
A25. Xiuzhou People's Government. (2011). *Details on the Paid Use of the Initial Pollution Allowances of Major Pollutants in Xiuzhou District (Trial Rules) [in Chinese].*
A26. Jiaxing Finance Bureau. (2010). *Management of the Finance of Paid Use of Pollution allowances [in Chinese].*
A27. General Office of the State Council of the People's Republic of China. (2014). *Instructions by the General Office of the State Council of the People's Republic of China on the Further Implementation of the Pilots on the Paid Use and Trading of Pollution Allowances [in Chinese].*
A28. Zhejiang People's Government. (2013). *Temporary Rules on the Paid Use and Trading of Pollution Allowances in Zhejiang Province [in Chinese].*
A29. Zhejiang People's Government. (2013). *Opinions by the General Office of Zhejiang People's Government on Strengthening the Quantitative Management of Environmental Resources Allocation to Promote Industrial Transformation [in Chinese].*
A30. Jiaxing Environmental Protection Bureau. (2014). *Reserve and Repurchase of Pollution Allowances in Jiaxing (Trail Rules) [in Chinese].*
A31. Jiaxing Environmental Protection Bureau. (2014). *Strengthening the Major Pollutants Approval and Supervision for Construction Projects [in Chinese].*
A32. Jiaxing Environmental Protection Bureau. (2014). *Technical Specifications on the Initial Pollution Allowances Approval and Pricing of Ammonia Nitrogen, Nitrogen Oxides in Jiaxing (Trail Rules) [in Chinese].*

A33. Jiaxing People's Government. (2014). *Deepening the Reform of Market-Oriented Allocation of Environmental Resource Elements in Jiaxing City [in Chinese].*

A34. Jiaxing People's Government. (2014). *Strengthening the Use of Performance Evaluation of Pollution Allowances to Promote Industrial Transformation and Upgrading in Jiaxing [in Chinese].*

A35. Haiyan People's Government. (2015). *Paid Use and Trading of Pollution Allowances in Haiyan County [in Chinese].*

A36. Haiyan People's Government. (2015). *Plan on Establishing the Market for Resources in Haiyan County [in Chinese].*

A37. Haiyan Environmental Protection Bureau. (2015). *Approval and Distribution of Pollution Allowances for the Major Pollutants in Haiyan County [in Chinese].*

A38. Haiyan Environmental Protection Bureau. (2016). *Paid Use of Pollution Allowances of the Major Pollutants in Haiyan County [in Chinese].*

A39. Jiashan Eco-civilization Office. (2014). *Strengthening the Use of Performance Evaluation of Pollution Allowances to Promote Industrial Transformation and Upgrading in Jiashan County (Trail Rules) [in Chinese].*

A40. Jiashan People's Government. (2015). *Reserve and Selling of Pollution Allowances Management in Jiashan (Trail Rules) [in Chinese].*

A41. Nanhu Committee of the Communist Party of China. (2014). *Plan on Accelerating the Reform of Market-Oriented Allocation of Environmental Resource Elements in Nanhu District [in Chinese].*

A42. Nanhu People's Government. (2015). *Deepening the Reform of Market-Oriented Allocation of Environmental Resource Elements in Nanhu District [in Chinese].*

A43. Nanhu People's Government. (2015). *Paid Use and Trading of Pollution Allowances in Nanhu District [in Chinese].*

A44. Nanhu People's Government. (2015). *Quantitative Management Approach of the Environmental Resource Elements in Nanhu District (Trail Rules) [in Chinese].*

A45. Nanhu People's Government. (2015). *Strengthening the Use of Performance Evaluation of Pollution Allowances to Promote Industrial Transformation and Upgrading in Nanhu District [in Chinese].*

A46. Pinghu. (2014). *Pollution Allowance Trading in Pinghu City [in Chinese].*

A47. Tongxiang. (2015). *Deepening the Reform of Market-Oriented Allocation of Environmental Resource Elements in Tongxiang [in Chinese].*

A48. Tongxiang. (2015). *Quantitative Management Approach of the Environmental Resource Elements in Tongxiang [in Chinese].*

A49. Nanhu People's Government. (2017). *Supplementary Rules on the Paid Use and Trading of Pollution Allowances in Nanhu District [in Chinese].*

A50. Jiashan People's Government. (2017). *Explanations of the Rules on the Comprehensive Evaluation of Industrial Enterprises in Jiashan County [in Chinese].* www.jiashan.gov.cn › art › art_7032_452608

Service levels for the four billion people with piped water on premises

Neil S. Grigg

ABSTRACT
While more than half of the people in the world have access to piped water on premises, no global assessment has been made of the availability, safety, reliability and affordability of service levels. The range of situations is challenging, but assessments based on performance can raise awareness and equip officials to promote higher and safer piped water service levels. The article explores explanatory variables for service levels and uses descriptive information and expert opinion to triangulate towards conclusions about how service levels vary. Research needs are identified for development of protocols and data classes for a piped water services rating system.

Introduction: assessing piped on-premises water supplies

According to the Joint Monitoring Programme for Water Supply and Sanitation (JMP, 2017a), in 2015 some 4.2 billion people, or about 58% of the global population, had access to piped water on premises. Piped water is one of the sources that will qualify as 'safely managed', according to the JMP's (2017b) revised system to assess access to safe drinking water. Other possibilities for safely managed drinking water on premises include boreholes, protected dug wells, protected springs, rainwater, bottled water and tank-truck water, depending on the effectiveness of their protection.

Within the category of piped water on premises, there is a large gap between the service levels for those with reliable, safe and affordable water and those who have services that are intermittent, of poor quality, and too expensive. Lumping the 4.2 billion people with access to piped water on premises together, there is no meaningful distinction between the service levels. This lack of a global estimate of levels of service for piped water is not because the problem is unimportant, but due to the complexity of the issue and division of responsibility to report service levels between health agencies and those representing utilities or regulators.

Drinking water service levels are embedded in the Sustainable Development Goals (SDGs) and are among many determinants of liveability in communities. They create a health and equity issue because poor-quality piped water can lead to disease outbreaks (Ercumen, Gruber, & Colford, 2014), and they create an economic issue because they affect business and industry and people without safe water face loss of work days. The global levels of piped water services are highly variable in cities and communities of various sizes with many different forms of water service arrangements.

Recognizing the need to assess piped water services more intensively, the JMP (2017b) now plans to study the service levels of people with piped water in more detail and to seek data from water providers and regulators, as well as from household surveys. This will be challenging, because global data on piped water services are not organized in any central way and the JMP must develop ways to access them. The JMP is working continually to develop better methods for overall assessments of drinking water access, which are explained in its introduction to the new 'safely managed' category (JMP, 2017b).

The JMP has previously based its data collection mainly on household surveys and censuses, but it will also seek information from administrative sources, including regulators. In parallel with these efforts, organizations that represent utilities are also evolving systems to assess the performance of water services. The term 'utility' is used here to mean any organized system to provide drinking water service to premises, such as local government or regional utilities, national and state utility companies, private companies, ancillary systems belonging to a business, community systems and water user associations.

One example of an assessment activity is the International Benchmarking Network for Water and Sanitation Utilities, which has operated since 1997 (Danilenko, van den Berg, Macheve, & Moffitt, 2014). Also, national water utility assessment programmes are evolving, as for example in the US (American Water Works Association, 2017). Despite these efforts, a global picture of the levels of service among piped water users is still lacking.

This article explores how to assess the service levels for the large share of the global population with piped water on premises. The study is exploratory because the data do not enable direct analyses and the evidence presented is largely circumstantial because it relies on inferences to connect conclusions to statements of fact. The article explains the types of organizations that provide piped water, the types of regulatory controls and how service levels vary across countries. Both quantitative and qualitative approaches are used, and different sources of data are used to triangulate towards conclusions about the quality of services available to the 4.2 billion people receiving piped water supplies. The intended contributions are to issues confronting attempts to assess levels of piped water service, to suggest how to improve the data and methods, and to identify research needs about the important underlying issues of safe piped water services for everyone.

Definitions and data elements

Assessing the global status of piped water services requires definitions and data classes, and developing these has proved to be a challenge due to the diversity of situations. The definitions began with the JMP (2017c) programme to define an improved source of drinking water, which 'by the nature of its construction and when properly used, adequately protects the source from outside contamination, particularly fecal matter'. This definition was used by JMP to report on progress towards the Millennium Development Goal (MDG) for drinking water.

In the JMP assessment system, the 'access' metric refers primarily to the distance the user must travel to obtain the water, which reflects the emphasis on non-piped services. Another metric implicit in the definition is whether the water is 'properly used', which injects an element of uncertainty into the data because it is impossible to know all conditions of water use. Taken together, the metrics for an improved source include construction quality, accessibility, and manner of use of the water. Now, the JMP (2017b) has acknowledged the limitations of indicators based on source type and introduced a revised indicator to consider the accessibility, availability and quality of drinking water. This will be a new rung at the top of the drinking water ladder called 'safely managed drinking water services', and will be used to assess progress towards the SDGs going forward. Safely managed drinking water requires an improved water source which is located on premises (piped or non-piped), available when needed and free of faecal and priority chemical contamination. These requirements deal with access, availability and quality of water. For off-premise supplies from improved sources, categories of 'basic' and 'limited' sources mean less than or more than 30 minutes collection times, respectively. 'Unimproved' sources are unprotected dug wells or springs, and 'no service' refers to collection from unprotected surface water sources. JMP will also be reporting on water status in schools and health care facilities.

The categories of drinking water access can be confusing. To clarify them, Figure 1 provides a mapping between the older and newer JMP systems. The two classes of improved water sources are merged into three categories shown for the new ladder, which are safely managed, basic and limited service. To qualify for the safely managed level, the water source must be on premises; some of these are piped and some are not piped. The unimproved categories remain the same, although with slight nuances in the way they are reported. The numbers on the diagram indicate percentages of the 2015 world population; for example 58% of 7.35 billion is 4.2 billion people with piped water on premises. There will be slight differences in the numbers due to rounding, of course.

By adding availability and quality to its criteria, the JMP is introducing new dimensions of performance assessment, but procedures will have to be worked out. Availability is measured by parameters such as the quantity of water available or used in a period, hours of service per day, and frequency of breakdowns, which are parameters that utilities track to assess their own performance. Piped systems that lose pressure are vulnerable to contamination, so availability and quality are related. Quality

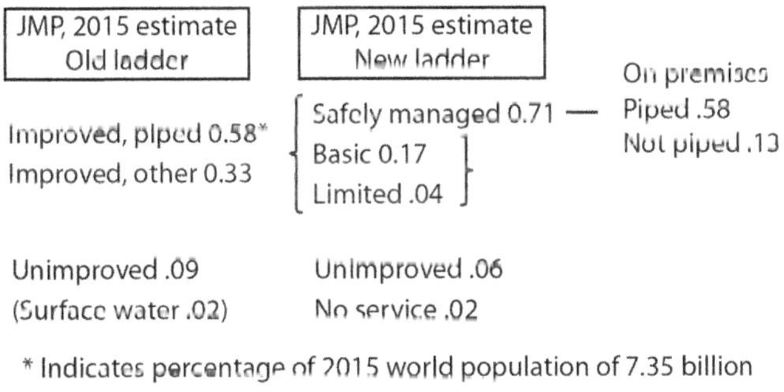

Figure 1. Mapping between old and new Joint Monitoring Programme rating systems.

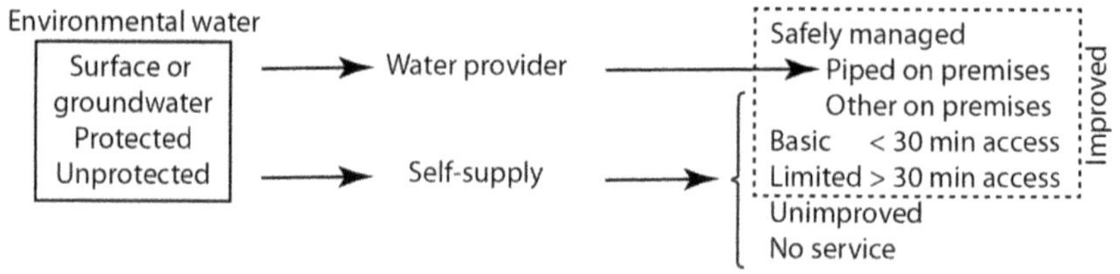

Figure 2. Water providers and piped water on premises.

would be measured by bacteriological tests during household surveys. In the case of piped water, while availability and water quality can be measured by utility performance data, high-quality water leaving a treatment plant is not always available to customers reliably because of the faulty distribution systems that afflict many communities.

How the new category of safely managed water relates to piped water is shown in Figure 2. On the left, the environmental water sources that are accessed by all are shown. People access this water either through water providers or by self-supply, which means that they are responsible for their own supplies and do not depend on a supplier. Water providers take on different forms and can range from local vendors to large public utilities. The term 'self-supply' is used by the US Geological Survey (2017) in its water use database programme, which has evolved over a number of years. It can apply whether the water source is a well on the premises or a surface source some distance away. Self-supply can also provide piped water on premises, but this would be infrequent.

Assessments of the performance of water providers focus on utility operations, such as revenue or water losses, and on water services to people, such as coverage, water quality and reliability of service. The concern addressed by this article focuses on the water services to people, which require metrics of availability (measured by the surrogate parameter of coverage), water quality and reliability. The JMP will attempt to survey these variables in the future, but it will take substantial effort to assemble reliable data.

Two perspectives of performance by water providers

The main reasons for the challenge in surveying the performance data of utilities include the complexity and large numbers of performance indicators. Assessing how well the providers meet expectations is approached through the management concept of performance assessment (Matichich et al., 2014). Studies of how to apply performance assessment to water service providers show that many indicators are needed to address these multiple expectations (Alegre et al., 2006; Cabrera, Dane, Haskins, & Theuretzbacher-Fritz, 2011).

The indicators fall into two broad categories dealing with utility performance and how well the needs of customers are met. This can be seen from a programme of effectively managed utilities, which has been developed by the US Environmental Protection Agency (USEPA) in cooperation with 10 non-governmental associations (Watereum, 2017). The programme uses 10 attributes of effectively managed water-sector utilities: product quality; employees and leadership; finances; operations;

enterprise sustainability; water resources; customers; infrastructure; community; and stakeholders. Of these, seven deal explicitly with how the utility is run, and three with external relations (customers, community and stakeholders). In the external relations category, customers or other stakeholders have multiple expectations of water service providers. Health professionals seeking to improve drinking water access for people could be considered in this category.

The two performance categories of how well the utility meets its business goals and how well it serves its customers are evident in other assessment systems. For example, the International Benchmarking Network for Water and Sanitation Utilities (IBNET) developed a metric called the IBNET Apgar, where the term 'Apgar' is taken from medicine as used to assess the overall health of a newborn infant (Danilenko et al., 2014). The Apgar score for water has five parameters: coverage, non-revenue water, collection period (for bills), operating cost coverage ratio, and affordability of water services. Of these, coverage and affordability address water services for people, but the score does not address water quality or reliability. Closely related, the Inter-American Development Bank and the International Water Association are developing an AquaRating system with 20 performance parameters to produce one score. These are both works in progress and will have a hard time being relevant to many types of countries and water service organizations, and use of the scores and indices will have to be taken in context (Danilenko et al., 2014).

While these two categories measure how well the utility is doing in its goals to serve customers and sustain its operations, an important element of water service is missing: how the utility's responsibility to serve all people is shaped. If a utility is a pure business enterprise, it may define its own responsibility to serve people who can pay. If it is a community organization, it may extend services to those who cannot pay and be sensitive to community needs in other ways as well. Behind these decisions lies the motivation of the provider and the forces driving its decisions. The distinctions between utility operations and service to users has been recognized by the International Organization for Standardization (2017) in two standards, one for the management of drinking water utilities and assessment of drinking water services, and one for the assessment and improvement of the service to users. In its standard for service to users, the organization wrote: 'The aim of water utilities is logically to offer services to everybody in the area of responsibility of the utility, and to provide users with a continuous supply of drinking water and the collection and treatment of wastewater, under economic and social conditions that are acceptable to the users and to the utility.' This implies a sense of public responsibility going beyond the business goals of a utility.

Concepts of performance management in business are well developed, and can be extended to operations of utilities. However, the side of performance measurement that assesses services to people is not well developed and is arguably more complex. It requires parameters of availability, water quality, reliability and affordability. Other terms also used include water safety, quantity, accessibility, continuity, equity, sufficiency and acceptability (Kayser, Moriarty, Fonseca, & Bartram, 2013; United Nations, 2017).

In summary, methods to assess the business performance of water utilities are well developed, but ways to assess the level of water services to people are less so. This opens the door to a series of important questions about the level of services that should be provided, how to pay for them, and about the responsibility to extend service to all people, no matter their ability to pay. These questions will be central to the emerging debates on the extent of the human right to water. As a utility must collect enough revenue to sustain its operations, it may not be able to address all the needs of the community. If it cannot, who has the responsibility? Perhaps the assessment of water services to people should not focus on the water provider alone but also on the responsibility of the community.

Types of utilities and other water service providers

The metrics of coverage, water quality and reliability will vary by types of piped water service providers in communities and cities. These range from small community organizations to large and sophisticated international corporations, and whether they are compelled to make reliable and high-quality water available will depend on their charters and governance. An unregulated private water provider may have no incentive other than profit, whereas a community organization may work under a social contract to provide water, no matter the economic consequences. The types will vary according to institutional frameworks, cultures, income and other socio-economic factors, as well as by size of communities. They can be started top-down, as when an authority intervenes to provide one, or bottom-up, as when citizens band together to form a community organization or utility. In some cases, an entrepreneur sees an opportunity and organizes a water service company. A national or regional government may empower a state organization to create a public utility. Sometimes, no organization is available, and people must obtain water from vendors. As the organizations develop, they respond to needs in their areas of service, and sometimes customers are satisfied. In other cases, difficulties may lead to affordability or other human rights issues.

No international inventory of water supply organizations is available, although the IBNET data set contains some information (Danilenko et al., 2014). While there are many organizational forms, service providers are generally public utilities, municipal organizations, private companies or cooperatives. They can also be mixed among these forms, such as a municipal organization with a private partner.

A few countries have organizational data on their water supply providers. Even with such national efforts, it is still difficult to characterize the organizational forms in use. In a late-1960s survey, the US Public Health Service found a 'bewildering array of large and small and independent and interconnected systems' (Harris, 1970). Since the Safe Drinking Water Act was passed in 1974, the USEPA has collected some organizational data through its Community Water Systems Survey (Temple, Baker, & Sloane, 1977). The most recent survey was in 2006, with information on water system size, water source, system ownership categories, and operational data for some 50,000 community water systems (US Environmental Protection Agency, 2017). Almost all utilities (97%) serve small communities of less than 10,000 in population. USEPA explained that the organizational types of these community water systems are evenly split between public and private ownership, with most publicly owned systems being the property of local city, county or special district governments. Privately owned systems include for-profit businesses and not-for-profit entities. Some 20% of all systems are ancillary, meaning their primary business is not water supply but water is provided as an integral part of their principal business.

To compare with the US approach, a brief examination of data was made from one high-income country (Germany) and three middle-income countries (South Africa, China and Colombia) where the writer has experience. No low-income countries were included because their statistics are unlikely to be well organized. Germany does not have a national water regulator like the USEPA or a national governmental water system data programme. Instead, its industry associations collect and publish water system statistics. They issued a recent report on the performance of the national water industry (Association of Drinking Water from Reservoirs et al., 2017), which includes utility statistics and lists types of organizations. These include the ancillary municipal enterprise (like the US municipal water utility as general government); the owner-operated municipal enterprise (like a separate authority under control of the city); the institution under public law (like a US special district); the autonomous company (a private company with city as shareholder); and the operations management model (public–private partnership).

The middle-income countries are in rapid transition, and none have detailed organizational data on water system characteristics. In South Africa, a centralized approach to water management is through the Department of Water Affairs (2013), which has been renamed the Department of Water and Sanitation. Drinking water systems are evolving, and there is little information on rural water systems. With its massive scale of ongoing urbanization, China faces many challenges in improving access to safe drinking water. Given the complexity of its administrative organization, responsibility for water is at multiple levels. Ongoing studies show one-quarter of cities being unable to provide adequate water pressure to many service areas. Statistics on safe water are difficult to obtain or nonexistent due to the decentralized nature of municipal infrastructure services, non-transparent sector management and relatively weak regulatory systems (Browder, 2007). Colombia is representative of the trend in Latin America to create integrated information systems at the national level. It is creating a unified information system for public services (Superintendencia de Servicios Públicos Domicilianos, 2017) which will draw data from diverse sources to make government more transparent. Due to past conflict and other issues, the data on water organizations are a work in progress.

These data systems focus on formal organizations, but many people are served by less formal community systems. Examples of such community-based models were identified in a study of five West African countries (Benin, Burkina Faso, Côte d'Ivoire, Ghana and Mali). Four models were water committees, water users' associations, water and sanitation committees, and water and sanitation development boards (Angbo, 1999). Obviously, to compile comprehensive data on their organization and performance would be very difficult.

Considering the variety of organizational types of water service providers, it would be challenging to create a comprehensive classification system. For subsequent discussion, they will be classified as public utilities, municipal organizations, private companies or cooperatives.

Roles of regulators in maintaining data on piped water

In addition to data from water providers, the JMP (2017b) will be seeking information from regulators. Effective water governance requires regulators to be independent and have the power to implement policies, laws and rules in a transparent way, while keeping policy, regulation and delivery separated (Rouse, 2013). These requirements would suggest that regulators should keep inventories of organizations, but, given the evolving status of national water supply agencies, how well these inventories are organized would range across a broad spectrum.

Organizing regulatory programmes involves different types of regulators, and which would have responsibility for data management is an important issue (Grigg, 2010). In the water sector, they regulate health and safety, water allocation, environmental water quality, fish and wildlife protection, rates and charges, and service access and quality. Here, we are mainly interested in health and safety, rates and charges, and service access and quality. In the US, the EPA issues rules to guide state government agencies to regulate health and safety. States have separate public utility commissions to regulate the rates of privately owned utilities, but normally not the rates of publicly owned utilities. Service access and quality are not regulated directly, but are controlled through political and legal processes, as are the rates of publicly owned utilities.

Among different countries, the regulatory arrangements will vary widely. The principles espoused by Rouse (2013) will not be evident in all places. As will be discussed later in the context of government effectiveness, regulation is related to governance, which depends on socio-economic and institutional parameters. Ideally, water regulators would maintain data on water service organizations and their levels of service, but this is not always the case. Data from regulators could help in studying levels of service, but data availability will be highly variable.

Rating piped water services

While the concepts of utility performance and services to people overlap, they are somewhat different. Rating water services to people must take into account customer experiences with access to services (percentage of people with piped services), reliability (water delivery), and water quality or safety. Given the importance of customer incomes, affordability is a very important issue as well.

Recognizing the many parameters and data limitations, a quantitative approach is not feasible, so a qualitative approach was implemented based on triangulation with three different types of data and observations. None of these are definitive, but by triangulating the three, a case can be made for the present levels of piped water services.

The first approach makes a hypothesis that average level of service is a function of government effectiveness (International Water Association, 2017), which depends largely on gross national income (GNI) per capita. This is an equity issue and seems intuitively correct in that people living in high-income countries normally get better public services than those in low-income countries (Feigenbaum & Hall, 2016). Rouse (2013) explained that governance depends on effective regulation with monitoring and reporting. World Bank (2017a) data include both government effectiveness and GNI per capita. Government effectiveness captures perceptions of the quality of public services, the quality of the civil service and the degree of its independence from political pressures, among other variables. Data on regulatory quality seem applicable, but they focus on the climate for private business, not environmental or health regulatory controls. Related indicators include rule of law, voice and accountability, and control of corruption. A correlation study of these indicators showed them to be highly correlated with government effectiveness, which is used here to measure the performance of utility officials and regulators.

Some 34 countries were selected to account for 80% of the global population, which showed a positive relationship between government effectiveness and GNI per capita. Intuition confirms this, as one senses that the governments of high-income countries like, say, Sweden, will run better than a low-income country that is approaching the status of a failed state. A check was made for correlation between the Gini coefficient of income inequality and access to improved water, but there was no apparent relationship.

Analysis showed that percentage of population with access to improved sources of drinking water is a function of GNI per capita, but improved water sources are not guaranteed to be safe. This means that the 2015 JMP estimates of access to safe water have a range of uncertainty due to the difficulty of obtaining global data (Bartram & Godfrey, 2015; JMP, 2017b; Martínez-Santos, 2017; Onda, LoBuglio, & Bartram, 2012). So, while the analysis shows that government effectiveness and access to improved sources are functions of GNI per capita, it does not prove that people with higher incomes are more likely to have good water services. It only shows that both are functions of GNI per capita.

Another data set is from World Bank's (2017b) Wikipedia Pilot Project, which uploaded descriptive summaries of national water and sanitation programmes. The programme closed in 2012 after some 90 overviews were prepared. Schiffler (2013) explained that the work was by volunteers, who spent substantial effort and sought to use primary data sources to build the summaries. While Wikipedia is not considered a reliable reference to cite in articles, the creators of the Pilot Project believed that the summaries would be authoritative and unlikely to be corrupted. The summaries provide what is arguably the best set of consolidated national water and sanitation information on the Internet.

The Wikipedia reports were studied for the same 34 countries, and each country was given a drinking water rating by noting information such as whether water supply is continuous or intermittent; whether an independent regulator is operational; whether there are frequent reports of waterborne sickness; and whether there is an organized water supply industry (an extensive number of utilities and other providers). As each summary is presented somewhat differently, it is impossible to be definitive on the criteria, so judgments had to be made. Countries were placed into three categories: safe and reliable water (essentially all people have high-level service); high level of piped water with lower service (most users have connections, but the water delivery and quality are unreliable); and low level of piped water with very low service (many users without connections and unreliable water delivery and quality).

Discussions of each criterion (continuous or intermittent service, independent regulator, reports of waterborne illness, and organized water supply industry) were evident in the postings, although some were more definitive than others. For some of the countries, less information was posted. To create scores from the four criteria, the following were noted:

- If a significant percentage of water providers had intermittent service, this was taken as evidence that water quality is compromised.
- An independent health regulator was a strong signal; a government ministry regulator without independence was weaker.

- Many utilities and participation by the private sector point to an organized water supply industry, but service mainly by a ministry with few other providers was judged weaker.
- As some postings included information on illness, but others did not mention it, this information was considered unreliable.

The analysis showed a relationship between national income and increasing levels of water services but does not prove a direct link. Wikipedia articles can be altered, and the information is complex and generalized. However, the evidence adds weight to the argument that water services do depend on national income to some extent.

This analysis could be corroborated to some extent by IBNET data, but this cross-analysis has not been performed. The approach would be to study the utilities in each of the 34 countries to note the coverage and affordability of water services. This would still not be definitive, because utility participation in IBNET is not uniform across the countries.

Water services seem easier to assess in low- and high-income countries, which are on opposite ends of the service spectrum. Service quality seems to vary more among middle-income countries. To probe this, I posed a question to a group of water industry academics, practitioners and research managers representing seven middle-income countries (Brazil, Indonesia, Iran, Mexico, South Africa, Peru and Turkey). The question was: is the water safe to drink anywhere in the country, safe only in a few zones of large cities, or generally unsafe in many places? Although it is an open-ended question, the group replied with definitive answers that showed convergence of opinion consistent with the World Bank postings and other literature:

- Water is sometimes safe in big cities, particularly in wealthier zones, but people often do not trust it and commonly treat or boil household water.
- While treatment systems may be effective, distribution systems are in poor condition, and water services may be intermittent.
- Sometimes water may be safe in specific zones or buildings, mainly due to point-of-entry treatment systems.
- The use of purchased water from unregulated vendors is common.
- Small communities have more difficulty than larger cities.

Additional experience factors can be obtained from national ministries, such as the US State Department (US Centers for Disease Control, 2017). They post travel warnings that are definitive in stating that high-income countries like Japan or the UK are like the US and you can trust the tap water. For most developing countries, travellers are advised not to drink the tap water. The general advice is: 'In most developing countries, tap water should probably not be drunk, even in cities. This includes swallowing water when showering or brushing your teeth. In some areas, it may be advisable to brush your teeth with bottled water.'

The experience of experts reflects some uncertainty. Water safety is explained as yes, no or maybe, but parameters such as reliability of service are not reported. Water is safe in developed countries and in some zones of developing countries where additional treatment may be provided. It is not safe in areas with weak or questionable services.

Travellers and people in general are left with uncertainty about water safety in many places.

From the World Bank socio-economic data, positive relationships are indicated of access to improved water sources with government effectiveness and per capita national income. The water-access data originated with the JMP, but are included in the World Bank data set. From the Wikipedia postings, water service and regulatory systems exhibit positive relationships with per capita national income of countries. The experts opined that water safety is better in developed countries and weaker in middle-income countries, although sometimes it can be acceptable in high-income zones of any country. All the evidence points in the same direction, which is intuitive and generally known: high-income people enjoy better and safer water services than low-income people. However, the data do not shed enough light on the issue to quantify the percentages.

This conclusion is repeated by IBNET staff (Danilenko et al., 2014):

> Despite the positive trend in the performance of utilities, there are large differences in Apgar scores among utilities, individual countries, and groups of countries. Utilities in low-income and lower middle-income countries tend to have lower Apgar scores than utilities in higher middle-income countries.... Utilities in low-income and lower middle-income countries tend to be more vulnerable to external shocks than utilities in middle-income countries.... Smaller utilities tend to have lower IBNET Apgar scores than larger utilities.

While the percentages cannot be quantified with any precision, global demographic and economic data indicate that the relationship between access to safe water and GNI per capita is one of poor or no services at the lower end and safe water at the upper end. For the middle-income countries, there is a gradually improving trend in services as income levels rise. At the higher end of the range, countries experience safe services like high-income countries, but those on the lower-end of the range are practically like low-income countries in their water services. There are exceptions, of course, and there will be zones with safe water even in the poorest countries. These phenomena were described by WHO director-general Tedros Adhanom Ghebreyesus: 'Safe water, sanitation and hygiene at home should not be a privilege of only those who are rich or live in urban centres' (World Health Organization, 2017).

To examine these ideas further, a few examples can be cited about types of water service organizations that are prevalent in a range of countries. Tanzania is a low-income country with a large water utility in the capitol city, the Dar es Salaam Water and Sewerage Authority. A World Bank (2012) project was launched to improve water supply and sanitation services in the service area through rehabilitation of facilities, extension of piped water supplies, and reorganization of operations. By the end of the project, only 25% of customers had access to 24-hour water service under adequate pressure. It was difficult to assess the results due to numerous data problems. One survey showed 82% piped water coverage, while another indicated 54%. Households and kiosks were connected through small independent piped networks that relied on boreholes. The majority consuming piped water (estimated at 45% in 2009) purchased it from neighbours, kiosks or water tankers. The water utility lacked data about its customer base. Revenues fell well short of covering operation and maintenance costs.

Bangladesh is just above the low-income line but provides an example in a populous country in Asia. Dhaka, the capital city, has water service provided through numerous paths, with the main utility being the Dhaka Water Supply and Sewerage Authority. A survey taken in 2005 showed 217,451 house connections, with a growth rate of about 4% annually. The utility reported that some 70% of house connections have access to 24-hour water supply. About one-third of households reported water quality as poor, mainly due to insects or dirt in the water, followed by smell of sewage and chemicals. Most domestic households reported that they boiled water before drinking. Of the population of about 12 million at the time, some 10 million lived in the utility service area, and the utility served 9 million people, or 80%, through house connections. Others are served by standpipes, and a smaller percentage by tankers. Some 200 bulk connections served about 100,000 people, mainly in the slum areas where NGOs handle installation, operation, maintenance and collection of revenues. An Asian Development Bank (2005, 2017) project aimed to strengthen the distribution network and capacity of the financial and operational management workforce. The project outcome was impressive, with rehabilitation of district metering areas and some 2,456 km of distribution network, along with construction of new house connections and tube wells. Still, Dhaka faces extreme challenges of poverty, new migrants and difficulty in providing service to all residents (World Bank, 2007).

At the lower end of middle-income countries, Kenya has a population of about 50 million, and in Nairobi, the main water supply system is operated by the Nairobi City Water & Sewerage Company. Service is afflicted by poor water quality and intermittent supply and by infrastructure challenges. The board members of the Nairobi Water Company were all fired in 2009. The city water and sewer department was spun off into the autonomous company in 2003. Only about 50% of the night time population of about 3 million have access to piped water; the others get water from kiosks, vendors, illegal connections or wells. Water supply to the slums is through kiosks, which are mainly privately owned, with only a few operated by community-based organizations or NGOs (Water Services Regulatory Board, 2016).

At the upper end of middle-income countries, Colombia has a population of about 50 million. Some large cities have high-performing utilities, but coverage in small cities and rural areas lags the larger cities. Also, large populations of low-income people lack access to safe water services. The nation has suffered decades of conflict that have prevented reforms from taking hold, but policies and institutional arrangements are in place to make continued improvements, even in small communities (Grigg et al., 2004).

The examples are limited in number, but they tend to confirm the general conclusions given above about variability of water services. The situations are changing, and improvements will depend on the effectiveness of governance at all levels. Meanwhile, assessing the level of water services to people remains challenging.

Conclusions

While some 4.2 billion people have piped water services, little attention has been given to assessing the levels of services they receive. Progress towards meeting the MDG and SDG targets for access to improved sources of drinking water has received more attention. This is appropriate because of the emphasis by the international community on alleviation of poverty and its effects on health and liveability. However, as the world continues to urbanize and more people are connected to piped water supplies, water safety, reliability and affordability will come under greater scrutiny.

The main current effort to assess drinking water services at the global level is by the JMP, which focuses on whether people have access to piped water. Under its new procedures, the JMP will attempt to assess reliability and safety, but its main source of data (from household surveys) cannot be considered very reliable on these points. Not only are people not trained in water quality, they may not always give care to their answers. The JMP recognizes these constraints, and continues to adjust its estimates and procedures. As a result, the JMP's future assessments could include more information on types of water providers and their service levels, but this would require much more extensive data than is available presently.

If more assessment information about service levels is made available it can enable benchmarking of safe water service conditions at different levels of national development and for management of regulatory programmes. This will require data about types of organizations and their service availability and reliability, as well as water quality and safety.

Development of a global assessment of piped water services can support programmes of service improvement and regulation towards the goal of higher-performing, equitable and affordable services. By defining and measuring levels of service more definitively, the issue of human rights to water will be clarified, and gaps will be exposed between good service levels and those of intermittent, poor-quality and expensive water. The challenge is that levels of service for piped water supplies are difficult to judge across many diverse situations.

It is intuitive that, for the most part, water services are better in high-income countries than in low- and middle-income countries, although problems and issues of equity occur everywhere. The analyses reported here of the relationships between water services, national income and governance effectiveness confirmed this by testing a hypothesis for the variation of service with national income with a triangulation approach. Further studies, supported by well-developed methods and quality-controlled data, could provide more insight into the relative levels of services for the 4.2 billion people with piped water on premises.

To create an assessment programme, a first step should be establishment of data standards and classes. The JMP's category of 'safely managed drinking water' currently lumps the services to 4.2 million people into one group, although with plans to assess reliability and quality, with an emphasis on household testing. A first step for data classes would be to develop a classification of type of provider for piped water on premises, with a starting point being expansion and refinement of the categories for public utilities, municipal organizations, private companies or cooperatives. A challenge is posed by the wide range of types of water supply providers, both in organizational forms and in scale. These vary widely in their attributes, as they range from small community organizations to large and sophisticated international corporations.

How services are organized is important, but depends on context. In high income countries with strong governance, water supply systems may be organized bottom-up, with local people taking responsibility rather than depending on higher government levels. Regulatory controls may be mandates or, in some cases, voluntary guidelines. For example, in the US and Canada, local governments are responsible for drinking water services. National-level drinking water rules are voluntary for provinces in Canada (Zubrycki, Roy, Venema, & Brooks, 2011) and mandatory in the US (US Environmental Protection Agency, 2017). For countries with weaker governance and institutions, donors and governments may attempt to organize systems top-down, like other development efforts. However, in many cases bottom-up community organizations based on self-help may work best. Organized utilities can be high-performing or low-performing. Small community organizations and cooperatives can also vary in performance, but they will generally lack the resources of larger organizations.

The work done by utility organizations on performance data classes can help, but it will not be feasible to apply the same criteria to a large and well-funded utility and a small cooperative. Going beyond rating the performance of business and system operations, it will also be important to assess the effectiveness of water services to people by considering access, reliability, quality and affordability. Normally, it will be easier to rate large, formal organizations than to rate small providers.

Assessing water services to people raises the question of how the responsibility of water providers to serve all people is shaped. Concepts for performance management of business operations are well developed, but those for assessing services to people are not. How is the responsibility of the provider to all people determined? Are customers only those the provider chooses? What is the community responsibility to provide water services, and at what level of service?

These questions about the level of the services also raise the issue of how to pay for them. A provider must collect enough revenue to sustain its operations and may not be able to address all needs of the community. If the provider works on an enterprise basis or with a profit-and-loss motive, how should the low-income customers be served? Is there a community responsibility to assist in mitigating problems of affordability?

Access to drinking water has in many cases emphasized government and donor roles, and some governments are more engaged in helping to improve drinking water than others. Government responsibility for drinking water is regarded as part of the social contract with people. The human rights movement addresses this duty of government, but private actors are also involved (Scanlon, Cassar, & Nemes, 2004). The issue was captured by this statement by the WHO director-general in speaking about access to safe water and sanitation: 'These are some of the most basic requirements for human health, and all countries have a responsibility to ensure that everyone can access them' (World Health Organization, 2017). The challenge will be in how the balance should be struck between rights and responsibility to pay.

There is a large gap in coverage between the role of the JMP in assessing all services and that of global utility assessment programmes such as IBNET and AquaRating. The gap occurs because the JMP is challenged to assess so many diverse situations for water and sanitation and the utility assessment programmes focus mainly on the business operations that are of most interest to the utilities. The gap is therefore in assessing how well people with piped water are actually served.

The JMP's new category of 'safely managed drinking water services' introduces the concept of availability to go along with access and use. As the JMP begins to collect data from utilities and regulators, it will be a change from its previous approach based on household surveys and modelling to make nationally representative estimates for international comparison (JMP, 2017b). How quickly this change will reach into distinguishing between levels of piped water services remains a question, however.

The study described here identified several categories of research needs, beginning with development of protocols and data classes for a piped water services rating system. Probing the effectiveness of the different pathways through which people get services will require organizational analysis and case study work. Testing the relationships between explanatory variables and service outcomes will be an expansion of the current focus on assessing business and operational performance of utilities. It can be considered an expansion of the category often labelled customer service, but interpreted in a broader way to include people who may have been excluded from service. To address the needs of the excluded people and how to meet those needs will require fundamental thinking about the public responsibility of water service providers and communities. As the percentage of people receiving piped water services increases, assessing the level of services they receive will become the key issue in global water supply assessment.

References

Alegre, H., Melo Baptista, J., Cabrera, E., Jr, Cubillo, F., Duarte, P., Hirner, W., ... Parena, R. (2006). *Performance indicators for water supply services*. London, UK: IWA Publishing.

American Water Works Association. (2017). *Partnership for safe water*. Retrieved August 4, 2017, from https://www.awwa.org/resources-tools/water-and-wastewater-utility-management/partnership-for-safe-water.aspx

Angbo, L. (1999). *Community-based management of piped water supply systems*. World Bank. Retrieved August 4, 2017, from http://documents.worldbank.org/curated/en/887031468151147488/pdf/719600WP0v20Bo0y0Systems0March01999.pdf

Asian Development Bank. (2005). *Dhaka water services survey*. Retrieved August 4, 2017, from https://web.archive.org/web/20100708002642/http://www.adb.org/Water/Tools/pdf/Dhaka-Assessment-FULL.pdf

Asian Development Bank. (2017). *Bangladesh: Dhaka water supply sector development program (Project Loan)*. Retrieved August 4, 2017, from https://www.adb.org/projects/39405-013/main

Association of Drinking Water from Reservoirs, German Association of Energy and Water Industries, German Alliance of Water Management Associations, German Technical and Scientific Association for Gas and Water, German Association for Water, Wastewater and Waste, German Association of Local Utilities. (2017). *Profile of the German water sector 2015*. Retrieved August 4, 2017, from http://www.dvgw.de/english-pages/drinking-water/profile-of-the-german-water-sector/

Bartram, J., & Godfrey, S. (2015). Drinking-water supply. In J. Bartram, R. Baum, P. A. Coclanis, D. M. Gute, D. Kay, S. McFadyen, ... M. J. Rouse (Eds.), *Routledge handbook of water and health*. London: Routledge.

Browder, G. (2007). *Stepping up: Improving the performance of China's urban water utilities*. World Bank. Retrieved August 4, 2017, from http://documents.worldbank.org/curated/en/840621468016215926/pdf/409640P0704130Public.pdf

Cabrera, E., Jr., Dane, P., Haskins, S., & Theuretzbacher-Fritz, H. (2011). *Benchmarking water services, guiding water utilities to excellence*. London, UK: IWA Publishing .

Danilenko, A., van den Berg, C., Macheve, B., & Moffitt, L. (2014). *The IBNET water supply and sanitation blue book 2014*. Washington, DC: World Bank.

Department of Water Affairs. (2013). Strategic overview of the water sector in South Africa. Retrieved August 4, 2017, from http://nepadwatercoe.org/wp-content/uploads/Strategic-Overview-of-the-Water-Sector-in-South-Africa-2013.pdf

Ercumen, A., Gruber, J. S., & Colford, J. M., Jr. (2014). Water distribution system deficiencies and gastrointestinal illness: A systematic review and meta-analysis, *Environ Health Perspect, 122*, 651–660.

Feigenbaum, J. J., & Hall, A. B. (2016). *How high-income areas receive more service from municipal government: Evidence from city administrative data* (Working paper). Retrieved August 4, 2017, from http://www.andrewbenjaminhall.com/Feigenbaum_Hall_Boston.pdf

Grigg, N. S. (2010). *Governance and management for sustainable water systems.* London: IWA Publishing.

Grigg, N. S., MacDonnell, L. J., Fontane, D. G., Howe, C. W., Livingston, M. L., Roesner, L. A., & Salas, J. D. (2004). *Integrated water management and law in Colombia: Institutional aspects of water management, allocation of uses, control of contamination, and urban drainage* (Report for the Ministry of Environment, Housing, and Regional Development). Fort Collins, CO: Colorado State University..

Harris, R. R. (1970, March). The water activities of the bureau of water hygiene. *Journal, American Water Works Association*, 155–164. doi:10.1002/j.1551-8833.1970.tb03878.x

International Organization for Standardization. (2017). *Activities relating to drinking water and wastewater services – Guidelines for the assessment and for the improvement of the service to users.* ISO 24510: 2007. Retrieved August 4, 2017, from https://www.iso.org/standard/37246.html

International Water Association. (2017). *Utility management.* Retrieved August 4, 2017, from http://www.iwa-network.org/news/utility-management/

Joint Monitoring Program for Water Supply and Sanitation. (2017a). *Progress on sanitation and drinking water 2015 update and MDG assessment.* Retrieved August 4, 2017, from http://www.who.int/water_sanitation_health/monitoring/jmp-2015-update/en/

Joint Monitoring Program for Water Supply and Sanitation. (2017b). *Safely managed drinking water services - JMP thematic report on drinking water.* Retrieved August 4, 2017, from https://washdata.org/

Joint Monitoring Program for Water Supply and Sanitation. (2017c). *Improved and unimproved water sources and sanitation facilities.* Retrieved August 4, 2017, from http://www.wssinfo.org/definitions-methods/watsan-categories/

Kayser, G. L., Moriarty, P., Fonseca, C., & Bartram, J. (2013). Domestic water service delivery indicators and frameworks for monitoring, evaluation, policy and planning: A review. *International Journal of Environmental Research and Public Health, 10*, 4812–4835.

Martínez-Santos, P. (2017). Does 91% of the world's population really have "sustainable access to safe drinking water"? *International Journal of Water Resources Development, 33*(4), 514–533.

Matichich, M., Hasit, Y. J., Yeager, F., Englesmith, J., Paralez, L., Francisco, W., & Greenwood, R. (2014). *Performance benchmarking for effectively managed water utilities.* Denver, CO: Water Research Foundation.

Onda, K., LoBuglio, J., & Bartram, J. (2012). Global access to safe water: Accounting for water quality and the resulting impact on MDG progress. *International Journal of Environmental Research and Public Health, 9*, 880–894.

Rouse, M. (2013). *Institutional governance and regulation of water services: The essential elements* (2nd ed.). London: IWA Publishing.

Scanlon, J., Cassar, A., & Nemes, N. (2004). *Water as a human right? IUCN environmental policy and law paper No. 51.* Retrieved August 4, 2017, from https://portals.iucn.org/library/sites/library/files/documents/EPLP-051.pdf

Schiffler, M. (2013). *Manuel Schiffler on KFW wikipedia project.* WaterCubeTV. Retrieved August 4, 2017, from https://www.youtube.com/watch?v=rBsRNoto0gU

Superintendencia de Servicios Públicos Domicilianos. (2017). *Servicio de Acueducto.* Retrieved August 4, 2017, from http://www.sui.gov.co/SUIAuth/logon.jsp

Temple, Baker & Sloane Inc. (1977). *Survey of operating and financial characteristics of community water systems.* USEPA.

United Nations. (2017). *International decade for action "Water for Life" 2005-2015.* Retrieved August 4, 2017, from http://www.un.org/waterforlifedecade/human_right_to_water.shtml

US Centers for Disease Control. (2017). *Traveler's health.* Retrieved August 4, 2017, from https://wwwnc.cdc.gov/travel/

US Environmental Protection Agency. (2017). *Community water systems survey.* Retrieved August 4, 2017, from https://www.epa.gov/dwstandardsregulations/community-water-system-survey

US Geological Survey. (2017). *Water use in the United states.* Retrieved August 4, 2017, from https://water.usgs.gov/watuse/

Water Services Regulatory Board. (2016). *A performance review of Kenya's water services sector: 2014 – 2015.* Retrieved August 4, 2017, from http://www.wasreb.go.ke/impact-reports

Watereum. (2017). *Ten attributes of effectively managed utilities.* Retrieved August 4, 2017, from http://www.watereum.org/resources/interactive-primer/ten-attributes/

World Bank. (2007). *Bangladesh, Dhaka: Improving living conditions for the urban poor* (Report No. 35824-BD). Retrieved August 4, 2017, from http://documents.worldbank.org/curated/en/587231468007834055/Bangladesh-Dhaka-Improving-living-conditions-for-the-urban-poor

World Bank. (2012). *Dar Es salaam water supply and sanitation project.* ICR Review. Retrieved August 4, 2017, from http://documents.worldbank.org/curated/en/512971474504993607/pdf/000020051-20140624203944.pdf

World Bank. (2017a). *Worldwide governance indicators.* Retrieved August 4, 2017, from http://data.worldbank.org/data-catalog/worldwide-governance-indicators

World Bank. (2017b). *The World Bank's wikipedia pilot project.* Retrieved August 4, 2017, from http://siteresources.worldbank.org/INTLACSUMSUSDEVDEP/Resources/WikipediaFlyer.pdf

World Health Organization. (2017). *2.1 billion people lack safe drinking water at home, more than twice as many lack safe sanitation.* Retrieved November 22, 2017, from. http://www.who.int/mediacentre/news/releases/2017/water-sanitation-hygiene/en/

Zubrycki, K., Roy, D., Venema, H. D., & Brooks, D. (2011). *Water security in Canada: Responsibilities of the federal government.* International Institute for Sustainable Development. Retrieved August 4, 2017, from http://www.iisd.org/pdf/2011/water_security_canada.pdf

Water and sanitation in Dhaka slums: access, quality, and informality in service provision

Sabrina Sharmin Haque, Monica Yanez-Pagans, Yurani Arias-Granada
and George Joseph

ABSTRACT

Slum populations are commonly characterized as having poorly developed water and sanitation systems and may access services through informal channels. However, there are limited representative profiles of water and sanitation services in slums, making it difficult to prioritize interventions that will make services safer for residents. This cross-sectional study examines the quality of and access to water and sanitation services in government-defined slums across Dhaka, Bangladesh. Access is generally high but is subject to quality issues related to safety, reliability and liability. Services are often operated by informal middlemen at various stages of provision.

Introduction

Bangladesh's rapid urbanization has propelled the country to be among the five fastest-growing economies in the world (World Bank, 2019) but has also placed stress on housing, infrastructure and overall quality of life in the cities, particularly for the poorest people. Low-income communities face unique challenges in accessing basic services, such as water, sanitation and hygiene (WASH). Unsafe WASH remains a leading cause of mortality globally, estimated to account for over 1.6 million deaths and over 84 million disability-adjusted life years in 2017 alone (Stanaway et al., 2018). Poor WASH can cause repeated cases of diarrhoea and asymptomatic enteric infections, especially in early childhood, which in turn have been linked to environmental enteropathy, growth faltering, and impaired cognitive functioning (Checkley et al. 2008; Guerrant et al. 2002; Humphrey 2009; Ngure et al. 2014). The WASH disease burden could be particularly significant in areas of high population density such as Bangladesh (Hathi et al., 2017).

Although policies have been changing, utilities traditionally have not been mandated to deliver services to slums due to their informal tenure. Instead, WASH services are often believed to be provided through informal markets operated by middlemen. On some accounts, residents have been observed to have illegal piped connections or private pumps

Supplemental data for this article can be accessed at https://doi.org/10.1080/02508060.2020.1786878.

(Hossain & Ahmed, 2015). Surface drainage and solid waste collection in slums are also neglected by city planners, with garbage and waste commonly found in the periphery of communities (Ahmed, 2014).

The lack of formalized services may undermine accountability systems that allow monitoring and enforcement of a minimum quality of service (Adams et al., 2015; Hanchett et al., 2003; Panday, 2017; UNICEF, 2010). Lack of regulation frameworks can also lead to poor cooperation and conflicts among different players across the service delivery channel (Banks, 2008). Further, underdeveloped WASH systems in slums ultimately contribute to widening intra-city housing disparities and may even weaken the 'urban health advantage' generated from better access to jobs and services compared to rural areas (Vlahov et al., 2005). For example, Bangladesh's slum population has the highest rates of child mortality and undernutrition in the country. Under-5 child mortality is 79% and 44% higher in slums areas compared to non-slum urban and rural areas, respectively (UNICEF, 2010), while nearly half of all children living in the slums of the country's major cities are stunted (NIPORT et al., 2015, 2016).

Although slum communities are often considered to have poor WASH, there are limited studies that measure a representative profile of WASH service delivery in slums. This is partly due to the difficulty of obtaining an accurate sampling frame of slum households. Though some surveys include slum households in sampling frames, many do not follow official government definitions of slum areas or offer disaggregated analysis of slum populations because they are not designed to be statistically representative of slum populations (Angeles et al., 2009; NIPORT et al., 2008, 2015). The lack of representative data limits the ability to prioritize WASH issues unique to this vulnerable population.

This cross-sectional study attempts to respond to data gaps by quantifying and qualifying WASH access in government-defined slums across Dhaka, Bangladesh. It presents statistics on access to WASH services representative of small, medium and large slums of Dhaka as defined by the government's 2014 Census of Slum Areas and Floating Population. The study additionally explores different dimensions related to the quality of and provision of WASH services. Dimensions related to quality of water service include types of infrastructure, reliability, accessibility, reported issues, sharing status and perceived quality. For sanitation, we examine types of infrastructure, sharing status, reliability, accessibility, faecal sludge management and barriers to investing in sanitation. Finally, we examine some of the mechanisms of service provision, including measurements of service providers for installation, day-to-day management, accountability mechanisms and payment collection schemes. These findings should inform opportunities for safe and sustainable interventions such as designing formalized mandates and clearer roles for utilities and non-state entities.

Methods

Cross-sectional data were collected in May 2016 and are representative of all the slums in the Dhaka North and South City Corporations.[1] The primary sampling units (PSUs) were slum communities. The sampling frame used was the Census of Slum Areas and Floating Population collected by the Bangladesh Bureau of Statistics (2014).[2] The census defines slums as compact settlements of five or more households, which grow unsystematically and haphazardly in unhealthy conditions in either government or privately owned vacant land. To operationalize the definition, the government uses six criteria: poor housing structure; high population density (one room for all household members; three or more structures per one decimal of land[3]); poor access to water and sanitation services (unsafe water; 15+ people sharing a toilet); inadequate lighting and road facilities; low socio-economic status (informal non-agricultural jobs); and conditions of land ownership (government; private vacant lots; abandoned building/houses; slopes of hills, or sides of rail-lines or roads). The census gave a sampling frame of 3360 slum communities with more than five households in Dhaka City Corporations. The census categorizes community size as small (5–10 households), medium (11–200 households) or large (more than 200 households).

PSUs were randomly selected and were equally allocated between small-and-medium-size slums and large slums using a combination of probability proportional to size and practical allocation. PSUs were allocated across strata using probability proportional to size, with the number of households used as the measure of size and rounding to account for the pre-determined cluster size of 10 households per PSU for medium-size slums and five households per PSU for small slum communities. The final sample includes 588 households, with 2.3% from small slums, 41% from medium-sized slums and 57% from large slums, which is closely aligned with the distribution of households from small, medium and large slums in the original sampling frame. In the second sampling stage, listings of households in selected communities were completed. Using these listings, 5–10 households were selected from each community using systematic equal probability sampling.

The administered survey measures basic demographic and housing characteristics and access to WASH following WHO/UNICEF Joint Monitoring Programme (JMP) definitions of tracking targets under Sustainable Development Goal 6.[4] The survey also asks questions on various dimensions of WASH service quality. For water services, this includes measures of reliability, accessibility, reported issues and perceived quality; and for sanitation services, this includes sharing status, reliability, accessibility, faecal sludge management and barriers to investing in sanitation. To understand the extent of informality of service providers, the survey additionally examines mechanisms of WASH service delivery, including household reported service providers for installation, day-to-day management, accountability mechanisms and payment collection schemes.

The analysis provides representative estimates by these three strata of slum size and additionally disaggregates by land ownership and household welfare levels. The concept of land ownership in slums in Bangladesh is complex and prone to measurement error.[5] To construct a measure of land ownership, the survey has a self-reported question on land ownership asked of household heads and then creates a community-level land ownership variable based on the mode of that response. To measure household per capita consumption, consumption aggregates are constructed and are fully consistent with the government's methodology for producing official poverty numbers using the Household Income and Expenditure Survey 2016/17.[6]

Results

Profile of slum households and residents

Table 1 summarizes the demographic and socio-economic characteristics of households by community size.[7] On average, households spend BDT4410 per capita (around USD53) per month on food, non-food, and housing. This corresponds to an average per capita consumption of USD1.76 per day. A typical household has 4.3 members, and 72% of them are dependents, as measured by the age dependency ratio.[8] Household heads are predominantly male (85.9%), on average 40 years old, and on average have lived in their current residence for 10 years. About 47% of adult slum residents report knowing how to read and write, but only about 18% completed primary school. Among school-age children (7–12 years old), 82% attend school on a regular basis. Nearly 5% of households report having benefited from a relief programme provided by the government or a non-government organization (NGO).

Overcrowding, poor housing conditions and fear of eviction are salient features faced by residents. One single room is normally shared by three people, and rooms are on average 12 square metres. Housing structures are mainly constructed of low-quality materials, including tin-shed (59% of housing structures) and semi-permanent materials (29% of housing structures). About 12% of housing structures are constructed of permanent materials. Over half of all residents report living in fear of being evicted from their housing structures. Slums also have dynamic rent housing markets, with 71% of residents reporting renting their housing structure. On average, 61% of slums in Dhaka are on land reported to be government owned, and 34% on land reported to be privately owned.

Main sources of drinking water

Overall, 97% of households have access to an improved water source as defined by JMP guidelines (Table 1). Access to piped water among households is nearly universal, at 95%. However, piped connections that run directly into each of the slum dwelling structures are rare (8%), with most households accessing piped water through a system connected to the slum compound, yard or plot (68%) or through a public standpipe (18%) (Table 2). The poorest households tend to share their water sources with a greater number of households (42 on average), compared to the richest households (23). Households are not generally too far from their main drinking water source (Figure S1 in the online supplementary material), spending seven minutes, on average, to reach it. Richer households tend to live closer to water sources compared to poorer households.

Water treatment and testing

Columns 8 and 9 in Table 2 show household water treatment practices and external testing for arsenic, iron, salinity and bacteria. Close to 75% of households do not do anything to their water to make it safer to drink. Among those who treat water, common practices include boiling (68%), straining through a cloth (17%), using a filter (10%), and adding bleach or chlorine (3%). When asked whether their main source of drinking water was ever tested by any parties for water quality issues, on average, 91% of households said either that the water was not tested or that they did not know whether it had been tested. This pattern does not

Table 1. Summary statistics by slum size.

	Small	Medium	Large	All	*T*-test, small–medium	*T*-test, medium–large
Monetary poverty						
Distribution of slum households (%)	2.3	40.8	57.0	100	-	-
Monthly per capita expenditure (BDT)	3,956	4,723	4,208	4,410		
Demographics and human capital						
Household size	4.3	4.2	4.3	4.3		
Dependency ratio (%)	80.8	76.1	68.9	72.1		
Age of household head (years)	39.2	40.2	40.3	40.2		
Female household head (%)	10.3	12.9	15.2	14.1		
Duration of residence household head (years)	7.8	8.3	12.6	10.7		*
Literacy rate for adults (%)	65.3	46.9	46.6	47.2	*	
Completed primary school for adults (%)	29.3	16.5	17.8	17.5	*	
School attendance, children 7–12 (%)	94.1	81.4	82.1	82	*	
Household received relief program (%)	3.4	3.8	5	4.5		
Labour markets						
Share of male adults who are earners (%)	94.4	89.9	88.2	89.0		
Share of female adults who are earners (%)	40.9	52.8	48.7	50.1		
Main source of income from industry (%)	6.9	4.2	15.1	10.5		***
Main source of income from services (%)	93.1	95.8	84.9	89.5		***
Housing						
Per capita rooms	0.32	0.32	0.33	0.33		
Room size (m^2)	12.1	11.8	11.7	11.8		
Share of renters (%)	96.6	82.2	61.2	70.6	**	***
Monthly rent (BDT)	2,849	2,692	2,390	2,548		**
Has separate kitchen (%)	55.2	87.9	63.4	73.2		***
Government-owned land (%)	17.2	27.1	86.6	60.8		***
Non-government and private owned land (%)	82.8	72.9	3.3	33.5		***
Other; do not know who owns land (%)	0.0	0.0	10.0	5.7		*
Fear of eviction (%)	17.2	21.8	71.6	50.1		***
Permanent dwelling structure (%)	24.1	12	11.4	11.9		
Semi-permanent dwelling structure (%)	37.9	36.7	22.8	28.8		
Tin-shed dwelling structure (%)	34.5	51.2	65.8	59.2		
Jhupri dwelling structure (%)	3.4	0	0	0.1		
Access to basic services						
Improved drinking water (%)	100	95.4	98.3	97.2		
Improved sanitation, JMP definition (%)	6.9	9.2	8.2	8.6		
Improved sanitation, including shared (%)	100	93.9	80.5	86.4		*
Electricity is main source of light (%)	100	99.2	93.6	96.1		

Notes: Data from Bangladesh Urban Informal Settlements Survey 2016. All summary statistics were computed using household weights except for mean monthly per capita expenditure (BDT), age, literacy rate for adults, completed primary school for adults, school attendance for children 7–12 , % of male adults who are earners, and % of female and male adults who are earners, which were computed using population weights. The *t*-test columns show the results of Wald tests of whether the difference in means/proportions between small and medium slums and medium and large slums are significant at $^*p < 0.1$, $^{**}p < 0.05$, or $^{***}p < 0.01$. The JMP defines access to improved sanitation as unshared facilities that are flushed to piped sewage systems, flushed to septic tanks, pit latrines with slabs, or ventilated improved pit latrines. The sanitation definition follows the same criteria as the JMP but includes as improved sanitation shared or unshared facilities that are flushed to piped sewage systems, flushed to septic tanks, pit latrines with slabs, and ventilated improved pit latrines.

seem to change based on the size of the community, land ownership, or the welfare status level of households.

Water quality issues

Table 3 presents a summary of the main quality issues identified by households. Although the results are subject to interviewer and response bias, these provide interesting insights

Table 2. Main source of drinking water, treatment and testing.

	Piped into dwelling (%)	Piped into compound, yard or plot (%)	Public tap/ standpipe (%)	Tubewell, Borehole (%)	Others (%)	Households who share water source (%)	Average number of households who share water source	Water treated (%)	Water tested (%)
	(1)	(2)	(3)	(4)	(5)	(6)	(7)	(8)	(9)
All	8.[illegible]	68.4	18.3	1.8	3.5	93.8	33.2	25.7	9.2
	By size of the slum								
Small	6.9	62.1	24.1	6.9	0.0	93.1	13.2	48.3	15.8
Medium	4.2	72.6	13.5	3.5	6.1	91.1	28.8	35.4	5.5
Large	10.8	65.7	21.4	0.4	1.7	95.8	37.0	17.9	11.4
T-test, small–medium							**		
T-test, medium–large	*							**	
	By land ownership								
Public	9.0	66.5	22.2	0.6	1.6	95.3	36.7	17.2	11.9
Private	5.4	77.9	5.0	4.3	7.5	92.3	21.3	45.5	4.8
Other	13.3	33.4	53.3	0.0	0.0	86.7	66.7	0.0	3.7
T-test, public–private			***					***	
	By per capita consumption expenditure								
Poorest	4.5	69.6	13.3	5.7	6.8	98.9	42.4	18.1	8.3
Second quintile	4.9	65.5	21.9	2.1	5.6	94.6	40.8	17.1	12.3
Third quintile	7.3	67.3	22.2	0.8	2.4	98.0	33.3	19.8	10.1
Fourth quintile	3.1	68.2	21.0	0.4	2.4	94.2	29.7	29.4	8.4
Richest	13.5	71.1	13.2	1	1.3	86.1	23.0	39.1	7.3
T-test, poorest–richest	*					***		***	

Notes: Data from Bangladesh Urban Informal Settlements Survey 2016. The *t*-test rows show the results of Wald tests of whether the difference in means/proportions are significant at $*p < 0.1$, $**p < 0.05$, or $***p < 0.01$. 'Other' includes bottled water and other categories not specified.

Table 3. Top issues reported related to water services.

	Bacterial contamination (%)	Funny smell, taste or colour (%)	Not enough water to meet all the demand (%)	Too expensive (%)	Other (%)
	(1)	(2)	(3)	(4)	(5)
All	7.4	49.6	7.5	1.7	33.8
	By size of the slum				
Small	3.4	72.4	0.0	0.0	24.1
Medium	7.7	56.7	3.8	0.4	31.4
Large	7.4	43.6	10.4	2.7	36.0
T-test, small–medium					
T-test, medium–large					
	By land ownership				
Public	7.6	46.7	7.2	2.8	35.6
Private	7.7	54.7	4.7	0.0	32.9
Other	3.3	50.0	26.7	0.0	20.0
T-test, public–private					
	By per capita consumption expenditure				
Poorest	6.5	38.2	6.7	4.6	44.1
Second quintile	5.6	49.4	7.8	2.0	35.1
Third quintile	5.2	49.6	9.5	0.0	35.7
Fourth quintile	10.8	54.4	7.9	0.9	26.0
Richest	8.2	53.4	5.8	1.5	31.2
T-test, poorest–richest		*			

Notes: Data from Bangladesh Urban Informal Settlements Survey 2016. The *t*-tests show the results of Wald tests of whether the difference in means/proportions are significant at $*p < 0.1$, $**p < 0.05$, or $***p < 0.01$. 'Other' includes arsenic or heavy metal contamination, salinity contamination, no challenges, do not know, and other categories not specified.

into the main water quality concerns of slum residents. Residents consistently report that the top issues are 'funny smell/colour/taste' (49.6%) and 'bacterial contamination' (7.4%) of drinking water. Other less prevalent issues include availability (7.5%) and high price of water (1.7%).

Reliability of water sources

Households were asked about the reliability of access to their main drinking water source. About 12% report service interruptions, with larger slums having significantly more service interruptions than small and medium slums (Table S3 in the online supplementary material). Richer households seem to be impacted less frequently by service interruptions than poorer households. More than half of households report flooding in the past year, with an average flood lasting around 16 days (Columns 1 and 2, Table 6). Almost a third of households are affected by limited supplies of drinking water during times of flooding (Column 3 in Table 6). On average, households that report being affected by flooding report being without access to a functional drinking water for 10 days (Column 4). Flooding affects larger slums more often than small and medium slums. However, the issues seem to be resolved more quickly in large slums compared to small and medium slums, as measured by the number of days that households need to wait to have a functional drinking water source back again. Lastly, richer households report being affected by flooding and longer repair wait times more often than poorer households.

Main sources of sanitation

On average, about 9% of households have access to an improved sanitation facility as defined by JMP guidelines (Table 1). If the definition is relaxed to consider shared facilities as improved, this number increases to 86%. Access to flush toilets connected to a piped sewer system or septic tank is limited (Table 4). About 8% of households have access to a flush or piped sewer system or septic tank. Most households access on-site sanitation, with 78% using improved pit latrines. Just over 10% of households have access to hanging toilets, which is a form of unimproved sanitation. The use of open pits or hanging toilets for sanitation is much more common in large slums and in slums on government-owned land. Richer households are slightly more likely to have access to a flush toilet, though the availability of those is quite limited.

Access to toilets exclusive to one household is rare. About 91% of households report sharing their sanitation facilities (Columns 9 and 10 in Table 4). On average, one toilet is shared with 16 households (around 70 people). Sharing toilet facilities is equally common across all slums independently of their size, though residents of smaller slums tend to share their sanitation facilities with fewer people, as expected. Households on publicly and privately owned land share their toilet facilities with other households just as frequently, but those on privately owned land share with a larger number of household on average: 19 versus 13 households, respectively. Richer households also tend to share their sanitation facilities slightly less frequently (83.9%, versus 95.4% among the poorest households).

Table 4. Access to sanitation facilities and households who share a toilet.

	Flush to piped sewer system, septic tank, (%)	Flush to somewhere else (%)	Pit latrine, ventilated (%)	Pit latrine with slab (%)	Pit latrine without slab / open pit (%)	Hanging toilet / hanging latrine (%)	Improved sanitation, JMP definition (excluding shared) (%)	Improved sanitation (including shared) (%)	Households that share a toilet (%)	Average number of households that share a toilet
	(1)	(2)	(3)	(4)	(5)	(6)	(7)	(8)	(9)	(10)
All	8.0	0.2	20.3	58.1	3.2	10.2	8.6	86.4	91.1	16.2
	By size of the slum									
Small	0	0	31	69	0	0	6.9	100	93.1	12.3
Medium	10.4	0.4	13.8	69.7	0	5.8	9.2	93.9	90.8	17.7
Large	6.7	0	24.5	49.4	5.7	13.8	8.2	80.5	91.2	15.2
T-test, small–medium	**									*
T-test, medium–large				*				*		
	By land ownership									
Public	6	0	20.1	54.6	5.3	14	8.7	80.7	90.7	12.4
Private	12.6	0.5	24	62.9	0	0	8.2	99.5	91.8	19.0
Other	3.3	0	0	66.6	0	30	10	70	90.0	39.3
T-test, public–private						***		***		**
	By per capita consumption expenditure									
Poorest	5.3	0.0	16.9	61	2.3	14.5	4.6	83.2	95.4	19.4
Second quintile	8.6	0.0	11.8	71.5	2	6.1	7.3	91.9	92.7	14.6
Third quintile	7.6	0.0	24.9	50	4	13.6	6	82.5	94	16.1
Fourth quintile	4.7	0.7	23.4	55.6	4.5	11	6.7	83.8	91.5	15.7
Richest	12.8	0.0	22.8	54.3	3.2	7	16.1	89.8	83.9	15.6
T-test, poorest–richest	*						**		**	

Notes: Data from Bangladesh Urban Informal Settlements Survey 2016. The *t*-test rows show the results of Wald tests of whether the difference in means/proportions are significant at $*p < 0.1$, $**p < 0.05$, or $***\ p < 0.01$.

Households wait on line for 7 minutes, on average, to use sanitation facilities. In small communities, households wait slightly less than in larger communities: 4.8 minutes in small slums on average, compared to 7.2 and 6.7 minutes in medium and large slums, respectively, though there is quite a lot of variability across households (Figure S2 in the online supplement). The waiting time is consistent across household per capita distribution and across land ownership.

Perceived satisfaction, safety and reliability

In terms of satisfaction with sanitation facilities, the majority of adult[9] male (89%) and female (84%) residents report feeling secure when using the facility at night (Table 5). Residents in smaller and privately owned slums feel slightly more secure in terms of using sanitation facilities at night, and more feel that they have good privacy compared to larger and government-owned-land slums, though the differences are not large.

Sanitation facilities are reported to be reliable for the most part; however around 10% of households report that facilities were temporarily unusable for at least one full day in the past two weeks. This number is substantially higher in small slums, where 28% of residents report this issue of reliability.

During times of flooding, 42% of households said that their sanitation facilities were inaccessible and reported an average of nine days without access to sanitation (Table 6). Consumption quintiles report varying levels of lack of access (32–47%), but the lack of access to sanitation during flooding seems to disproportionately affect better-off households.

Faecal sludge management

Table 7 shows statistics on faecal sludge management and on where the sanitation facilities used by households empty to. Residents were asked to report where contents of their sanitation facilities emptied to. Responses were coded as 'safe' or 'unsafe', where faecal sludge management was 'safe' if contents emptied either directly to a piped sewerage network, a septic tank with no outlet, or a lined pit with no outlet. Overall, only 8% of households could be classified as practicing safe faecal sludge management. Among households with access to improved sanitation (unshared), only 2% practiced safe faecal sludge management. This number most closely represents access to 'safely managed sanitation', the new sanitation indicator developed by the JMP for meeting Sustainable Development Goal 6. The large majority report that contents are emptied to drains either through lined pits (42.9%) or a septic tank connected to a drain (9.8%). Moreover, 13.7% of all households report that their sanitation facilities are emptied directly to a nearby lake, pond or river, and 6.1% to a septic tank connected to open ground. Among owners of pit latrines or sanitation facilities that link to septic tanks, about 68% report never having emptied their pit or septic tank. This does not necessarily tell us about safe faecal sludge management, but at least sheds light on pit and septic tank emptying practices.

Table 5. Satisfaction with sanitation facilities available.

	Feels safe using facility at night, adult male (%)	Feels safe using facility at night, adult female (%)	Satisfied with privacy, adult male (%)	Satisfied with privacy, adult female (%)	Facility unusable for full day in past two weeks (%)
	(1)	(2)	(3)	(4)	(5)
All	89	84.4	73.9	68.9	9.2
	By size of the slum				
Small	96.6	96.6	82.8	82.8	27.6
Medium	87.3	83.1	65.8	59.1	8.9
Large	89.8	84.8	79.2	75.2	8.7
T-test, small–medium		**			
T-test, medium–large			*	**	
	By land ownership				
Public	87.1	84.5	78.2	74.3	9.6
Private	92.1	88.3	67.8	61.2	8.4
Other	90.0	60.0	63.3	56.7	10.0
T-test, public–private					
	By per capita consumption expenditure				
Poorest	86.8	86.8	67.6	68.2	11.1
Second quintile	85.7	80.2	65.7	62.2	4.4
Third quintile	87.4	82.6	73.1	68.3	11.0
Fourth quintile	88.4	86.2	76.0	68.3	10.7
Fifth quintile	94.7	85.7	83.0	75.2	8.8
T-test, poorest–richest	**		**		

Notes: Data from Bangladesh Urban Informal Settlements Survey 2016. 'Satisfied with privacy of sanitation facilities' corresponds to 'very satisfied' and 'satisfied' responses of households. The t-test rows show the results of Wald tests of whether the difference in means/proportions are significant at $^{*}p < 0.1$, $^{**}p < 0.05$, or $^{***}p < 0.01$.

Table 6. Accessibility of water and sanitation facilities during times of flooding.

	Households that experienced flooding (%)	Number of days of flooding	Households with limited water availability during flooding (%)	Number of days with a non-functional water source during flooding	Households with a non-functional toilet facility during flooding (%)	Number of days with a non-functional toilet facility during flooding
	(1)	(2)	(3)	(4)	(5)	(6)
All	52.2	15.6	28.9	10.4	42.1	9.4
				By size of the slum		
Small	44	5.0	45.5	3	45.5	4.4
Medium	47.7	11.4	30.8	16.7	41.7	14.6
Large	55.7	18.4	27.3	6.4	42.2	6.5
T-test, small–medium		*		**		**
T-test, medium–large				*		**
				By land ownership		
Public	58.2	19.0	36.8	11.4	50.7	10.1
Private	39.5	8.4	16.2	4.4	31.1	6.3
Other	62.0	9.3	0.0	-	-	-
T-test, public–private		*	**	**		
				By per capita consumption expenditure		
Poorest	43.6	13.0	26.4	4.2	31.8	5.4
Second quintile	47.3	21.5	25.4	10.5	46.1	11.0
Third quintile	53.4	12.8	32.3	11.5	46.0	9.0
Fourth quintile	54.9	17.1	37.9	10.9	47.0	9.1
Fifth quintile	58.3	14.0	22.4	12.1	37.7	10.7
T-test, poorest–richest				**		**

Notes: Data from Bangladesh Urban Informal Settlements Survey 2016. The t-test rows show the results of Wald tests of whether the difference in means/proportions are significant at $^{*}p < 0.1$, $^{**}p < 0.05$, or $^{***}p < 0.01$.

Table 7. Faecal sludge management.

	Lined pit with overflow to drain elsewhere (%)	Directly to drain/ ditch (%)	Directly to lake/pond/ river (%)	Septic tank connected to drain (%)	Septic tank connected to opened ground (%)	Others (%)
	(1)	(2)	(3)	(4)	(5)	(6)
All	42.9	13.9	13.7	9.8	6.1	13.6
	By size of the slum					
Small	44.8	0.0	13.8	37.9	0.0	3.4
Medium	48.4	11.9	3.1	12.7	8.0	15.8
Large	38.8	15.9	21.2	6.7	5.0	12.4
T-test, small–medium		**				*
T-test, medium–large			***			
	By land ownership					
Public	35.8	19.8	18.6	6.3	5.0	14.6
Private	51.6	5.6	1.4	18.0	9.3	14.1
Other	66.6	0.0	33.4	0.0	0.0	0.0
T-test, public–private		**	***			
	By per capita consumption expenditure					
Poorest	36.5	11.6	15.6	13.8	9.8	12.7
Second quintile	40.2	14.2	10.8	11.5	10.3	13.0
Third quintile	40.0	16.5	14.3	11.0	5.4	12.8
Fourth quintile	48.6	13.9	12.8	7.8	3.4	13.6
Richest	46.6	13.2	14.8	6.8	3.4	15.2
T-test, poorest–richest						

Notes: Data from Bangladesh Urban Informal Settlements Survey 2016. The *t*-tests show the results of Wald tests of whether the difference in means/proportions are significant at $*p < 0.1$, $**p < 0.05$, $***p < 0.01$. 'Other' includes directly to piped sewer system, septic tank connected to piped sewer system, septic tank with no outlet, lined pit with no outlet, unlined pit, and other categories not specified.

Barriers to upgrading sanitation

Households reported that the main constraints to improving their sanitation facilities are lack of finance and lack of interest by landlords to invest (Figure S3 in the online supplement). In large slums, lack of space to build the facilities is also frequently reported as an important issue. Even among richer households lack of space is a reported constraint, which suggests that households who might have the financial means to invest in improving sanitation merely do not have the space to do so.

Mechanisms of water and sanitation service delivery

Installation, management and accountability

Tables S4, S5 and S6 in the online supplement summarize statistics on the ownership of drinking water sources, management structure, and accountability mechanisms for escalating complaints. Water infrastructure is mainly installed by landlords (56.8%); D-WASA, the local utility company responsible for water distribution in Dhaka (17.3%); and NGOs (12.9%). Although landlords are the owners of virtually all drinking water sources in small and medium slums (100% and 81%, respectively), NGOs own a non-trivial share of water sources in large slums (21%). Landlords are reported to instal a mix of tubewells and piped water sources, while D-WASA is reported to instal a portion of only piped water sources. Landlords are the main owners of water sources used by slum residents on privately owned land (90%) while in large slums D-WASA and NGOs own on average 27% and 21% of drinking water sources, respectively.

Sanitation facilities are mainly installed by landlords or middlmen (63%), NGOs (20%), and private companies or residents (15%) (Table S7 in the online supplement). Although landlords are almost the exclusive providers of sanitation facilities in small and medium slums, NGOs have been supporting several of these investments in large slums. Compared to water services, there is a much larger role for household investments in sanitation infrastructure. NGOs and households themselves have been much more proactive in investing in sanitation infrastructure on government-owned land compared to slums on privately owned land. In contrast, in communities where land ownership is unclear or it is not owned by the government, investments in sanitation facilities have mainly come from NGOs. Landlords rely heavily on community leaders for the day-to-day management of both water and sanitation facilities across all types of slums (Tables S6 and S9 in the online supplement).

Payments for water and sanitation services

Landlords also seem to depend heavily on community leaders and private individuals to collect water payments from households (Table 8). On average, 46% of payments for water services in medium slums are channelled through community leaders, while in large slums NGOs take some of the payment collection responsibility (17%). The use of community leaders to collect water payment is slightly more prevalent in slums on privately owned land (around 40.3%) compared to those on government-owned land (35%). Interestingly, community leaders are responsible for collecting the water services

Table 8. Water and sanitation payments: who do you pay?

	Landlord / middlemen (%)		D-WASA (central government) (%)		Private company / person (%)		NGO (%)		Community leaders (%)		Others (%)	
	Water (1)	Sanitation (2)	Water (3)	Sanitation (4)	Water (5)	Sanitation (6)	Water (7)	Sanitation (8)	Water (9)	Sanitation (10)	Water (11)	Sanitation (12)
All	13.3	17.9	5.5	-	24.5	2.0	13.6	1.6	35.0	77.0	8.2	1.5
	By size of the slum											
Small	-	0.0	-	-	-	75.0	-	0.0	-	25.0	-	0.0
Medium	12.5	32.2	8.3	-	16.7	0.0	0.0	0.0	46.0	64.4	16.5	3.3
Large	13.5	6.2	4.8	-	26.3	0.0	16.8	3.1	32.3	90.6	6.2	0.0
T-test, small–medium	-		-	-	-		-		-		-	
T-test, medium–large				-			**					
	By land ownership											
Public	14.4	4.5	4.2	-	26.6	0.0	14.8	4.5	34.5	86.9	5.5	4.1
Private	0.0	40.6	19.9	-	0.0	5.5	0.0	0.0	40.3	53.8	39.8	0.0
Other	-	5.9	-	-	-	0.0	-	0.0	-	94.1	-	0.0
T-test, public–private	***			-	**		**					
	By per capita consumption expenditure											
Poorest	0.0	25.8	0.0	-	37.2	4.3	0.0	5.3	56.4	64.6	6.4	0.0
Second quintile	17.5	25.3	4.4	-	30.6	0.0	9.6	0.0	37.9	74.7	0.0	0.0
Third quintile	15.0	8.9	16.0	-	20.8	0.0	5.7	0.0	26.4	81.1	16.1	10.0
Fourth quintile	14.8	11.5	0.0	-	24.6	2.9	27.7	0.0	22.9	85.6	10.0	0.0
Richest	15.1	15.8	6.8	-	14.2	0.0	19.5	0.0	34.6	84.2	9.8	0.0
T-test, poorest–richest	*		**	-								

Note: Data from Bangladesh Urban Informal Settlements Survey 2016. The t-tests show the results of Wald tests of whether the difference in means/proportions are significant at $*p < 0.1$, $**p < 0.05$, $***p < 0.01$. 'Other' for water includes tanker truck manager, water vendor and other categories not specified; for sanitation, it includes other categories not specified.

payments from the poorest households, while richer households tend to pay directly to the landlords or NGOs operating in the slum.

On average, 71% of households rent their housing structure. Among those households who rent, on average 89% have their water utilities included as part of the rent (Table S10, Column 2, in the online supplement). Therefore, rent inclusive of water services is a quite common arrangement in slums.[10] In Column 3, the average payment for water services among slum households who report owning their housing structures is given. Although just a third of households report owning their homes, these estimates provide a broad approximation that can serve as a reference on how much slum households pay to access water services. Slum households owning their housing structure pay on average BDT169 (around USD2) monthly for water services. Access to water services is cheaper on average in larger slums, at around BDT155 per month, compared to medium slums, where households pay BDT308 monthly for these services on average; for small slums, no household reports owning their housing structure. These estimates should be interpreted with caution, however, as they are based on small samples. Lastly, as a share of total per capita consumption (food, non-food and housing, Column 4), households spend around 1% of their expenditures on water services.

Figure S4 in the online supplement shows how frequently payments for water services are collected from households who own their housing structure or pay for water services separately from their rent. Most households are charged monthly or weekly. The poorest households tend to pay per use or are charged weekly, while richer households are almost always charged monthly. Similarly, households in private slums are billed for water services on a weekly basis exclusively. This suggests that landlords provide different payment frequencies to potential renters, perhaps based on their ability to pay.

Landlords also rely largely on community leaders to collect payments for the use of the sanitation facilities. On average, 77% of residents pay to community leaders for sanitation services (Table 8). While private companies frequently act as collector agencies for the provision of water services (25%), community leaders are almost exclusively responsible for collecting payments for the use of sanitation facilities. This does not vary across the different types of community size, land ownership, or socio-economic status of households.

Unlike water services, 6% of residents who rent their dwelling report having the use of sanitation facilities included as part of their rent (Table S11 in the online supplement). In small slums, 18% of renters report having the use of sanitation facilities included in the rent, which is consistent with the higher average rents paid by residents in these slums compared to large slums. In contrast, 8% of renters in large slums report having the use of sanitation facilities included as part of their rent.

Table S12 in the online supplement shows different measures of the magnitude of payments made by residents for sanitation services. Although sample sizes are generally small, these numbers suggest that residents pay BDT36 per month on average for use of the sanitation facilities available in slums. The amount spent on sanitation services is substantially lower than the amount spent on water services. Accessing water services costs households on average five times more than sanitation facilities. Access to sanitation facilities costs almost twice as much in small slums compared to large slums, and richer households tend to spend slightly more on average to access these services. Lastly, payments for use of sanitation services are billed to households monthly as a fixed amount, though in some cases, such as in large and publicly owned slums, residents can pay weekly (Figure S5 in the online supplement).

Provision of water services to slums by D-WASA

Though middlemen are the predominant service provider, the local utility, D-WASA, is sometimes reported to play a role in water service provision, whether it be in installation, accountability, management, or payment, particularly in large slums. This might suggest that D-WASA is providing water points at the border of slums or even in some cases inside the slums, despite the lack of a formal mandate. To explore this hypothesis, we look at the location of the main drinking water source for households in large slums who report receiving their water services directly from D-WASA. Table S13 in the online supplement shows these statistics estimated separately for four groups of slum residents: those who report that D-WASA installed their water source; those who report that D-WASA manages their water source; those who report reporting issues with the water source to D-WASA; and those who report paying D-WASA for water service.

Fifty to sixty-five per cent of residents who report receiving water services from D-WASA access this water source directly from their own dwelling, yard, or plot Among those who pay D-WASA directly for water services, more than 82% have the water source installed directly in their own yards or plots. Overall, these estimates suggest that D-WASA is not only installing water sources in the border of slums but might be providing direct services to some slum households, though the share benefiting from this direct service is very small (5.5–17.5% of residents).

Discussion

This article focuses on measuring access levels to water and sanitation and the 'informality' of service markets in the slums of Dhaka, Bangladesh. Given the traditional lack of formal mandates for public utilities to serve slum areas (although this is changing in some areas of Bangladesh), overall access to water and sanitation infrastructure is high. Access to improved water is nearly universal, at 97%, and 86% of households use shared or unshared improved sanitation technology. These results seem to be consistent with findings from other surveys that measure access in slums in Bangladesh (NIPORT et al., 2008, 2015; UNICEF, 2015). However, indicators of improved infrastructure access may mask important dimensions of safety. This study is limited in that it could not measure updated JMP indicators for safely managed water, which include indicators of functionality and water quality (WHO/UNICEF, 2019). We find that WASH services are mostly provided by non-regulated providers, with evidence to support claims of 'informality' and strong presence of middlemen providers. The data suggest that local landlords or middlemen play a large role in installation and management, while the local utility seems to have a limited role in the overall spectrum of service delivery.

Informal service provision may reduce accountability for meeting minimum standards of quality. For example, although nearly all households use piped water connections, many report issues of reliability, functionality and accessibility, especially during monsoon seasons. Service interruption can indicate leaks and malfunctioning pipes that compromise water quality and heighten risks to public health and safety (Ercumen et al.,

2015; Shams et al., 2016). Over half of study households reported issues related to water quality. Other studies have found that piped water supply is among one of the most heavily contaminated types of water sources in Bangladesh, most likely due to leaks, intermittent supply and unsafe handling (Pickering et al., 2019; UNICEF, 2015; World Bank, 2018). In our study, just a quarter of slum households reported treating their water. As middlemen such as landlords are the primary service providers, point-of-tap water treatment, such as water filters or automated chlorination systems, could be a viable short-term solution (Pickering et al., 2015) until the formal capacity to monitor and enforce quality standards of piped water is developed.

On sanitation, open defaecation and use of unimproved sanitation technology is minimal in Dhaka slums. However, there is a high dependence on shared or public toilets. Over 90% of households share sanitation facilities, and the average ratio of households to sanitation facilities is 16 to 1. Using JMP standards to exclude improved sanitation that is shared between two or more households, only 9% have access. These findings suggest that there is need to reduce shared sanitation, or at least reduce the sharing ratio and improve the management of public toilets. When considering faecal sludge management, we estimate that 2% have access to the JMP's conceptualization of 'safely managed sanitation'. The issue of safe faecal sludge management is not limited to slum communities. One study estimates that 2% of the human excreta in Dhaka is safely managed, while the rest is discharged into open water. This is mostly due to the lack of regular maintenance of on-site sanitation facilities such as pit latrines and septic tanks (Blackett et al., 2014). Policy makers have urged better faecal sludge management by implementing city-wide sanitation initiatives that include both on-site and off-site sanitation technologies. It is vital that slum communities are also included in these strategies and that formal, regulated faecal sludge management services are made available to residents.

Finally, this study responds to knowledge gaps on the basic characteristics of slum populations. Slums are not a transitory place for migrants; on average, household heads report about 11 years of residence in the community. There also seems to be a mix of consumption patterns, illustrating that not every household is living under the national poverty line and that households are making conscious trade-offs between financial gains and living conditions. The findings additionally highlight inequalities across slums based on the characteristics of the community itself, such as the size of the slum and whether the slum is on private or public land. For instance, households in larger slums are more likely to face threats of eviction and experience more crowding. On the other hand, they are also more likely to obtain services from NGOs and utilities compared to small and most medium slums. This heterogeneity demonstrates the need for interventions tailored to different types of slum contexts.

Given the trends of rapid urbanization and growing urban poverty in Bangladesh, informal settlements will need to be a focal population in sustainable development strategies, not only for poverty reduction but also for improving health, equitable access to basic services, and economic, social and environmental resilience. This will require better monitoring and research efforts that can better describe the unique problems these communities face, as well as the institutions and stakeholders that impact these communities.

Notes

1. The micro-data from Bangladesh Urban Informal Settlements 2016 are publicly available at https://microdata.worldbank.org/index.php/catalogue/2864/.
2. The sampling frame included a total of 3360 urban slums. We excluded 34 slums originally included in the Census of Slums and Floating Population 2014 from the sampling frame because they had less than five households, per the official government definition of slums.
3. 1 decimal equals 436 square feet or 40.5057 m^2.
4. Unlike the JMP, we do not include chemical or microbial testing of priority water contaminants.
5. There are multiple official agencies in charge of land records in Bangladesh, so it is quite complex to find out information on land tenure for many records, even those on privately owned land.
6. For details on how the official poverty numbers are produced in Bangladesh, see Ahmed et al. (2017).
7. Tables S1 and S2 in the online supplementary material present similar statistics by slum land ownership and quintiles of per capita household expenditure.
8. The dependency ratio is defined as the number of household members aged zero to 14 or over 55, divided by the number of household members aged 15 to 55.
9. Defined as over 15 years of age.
10. Table S12 in the online supplementary material presents a summary of the utilities that are reported to be included as part of the housing rent payment by slum households who are renters. Results suggest that almost all renters have their water and electricity utilities automatically included in their rent as a fixed payment. In small and medium slums water is almost always included, while in larger slums some renters pay for their water services separately.

Acknowledgments

This study was a part of the larger WASH Poverty Diagnostic initiative of the World Bank, supported by the Swedish International Development Cooperation Agency and the Japan Trust Fund for Scaling Up Nutrition.

Disclosure statement

No potential conflict of interest was reported by the authors.

ORCID

Yurani Arias-Granada http://orcid.org/0000-0001-6382-5366

References

Adams, A. M., Islam, R., & Ahmed, T. (2015). Who serves the urban poor? 2015. A geospatial and descriptive analysis of health services in slum settlements in Dhaka, Bangladesh. *Health Policy Plan*, *30*(Suppl 1), 32–45. https://doi.org/10.1093/heapol/czu094

Ahmed, F., Arias-Granada, Y., Eugenia-Genoni, M., Yanez-Pagans, M., Yoshida, N., Roy, D., & Latif, A. (2017) Description of the official methodology used for poverty estimation in Bangladesh for 2016/17. In *Household income and expenditure survey 2016/17*. World Bank. http://catalog.ihsn.org/index.php/catalog/7399.

Ahmed, I. (2014). Factors in building resilience in urban slums of Dhaka, Bangladesh. *Procedia Economics and Finance*, *18*, 745–753. https://doi.org/10.1016/S2212-5671(14)00998-8

Angeles, G., Lance, P., Barden-O'Fallon, J., Islam, N., Mahbub, A. Q. M., & Nazem, N. I. (2009). The 2005 census and mapping of slums in Bangladesh: Design, select results and application. *International Journal of Health Geographics, 8*(1), 32. https://doi.org/10.1186/1476-072X-8-32
Bangladesh Bureau of Statistics. 2014. *Census of slum areas and floating population.* Final report. Statistics and Information Division, Ministry of Planning.
Banks, N. (2008). A tale of two wards: Political participation and the urban poor in Dhaka city. *Environment and Urbanization, 20*(2), 361–376. https://doi.org/10.1177/0956247808096116
Blackett, I. C., Hawkins, P. M., & Heymans, C.2014. The missing link in sanitation service delivery: A review of fecal sludge management in 12 cities. Working Paper 88258, WSP, World Bank, Washington, DC.
Checkley, W., Buckley, G., Gilman, R. H., Assis, A. M., Guerrant, R. L., Morris, S. S., Mølbak, K., Valentiner-Branth, P., Lanata, C. F., Black, R. E., & Childhood Malnutrition and Infection Network. (2008). Multi-country analysis of the effects of diarrhoea on childhood stunting. *International Journal of Epidemiology, 37*(4), 816–830. https://doi.org/10.1093/ije/dyn099
Ercumen, A., Arnold, B. F., Kumpel, E., Burt, Z., Ray, I., Nelson, K., & Colford, J. M. (2015). Upgrading a piped water supply from intermittent to continuous delivery and association with waterborne illness: A matched cohort study in urban India. *PLoS Medicine, 12*(10), e1001892. https://doi.org/10.1371/journal.pmed.1001892
Guerrant, R. L., Kosek, M., Moore, S., Lorntz, B., Brantley, R., & Lima, A. A. (2002). Magnitude and impact of diarrheal diseases. *Archives of Medical Research, 33*(4), 351–355. https://doi.org/10.1016/S0188-4409(02)00379-X
Hanchett, S., Akhter, S., & Khan, M. (2003). Water, sanitation and hygiene in Bangladeshi slums: An evaluation of the Wateraid-Bangladesh urban programme. *Environment and Urbanization, 15*(2), 43–55. https://doi.org/10.1177%2F095624780301500219
Hathi, P., Haque, S., Pant, L., Coffey, D., & Spears, D. (2017). Place and child health: The interaction of population density and sanitation in developing countries. *Demography, 54*(1), 337. https://doi.org/10.1007/s13524-016-0538-y
Hossain, K., & Ahmed, S. (2015). Non-conventional public-private partnerships for water supply to urban slums. *Urban Water Journal, 12*(7), 570–580. https://doi.org/10.1080/1573062X.2014.923473
Humphrey, J. H. (2009). Child undernutrition, tropical enteropathy, toilets, and handwashing. *The Lancet, 374*(9694), 1032–1035. https://doi.org/10.1016/S0140-6736(09)60950-8
Ngure, F. M., Reid, B. M., Humphrey, J. H., Mbuya, M. N., Pelto, G., & Stoltzfus, R. J. (2014). Water, sanitation, and hygiene (WASH), environmental enteropathy, nutrition, and early child development: Making the links. *Annals of the New York Academy of Sciences, 1308*(1), 118–128. https://doi.org/10.1111/nyas.12330
NIPORT, ICDDRB, & MEASURE Evaluation. (2015). *Final report for the Bangladesh urban health survey 2013.*
NIPORT, ICDDRB, MEASURE Evaluation, & ACPR. (2008). *Final report for the Bangladesh urban health survey 2006.*
NIPORT, Mitra and Associates, & ICF International. (2016). *Bangladesh Demographic and Health Survey (DHS) 2014.*
Panday, P. (2017). *Reforming urban governance in Bangladesh: The city corporation.* Palgrave McMillan.
Pickering, A. J., Crider, Y., Amin, N., Bauza, V., Unicomb, L., Davis, J., Luby, S. P., & Carpenter, D. O. (2015). Differences in field effectiveness and adoption between a novel automated chlorination system and household manual chlorination of drinking water in Dhaka, Bangladesh: A randomized controlled trial. *PloS One, 10*(3), e0118397. https://doi.org/10.1371/journal.pone.0118397
Shams, S., Ahsan, A., Al-Mamun, A., & Arunkumar, T. (2016). Physical risk assessment for urban water supply in a developing country: A case of Mega City Dhaka. *Engineering Journal, 20*(3), 3. https://doi.org/10.4186/ej.2016.20.3.23
Stanaway, J. D., Afshin, A., Gakidou, E., Lim, S. S., Abate, D., Abate, K. H., … Abdela, J. (2018). Global, regional, and national comparative risk assessment of 84 behavioural, environmental and occupational, and metabolic risks or clusters of risks for 195 countries and territories, 1990–2017:

A systematic analysis for the Global Burden of Disease Study 2017. *The Lancet*, *392*(10159), 1923–1994. https://doi.org/10.1016/S0140-6736(18)32225-6

UNICEF. (2010). *Understanding urban inequalities in Bangladesh: A prerequisite for achieving vision 2021.*

UNICEF & Bangladesh Bureau of Statistics. (2015). *Bangladesh Multiple Indicator Cluster Survey (MICS) 2012–2013.* UNICE.

UNICEF & Bangladesh Bureau of Statistics. (2019). *Bangladesh Multiple Indicator Cluster Survey (MICS) 2019.* UNICE.

Vlahov, D., Galea, S., & Freudenberg, N. (2005). The urban health “advantage”. *Journal of Urban Health*, *82*(1), 1–4. https://doi.org/10.1093/jurban/jti001

WHO/UNICEF. (2019). *Progress on drinking water, sanitation and hygiene: 2019 update and SDG baselines.* World Health Organization.

World Bank. (2018). *Precarious progress: A diagnostic of water supply, sanitation, hygiene, and poverty in Bangladesh. WASH poverty diagnostic.*

World Bank. (2019). *Bangladesh development update, April 2019: Towards regulatory predictability.*

Poor water service quality in developed countries may have a greater impact on lower-income households

Anna Robak and Henning Bjornlund

ABSTRACT

This article analyzes the relationship between perceived service quality and averting behaviours and averting expenditures across prominent water service attributes, using revealed preference survey data from New Zealand water supply customers. It finds that nearly 50% of consumers undertake averting behaviours, investing substantial amounts in improving their water service quality. Unexpectedly, lower-income households were more likely to undertake averting behaviours for the same perceived service quality, and spent more in their averting behaviours. This suggests that the burden of low-quality service is greater on lower-income households, not only relative to income but in absolute terms.

Introduction

Providers of urban infrastructure services across the developed world are facing increasing pressure from regulators and customers to provide affordable services (Bontemps & Nauges, 2006; OECD, 2010). The large number of willingness-to-pay studies around urban infrastructure suggests that infrastructure providers and researchers interpret an affordable service as one that has a price that customers are willing to pay. However, revealed preference studies have shown that infrastructure service users invest time and money in improving the quality of their service – particularly in a developing country, or following a contamination event – by undertaking behaviours, such as travelling for water and boiling, and purchasing equipment (Dupont, Adamowicz, & Krupnick, 2009). Household behaviours are directed at improving taste, smell, reliability and many other service quality attributes. If individuals invest time and money, then researchers considering affordability should include the costs of these actions and purchases. Altaf (1994) urged that the public planning process consider household responses in a developing country, and we argue that this consideration is just as warranted in a developed country. Two main questions therefore arise with respect to affordability and equity: First, could it be more economically efficient for households to undertake behaviours and purchases to attain their desired quality of service rather than service providers investing to provide the desired quality to all? And second, is there a difference in the extent to which richer and poorer households are burdened by levels of service that are lower than desired? The answers to these questions are critical for urban infrastructure providers when making investment decisions under tight budget constraints.

The bigger question, which remains largely unanswered, is whether or not there are significant averting behaviours undertaken in a developed country outside of contamination events.

This article provides insight into the trade-off between household averting behaviours and centrally supplied quality by investigating whether the household averting behaviours and purchases differ according to perceived quality of service, based on the following research questions: (1) Is a household that is less satisfied with their quality of service more likely to undertake an averting behaviour? (2) Does a household that is less satisfied with their quality of service spend more time and/or money customizing it? (3) Are there significant and non-negligible marginal benefits from improving the quality of service? And it explores the second question of relative affordability by exploring the different averting behaviour responses and purchases of lower-income households using the following research questions: (4) Are lower-income households less likely to undertake averting behaviours? (5) Do lower-income households spend less than higher-income households customizing their quality of service through averting behaviours?

A service has a number of attributes that individuals use to assess its quality. For each attribute, an economically efficient and equitable level of service is one that individuals are willing to pay to achieve when they have complete knowledge about the service and its impacts. This study identifies and monetizes averting behaviours to estimate the cost incurred by households that undertake behaviours and purchases to achieve the quality of service they desire – when the one provided is less than desired. This study focuses on private investments, as there is no evidence to suggest that other coping mechanisms, such as the emotional stress and lost wages identified by Wutich and Ragsdale (2008) in a developing country, would be relevant in a developed country.

Among infrastructure services, water supply is arguably the most critical, as water is a basic human need. Water suppliers worldwide are increasingly struggling to maintain quality of service while keeping the service affordable because of ageing infrastructure, increasingly stringent legislation, and rising pollution levels (New Zealand Treasury, 2009; Renzetti, 1999; USEPA, 2010). To achieve affordability, most regulators and governments would allow service quality reductions, provided they are supported by a rigorous cost–benefit analysis (European Commission, 2008; Hajkowicz, Young, Wheeler, Hatton MacDonald, & Young, 2000; New Zealand Treasury, 2009). This study therefore uses the case of municipal water supply infrastructure to answer questions about what constitutes an economically efficient and equitable quality of service.

When assessing quality of service, water suppliers consider a number of attributes, such as the frequency and duration of service interruptions, aesthetic considerations, and protection from illness (Alegre et al., 2006). They therefore have to consider the economically efficient and equitable quality of service for each of these attributes.

While a large number of studies have identified which attributes are most important to households through revealed and stated preference studies, they have done so by asking households to rank or value a *predetermined list of attributes.* Hatton MacDonald, Morrison, and Barnes (2010) specified the frequency of, and notification about, service interruptions; Genius et al. (2008) specified water supply security and quality; Willis, Scarpa, and Acutt (2005) specified 14 predetermined attributes; and

Pattanayak, Yang, Whittington, and Bal Kumar (2005) specified taste, colour, health risk, and reliability of the water supply. It is well documented in the literature, however, that what people say they do or want is not always reflected in their actions. This phenomenon occurs due to cognitive dissonance, hypothetical bias, and even question ordering (Ajzen, Brown, & Carvajal, 2004; Bertrand & Mullainathan, 2001).

To ensure that the most important attributes are considered, this study asked people open-ended questions about the full range of behaviours and equipment purchases related to their water service. We asked *why* they undertook these behaviours and purchases, to identify and evaluate what was most important to them, given their current perceived level of service. A search of the literature suggests that this is the first study to identify the most important attributes using a revealed preference technique and allowing respondents to specify attributes that may not be mentioned in the literature.

This study uses the water industry in New Zealand to explore the extent of household-level averting behaviours and expenditures in a developed nation, and the existence of any significant private investment burden on lower-income households. While water supply quality in developing countries is likely to have far more serious health and livelihood impacts, the economic impacts of the full range of water supply quality attributes have not yet been documented in a developed nation. Further, as climate change affects water availability, researchers will be looking at developing-country impacts to predict the impacts of a lower quality of service, and economists will use these predictions to price preventive investments in infrastructure. All other identified averting expenditure studies have either focused on one or two attributes, such as safety or taste, excluding other concerns (Beaumais, Briand, Millock, & Nauges, 2010; Dupont et al., 2009), or been based in developing countries (Aziz, 2007; Haq, Usman Mustafa, & Ahmad, 2008; McConnell & Rosado, 2000; Pattanayak et al., 2005). They also have not investigated the extent of burden on lower-income households. This study seeks to identify all attributes for which households undertake averting expenditures.

Water-service-related averting expenditures

Averting expenditures include the purchase and time costs that households incur to avoid the adverse impacts of some phenomenon (Birol, Koundouri, & Kountouris, 2008) – typically, in response to perceived environmental and health risks (Abrahams, Hubbell, & Jordan, 2000). In developed countries, for example, households purchase bottled or tanker water, or filter or boil their tap water (Janmaat, 2007; Laughland, Musser, Shortle, & Musser, 1996). In developing countries, in addition to filtering or boiling, or purchasing bottled water, it is common to travel to collect water for domestic purposes (Hahn, Riederer, & Foster, 2009; Larson, Minten, & Razafindralambo, 2006). In a developed country, travelling for water is not as burdensome as in developing countries, as it typically involves driving to a spring or to purchase large containers. The value of upgrading a potable water supply system (PWSS) can therefore be equated to consumers' averting purchase and time investments.

Averting expenditures are widely used by economists because the method is grounded in the well-developed field of household production theory (Laughland, Musser, Musser, & Shortle, 1993). In the PWSS context, the household can be considered to *produce* good-quality water by investing their own time and money in purchasing a substitute good such as bottled water or by improving the current water quality by buying a filter. In the field of water supply, the averting expenditure method is the most common revealed preferences (RP) approach, and is an appropriate measure of willingness to pay when the cost of averting behaviour/expenditure exactly addresses the perceived problem (Whitehead & Van Houtven, 1997).

RP techniques, and averting expenditures in particular, have some limitations. Beshears, Choi, Laibson, and Madrian (2008) argue that RP do not always represent an individual's true interests; RP are confounded by misinformation, miscalculation and impulsiveness. Individuals may accept default options or avoid taking action or making purchases through procrastination, or delay taking action or making purchases because of the complexity of choices or distance of the consequences. In addition, their choices depend on personal experience, and they are willing to pay more for brands that have more effective marketing. While the authors recognize that these factors mean that consumers' valuation of a higher level of service may not manifest itself in purchases or behaviours, therefore leaving some residual value unmonetized, the main purpose of this research is to uncover consumers' total spending, their behaviours, and the motivations for their actions.

The averting expenditure literature on PWSS focuses largely on the aesthetic and health-related water quality performance attributes, as summarized in Table 1; in developing countries in particular, there is also a body of averting behaviour literature on unreliable or intermittent water supplies (Majuru, Suhrcke, & Hunter, 2016). Averting expenditures related to other performance attributes, such as frequency of service interruptions, water restrictions and mains pressure are not quantified in the literature, but behaviours and equipment are identified. The literature review reveals only one averting expenditure study of water 'reliability' (Pattanayak et al., 2005), in a developing country. The research related to each performance attribute, as well as any perceptions related to expenditures, is outlined in the following subsections.

Water quality

The averting expenditure literature on PWSS is frequently ambiguous on whether behaviours and purchases are undertaken to avoid bad taste, smell and appearance, to reduce risk of illness, or to avoid some other undesirable attribute. In this section, therefore, averting expenditures related to all types of water quality attributes are discussed. The drivers for these expenditures, the types of behaviours and purchases, the explanatory variables, and the country in which the study was conducted are summarized in Table 1.

These studies consider averting expenditures such as buying bottled water, boiling water, carting water, household filtration and treatment. Consumers typically invest time and money on these behaviours and purchases when they experience problems with taste, smell and/or appearance or are concerned about microbiological or chemical contamination.

Abdalla, Roach, and Epp (1992), for example, found that respondents who perceive greater health risks are more likely to make averting expenditures to protect against chemical contamination. Similarly, Larson and Gnedenko (1999) found that

Table 1. Summary of averting expenditure literature.

Averting due to	Averting expenditures	Explanatory variables		Reference
Event: chemical contamination	Bottled water; home water treatment, hauling, boiling	Higher perceived health risks, children in home	Pennsylvania, US	Abdalla et al. (1992)
Event: contamination by microorganism (*Giardia lamblia*)	Boiling, carting, purchasing water	n/a	Pennsylvania, US	Harrington, Krupnick, and Spofford (1989)
Event: contamination by microorganism (*Giardia lamblia*)	Boiling, carting, purchasing water	n/a	Pennsylvania, US	Laughland et al. (1996)
Perceived water quality	Filtering, boiling; bottled water	Women, 'better' occupations, higher incomes, high school training of household head, people with children under five	Brazil	McConnell and Rosado (2000)
Protection from arsenic poisoning		Children in household	Bangladesh	Aziz (2007)
Perceived water quality	Boiling, filtering, chemical treatment	n/a	Abbottabad, Pakistan	Haq et al. (2008)
Perceived water quality	Boiling, settling, filtering; bottled water	Perceived water quality	Moscow	Larson and Gnedenko (1999)
Implied perceived water quality	Bottled water and soft drinks	n/a	France	Carpentier and Vermersch (1997)
Perceived raw water quality	Filtering, bottled water; finding own sources	Greater perceived pollution levels	Korea	Um et al. (2002)
Water quality		Lower education	Canada	Janmaat (2007)
Raw water quality	Bottled water	Lower income, household head retired, living in urban areas	France	Bontemps and Nauges (2006)
Safety and taste	Bottled water, household filtration	Higher income, unpleasant taste experiences with tap water, non-French-speaking, being a male with children in one's household, tap water health concerns, belief that bottled water is safer than tap water	Canada	Dupont et al. (2009)
Safety and taste	Bottled water, boiling, household filtration	n/a	Canada	Dupont (2005)
Safety and taste	Bottled water, filtered tap	Perceived unsafe tap water, dissatisfied with taste, odour, appearance, non-white, income, younger, information on current or prior tap water problems	Georgia, US	Abrahams et al. (2000)
Taste and odour	Bottled water or home treatment devices	Dissatisfaction with taste	Quebec, Canada	Levallois et al. (1999)
Taste and odour and safety	Bottled water, household treatment devices, boiling	Aesthetic concerns	Toronto, Canada	Auslander and Langlois (1993)

respondents perceiving lower drinking water quality are more likely to make averting expenditures. These perceptions should therefore be incorporated into averting expenditure models.

Although in the developed-country studies water suppliers delivered a quality of water that is considered safe,[1] between 40% and 86% of PWSS consumers in Canada and France purchase bottled water rather than consuming tap water (Bontemps & Nauges, 2006; Levallois, Grondin, & Gingras, 1999) and/or boil or treat their tap water (Auslander & Langlois, 1993; Dupont, 2005; Janmaat, 2007). Most of these studies find that consumers are more likely to purchase bottled water or boil or filter their water if they perceive a bad taste, smell, or appearance of the water, or if they perceive a health risk.

Despite a lack of scientific evidence, people are likely to believe water is unsafe when its taste, odour, or appearance is poor (de Franca Doria, 2010; Jardine, Gibson, & Hrudey, 1999). De Franca Doria (2010) found that consumers' attitudes towards chemicals, familiarity with water quality properties, trust in suppliers, past water quality problems, and information obtained through the media and through personal networks also influence their perception of water quality. These perceptions have therefore been incorporated into the survey used for this article, with the exception of the 'information' parameter. This parameter was not included because it did not arise in the 'reasons for undertaking the averting behaviour' and would have required an additional level of prompting across all service quality attributes. The information questions would be better asked in a targeted survey on a single service quality attribute.

Although travelling for water is not commonly reported in the developed-country literature, anecdotal evidence from the interview stage of this research suggests that some New Zealanders travel for better tasting or 'healthier' water. Travelling for water is therefore included as an averting expenditure in this study, based on reported time and distance travelled.

Water pressure

A review of the literature revealed no studies examining averting expenditures to adjust household water pressure. Many stated-preference studies, however, include water pressure. Thus, we have included it here. We asked about both water pressure that is too high and too low, as equipment exists for both problems.

Service interruptions and water restrictions

In one of the few studies on averting expenditures due to lack of water service reliability, Pattanayak et al. (2005) examined the 'coping costs' of Kathmandu households. The costs are reported in terms of five behaviours (collecting, pumping, treating, storing and purchasing); however, the reasons for these behaviours are not reported. The review did not identify any other averting studies related to improvements in service interruptions or water restrictions.

In this study, consumers were asked to rate the taste, smell and appearance of their water, their perceived level of protection from waterborne microorganisms, pressure, and hardness, as well as the number of service interruptions and water restrictions they recalled. These perceptions help identify who is likely to undertake averting expenditures.

The averting expenditures and their drivers identified in the literature were also used in the survey for this study. While the literature focuses on averting expenditures related to water quality, this research examines people's expenditures related to the many other aspects of PWSS performance.

Methodology

To estimate willingness to pay for PWSS impacts and to predict which perceptions have the strongest influence on averting behaviours, three steps were followed: identifying averting behaviours and equipment purchases and the reasons for each; estimating the cost of each averting behaviour and purchase (and therefore monetize each 'reason'/impact); and undertaking econometric analysis to understand which perceptions most influence behaviours and purchases.

Respondents were recruited for a New Zealand–wide internet panel via invitation email from the panel company, with a link to the survey. While the internet panel company had warned the researchers that women 24–35 might be overrepresented, the descriptive statistics showed that this was not the case.

The opening screen of the survey described the aim of the research and provided contact details for the researcher, as well as the University of South Australia's Ethics Committee, who had approved the survey and recruitment methodology.

The online survey format allowed customized follow-up questions, allowing us to ask *why* respondents had each piece of equipment (Fleming & Bowden, 2009; Olsen, 2009). The online format may preclude some respondents from participating, but 80% of New Zealanders have internet access (Statistics New Zealand, 2013b). Using an internet panel further restricted participants to individuals who had already registered to participate in surveys. They may also share characteristics that are not evident in the demographic data. An additional potential source of bias was that respondents who were dissatisfied with their service, or had undertaken averting behaviours, may have been more likely respond. To mitigate this type of bias, the survey title was neutral, 'Water Use in Your Home'; and the responses were not more negative than those reported in other studies. In terms of the people who responded to the survey, our analysis of completions compared to non-completions showed no statistical difference in perception of water service quality or averting behaviour. Further, demographically as well as geographically, the statistics proved to be representative of national demographics (Table 2).

The internet survey was sent to 15,600 panel members, and 2300 responses were received, for a response rate of 15%. This rate was three times higher than what the internet panel company had expected, suggesting a high level of interest in the topic. Of these respondents, 89% were connected to a town supply and therefore relevant to this article. A further 130 responses were not included as they were considered not genuine, due to very short completion times, showing predictable patterns such as always choosing the first option, completing the survey more than once, providing implausible responses (e.g., very high water rates or frequency of water restrictions), or providing

Table 2. Descriptive statistics for all respondents.

Variable	Description	*Obs.*	*Mean*	*Std. Dev.*	*Min*	*Max*	*Nat'l avg.**
Perception variables							
taste_rate	Taste, smell and appearance of water rated as						
0	Terrible	1917	0.03	0.16	0	1	
1	Bad	1917	0.05	0.23	0	1	
2	Could be better	1917	0.50	0.50	0	1	
3	Excellent	1917	0.52	0.49	0	1	
protectHow protected you feel	from microorganisms in water						
0	Not at all	1917	0.03	0.18	0	1	
1	Not well	1917	0.16	0.37	0	1	
2	Mostly	1917	0.60	0.49	0	1	
3	Completely	1917	0.20	0.40	0	1	
pressure	Mains pressure is...						
0	Too low	1917	0.03	0.16	0	1	
1	Acceptable	1917	0.81	0.39	0	1	
2	Too high	1917	0.16	0.36	0	1	
Restri_dur	Duration of longest water restriction (weeks)	1264	7.01	4.96	0	36	
Restri_freq	Water restrictions over 30-year period (no.)	1917	6.24	8.60	0	50	
Fluoride – belief about current status	0 No fluoride	1917	0.18	0.37	0	1	
	1 Fluoride	1917	0.43	0.33	0	1	
	2 Don't know	1917	0.39	0.30	0	1	
Fl_dis	0 Satisfied with presence or absence of fluoride /1 dissatisfied	1917	0.15	0.35	0	1	
Interr	Annual interruptions lasting over 4 hours (no.)	1905	1.07	2.96	0	50	
Hard	Perception that water is...						
	0 Not hard	1917	0.59	0.49	0	1	
	1 Somewhat hard	1917	0.15	0.36	0	1	
	2 Hard	1917	0.10	0.30	0	1	
	3 Don't know	1917	0.17	0.37	0	1	
Socio-demographic variables							
Male	1 Male / 0 female	1917	0.42	0.49	0	1	0.49
Income	1 Personal income < $15k; 2 15–25; 3 25–35; 4 35–45; 5 45–55; 6 55–65; 7 65–75; 8 75–85; 9 85–95; 10 95–105; 11 105–115; 12 115–125; 13 125–135; 14 135K+	1882	4	2.81	1	14	4.1
Poor	1 Personal income < $45,000; 0 otherwise	1882	0.60	0.49	0	1	0.49
rates_spend	Annual rates (NZD)	1917	270	138	10	1100	
own_property	1 Own / 0 Do not own property of residence	1913	0.62	0.49	0	1	0.53
ethnicity	1 Caucasian	1917	0.76	0.43	0	1	0.74
	2 Maori	1917	0.07	0.26	0	1	0.15
	3 Pacific Island	1917	0.05	0.21	0	1	0.07
	4 Asian	1917	0.04	0.19	0	1	0.12
	5 African	1917	0.01	0.09	0	1	-
	6 Other	1917	0.07	0.26	0	1	-
D_kids	1 Children living in household / 0 No children living in household	1917	0.38	0.49	0	1	0.33
Num_hh	No. of people in household	1917	2.85	1.37	1	6	2.7

* National average. Source: http://www.stats.govt.nz/browse_for_stats/people_and_communities/geographic-areas/urban-rural-profile-update.aspx. n/a = not available / not applicable.

conflicting responses. After eliminating the 130 'inconsistent' responses, there were 1917 useable responses.

Respondents were asked to identify any equipment they had at their residence, or any behaviours they undertook, such as boiling water or travelling to collect water, to improve or avoid using their household water supply. A list of possible responses was provided based on a pilot, and respondents were able to specify 'other' behaviours and equipment purchases.

For each piece of equipment selected, respondents were asked to provide the main reason for owning it. The 'equipment reason' helped verify that the equipment's purpose was to improve the quality of their water supply, and identify the reasons why consumers spend money to improve the quality of their water supply and how much they spend. This approach helped distinguish between main concerns being about taste, smell and appearance versus concerns about water safety.

If respondents indicated they travelled for water or had tanker, container, or bottled water, they were asked how much money they spent on this and the distance they travelled to collect it, on a weekly basis.

Descriptive statistics were used to estimate the proportion of the respondents using averting behaviours or purchases, the main reasons for these behaviours or purchases, and the annual amount spent on them.

A cost estimate of averting behaviours and purchases was used to monetize the impacts of PWSS performance, confirming people's willingness to pay for the most prominent impacts. Research on household potable water supply equipment shows that people's recollections about purchase and installation costs differ significantly from market values. For this reason, Janmaat (2007) relied on market prices; this study also takes this approach.

Costs were estimated based on the capital, operating and maintenance costs proposed by Jagals and Rietveld (2011): the cost to develop the source and the cost of storage, treatment and pumping.[2] In the latter three cost categories, both capital and recurrent costs (e.g., operating, maintenance and monitoring) are considered. For recurrent costs, we included the costs of replacement parts, power,[3] and time incurred.[4] To estimate the costs for storage and pumps, minimum capacities were assumed. Initial and follow-on costs for each piece of equipment and underlying assumptions are fully detailed in Robak (2013). Expenditures were estimated based on New Zealand pricing information published on the internet. Details of these prices are available in Robak (2013). Disposal costs are not considered, due to the small size and low cost of disposing of most household-related water equipment. These cost estimates assume a 10-year life, which is the expected life of most household equipment (InterNACHI, n.d.).

The amortized value was based on a weighted average cost of capital of 8%, as recommended by the New Zealand Treasury (2008) for water and other infrastructure projects, and annual replacement of major parts.

Sensitivity tests were carried out to test the effects of uncertainty in cost estimates with respect to discount rate, analysis period, people's value of time, purchase costs, the proportion of the population having the reported equipment, equipment life, and discount rates. Results of the sensitivity tests, reported in full in Robak (2013), showed that the RP cost estimates are accurate to ±60%. The estimates are most sensitive to the analysis period and assumptions about the value of time.

Logistic regression was used to provide an understanding of which perception and demographic factors made consumers more likely to employ averting behaviours and expenditures. The main aim was to test the strength of relationships between perceived service levels and averting behaviours and purchases. The model for each major averting behaviour or purchase (i.e., treat, source, store, increase pressure, reduce pressure, boil water, filter water, undertake any averting behaviour) was specified as:

$$
\begin{aligned}
U_i = {} & \beta_0 + \sum\nolimits_{j=0-3} \beta_{1,j} \times \text{taste}_j \\
& + \sum\nolimits_{k=0-3} \beta_{2,k} \times \text{protect}_k + \beta_3 \times \text{low\_pressure}_i + \beta_4 \times \text{high\_pressure}_i \\
& + \beta_5 \times \text{restri\_dur}_i + \beta_6 \times \text{restri\_freq}_i + \beta_7 \times \text{fl\_dis}_i \\
& + \beta_8 \times \text{interr}_i \quad + \beta_9 \times \text{male}_i + \beta_{10} \times \text{interr}_i \,\&\, \text{male}_i + \ \beta_{11} \times \text{income}_i \\
& + \beta_{12} \times \text{own\_prop}_i \ \ + \ + \sum\nolimits_{l=1-6} \beta_{13,l} \times \text{ethinicity}_l + \ \beta_{14} \times \text{D\_kids}_i \\
& + \ \beta_{15} \times \text{num\_hh}_i
\end{aligned} \tag{1}
$$

where U is the probability of undertaking averting behaviour i. The variables and categories (j = 0 to 3; k = 0 to 3; l = 1 to 6) are as described in Table 2, and each of the following was tested as the dependent variable i:

- *do_sth*: undertake any averting behaviour or purchase
- *treat*: treat water in any way, using filters, boiling, etc.
- *source*: source additional or alternative water (e.g., tankering, rainwater tanks)
- *store*: have water storage equipment
- *press_up*: have pressure-enhancing equipment
- *press_down*: have pressure-reducing equipment
- *boil*: boil water
- *filter*: have some form of filtration
- *movableonly*: only undertake averting behaviours that are 'movable' from one residence to another, including boiling water, jug filters, travelling for water, and purchasing bottled or container water.

Results

Descriptive statistics

The survey produced usable responses from 1917 potable water consumers from across New Zealand. Respondents who reported implausible values for water rates (>$1200) and frequencies and durations of water restrictions (>60 in 30 years; >50 weeks per year) and service interruptions (>52 per year) have been removed from the analysis, as their perception responses cast doubt on all of their answers. The respondents are reasonably representative of the national average (Table 2): male respondents were 42% of the sample, compared to a national rate of 49%; Caucasian respondents were 76% of the sample, compared to a national rate of 65%; property owners were 62% of the sample, compared to a national rate of 53%; and the sample had an average personal income in the range of $35,000–45,000, compared to a national average of $48,800 per year.

Many respondents reported more than one type of averting behaviour or purchase. Nearly half of the respondents (48%) on town supply undertake behaviour or have equipment in their homes to improve the performance of their PWSS, and over 30% treat their water by filtering or boiling (Table 3). This proportion is comparable to

Table 3. Averting behaviours and cost estimates.

Behaviour	% undertaking this behaviour	Estimated time and money expenditure per household (amortised $/y)		Main reason for undertaking this behaviour/purchase (% of respondents)				
		Across population	Across consumers who undertake this behaviour or purchase	Taste & odour	Concerns about microbes	Concerns about chemicals	To protect fitting and equipment	For babies
Boil	10	74	711	35	27	15	3	4
Filter	23	44	150	51	18	11	1	0
Store	4	8	104	14	1	2	0	0
Source	9	52	684	15	0	2	0	0
Pressure	6	6	97	0	0	0	23	0
*Total**	48	184	-					

To soften water	Conserve water	Reduce rates	Irrigation	Protect against water restrictions	Comfortable shower	Other	Not my decision: it was there when I moved in
1	0	0	0	0	0	15	0
1	0	0	0	0	0	10	9
0	8	6	11	7	2	28	7
0	9	13	24	8	0	20	13
0	20	7	2	1	21	18	9

* Totals may not be the sum of the cells above, as some respondents undertake multiple averting behaviours.

Canadian studies (Auslander & Langlois, 1993; Dupont, 2005; Janmaat, 2007) – the only other developed-country studies in which these averting behaviours were quantified outside of contamination events – which find that about half of respondents improve or avoid using their tap water.

Cost estimate

The total amortized estimated spending by the average New Zealand household on water-service-related averting behaviours is $184 per household per year (Table 3).[5] Considering that there are approximately 1.5 million households in New Zealand (Statistics New Zealand, 2013a), total amortized expenditures are on the order of $280 million annually. The average New Zealand household spends $74 per year boiling water, $44 filtering it, $52 storing it, $8 sourcing better water, and $6 adjusting the pressure.

A number of reasons for investing time and money in household water supply improvements were identified, some of which receive little attention in the literature:

- To secure water to irrigate the garden
- To have control over the water bill
- To conserve water
- To protect against water restrictions[6]
- To protect fittings and household equipment (e.g., hot water cylinders)

Due to space constraints, the reasons for each averting expenditure are not reported. However, they are available in Robak (2013). These behaviours and purchases should be accounted for in future cost–benefit analyses.

To estimate *how much benefit water suppliers could achieve by improving the quality of service centrally*, there must be a difference between how much people spend when they are satisfied with their service and how much they spend when they are dissatisfied. This study examines the differences between households' investments based on their current perceived level of service. Households were grouped by perception for each service quality attribute. Based on reported averting behaviours and purchases, averting costs were estimated for each perception cohort.

In testing whether respondents reporting a higher or lower perceived quality of service are significantly less or more likely to undertake averting behaviours or purchases, we find that this is true with $p < 0.05$ for four types of attributes. Respondents are significantly less likely to undertake averting behaviours who report 'excellent' water aesthetics; who report feeling 'completely' protected from microorganisms; who report feeling 'completely' protected from chemicals; and who report that their water is 'not hard' (Table 4). One finding is surprising and warrants greater investigation: that the lowest rate of averting behaviour directed at protection from microorganisms occurs when respondents feel 'not at all' protected. This result may be due to the low sample size, with only 61 respondents feeling 'not at all' protected from microorganisms. Alternatively, respondents may believe that the situation will be resolved by their water supplier in the near term.

Table 4. Average annual averting expenditure, by perceived quality of service.

Attribute	Perception	Annual household averting expenditure with this perception	% of respondents who undertake an averting behaviour or purchase
Taste, smell & appearance	Terrible	$207	38%
	Bad	$189	35%
	Could be better	$180	24%
	Excellent	$201	10%***
Protection from microorganisms	Not at all	$128	1.5%***
	Somewhat or mostly	$301	8.2%***
	Completely	$375	4.3%***
Protection from chemicals	Not at all	$323	15%**
	Somewhat or mostly	$288	5.4%**
	Completely	$259	1.8%***
Hardness	Not hard	$157	0.3%***
	Hard	$272	3.2%***

$^{**}p < 0.01$, $^{***}p < 0.001$.

To estimate the relative value of improving the taste of water, smell and appearance to 'excellent', the cost of improvement was multiplied by the rate of averting behaviours and purchases within each perception cohort. Depending on whether households perceive the current water supply to taste, smell and appear 'terrible', 'bad', or 'could be better', the marginal benefits of providing water with an 'excellent' taste, smell and appearance are $59, $46, or $23 per household per year (Table 4).[7] Similarly, improving perceived protection from chemicals from 'not at all' to 'completely' protected could offer marginal benefits of $44 per household per year. Improving to 'somewhat or mostly' protected would have benefits of $33 per year. If the water utility can make these improvements *or change these perceptions* for less than the cost of the averting behaviours and purchases, then it is more economically efficient for the water utility to do so than households. The results suggest that people place the greatest value on improvements in taste, smell and appearance, and in protection from chemicals. It was hypothesized that lower-income households are less likely to improve taste, smell and appearance, as they are would be less likely to value costly interventions that improve only the aesthetics of their drinking water. To test this, the sample was divided into two sets: respondents with incomes at or below the national median, and respondents with incomes above the median.

The major difference between lower- and higher-income respondents is that the rate of lower-income respondents reporting water that tastes, smells or appears 'terrible' is twice that of higher-income respondents (Table 5). Lower-income households who

Table 5. Differences in perception, spending, and rates of averting behaviour for respondents with lower and higher incomes.

	% having this perception		Spending per household that undertakes this behaviour		% of respondents undertaking this behaviour	
	Income ≤ median	Income > median	Income ≤ median	Income > median	Income ≤ median	Income > median
Taste, smell and appearance						
Terrible	3.4%	1.7%	$ 194	$ 134	49%	8%***
Bad	5%	6%	$ 207	$ 160	33%	37%
Could be better	52%	48%	$ 189	$ 167	23%	26%
Excellent	40%	45%	$ 219	$ 169	11%	9%

$^{***}p < 0.001$.

engage in averting behaviours spend more to improve the taste, smell and appearance of their drinking water; and lower-income households are significantly more likely to undertake averting behaviours and purchases to improve the taste, smell and appearance of their drinking water. This last finding is significant to $p < 0.05$. It is surprising that very few of high-income respondents reporting water with terrible taste, smell and appearance do anything to improve their water's aesthetics. There are several possible explanations. First, the number of higher-income respondents reporting 'terrible' water aesthetics was very low: just 13. Also, it is possible that higher-income respondents have higher expectations of their water supplier, and demand that their water supplier provide these improvements.

The high rate of lower-income respondents reporting water with a terrible taste, smell or appearance is not surprising. The reason has not been tested, but it is plausible that lower-income respondents receive water from smaller water supply systems that may have a lower degree of treatment, or that their homes are located at far ends of the water supply system, where water may become stagnant. Delpla, Benmarhnia, Lebel, Levallois, and Rodriguez (2015) report that few studies have investigated whether poorer households receive poorer-quality drinking water; they found that poorer households were more likely to have high lead concentration (indicating poorer water quality), but lower trihalomethanes (indicating better water quality). Turgeon, Rodriguez, Theriault, and Levallois (2004) found that *perceived* taste and health risks were significantly related to proximity to water treatment plants, reservoirs and water supply system extremities. Further, Turgeon et al. (2004) found definite patterns in how low-income households were distributed relative to these parts of the water supply system – but the patterns varied from city to city.

The other two findings are somewhat surprising: that lower-income households choose more costly averting behaviours, and are more likely to undertake averting behaviours. Lower-income respondents relied more on boiling their water or purchasing tanker, container or bottled water than higher-income respondents. Boiling and purchasing bottled water have higher costs than the other alternatives, but their upfront costs are lower. The authors postulate that lower-income respondents choose these 'higher-cost' alternatives because other forms of aesthetic improvement, such as filtration devices, have costlier initial investments. This finding suggests that lower-income households are more burdened by low levels of service than are higher-income households. Another possible explanation may be that lower-income respondents are more likely to rent their homes than higher-income respondents and thus wish to avoid an investment that cannot be relocated; however, our analysis showed that only 8% of respondents undertook 'movable only' averting behaviours.

Logistic models – reasons for averting behaviours and purchases

Logit models were run for each type of averting behaviour and purchase, based on Eq. (1). The pressure-enhancing and pressure-reducing models were not significant and are not discussed further. Also, a number of variables proved to have low significance in all models. Perception of the water supplier's investment efficiency and being born in New Zealand were therefore removed from the models. Furthermore, in the initial model runs, the variables education, age and business ownership were included because the literature indicated their significance; however, variance inflation factors showed that these variables were highly correlated with the perception variables. We retained the perception variables in the model rather than the socio-demographic variables, as these were more significant in explaining the averting behaviours.

Adamowicz et al. (2005) note the importance of household dynamics in RP data as revealed preferences are a function of collective decisions, which may not be reflected in individual survey respondents' characteristics. In this study, the two significant socio-demographic variables were *individual* income and ethnicity; household variables were not collected, but may have resulted in greater significance.

In the models, 'excellent taste (smell and appearance)' is the base taste, smell and appearance; 'acceptable pressure' is the base water main pressure, and 'completely protected' is the base level of protection from microorganisms. The Wald tests (chi-squared statistics and model significance) show that all models are significant, and the main findings are as follows (Table 6).

Who undertakes any type of averting behaviour or purchase related to household tap water? Respondents who own their residence, dislike the taste, smell, or appearance of their water, feel not 'completely protected', identify as Maori, and experience more frequent water restrictions were most likely to engage in averting behaviours or purchases. It is unsurprising that property owners are more likely to engage in averting behaviour, as many averting purchases are costly and can be difficult to move from one home to the next. And renting residents might not need the equipment at their next home. Um, Kwak, and Kim (2002) is the only study that has examined the influence of home ownership on averting behaviour, but found little significant influence. This study is therefore the first to demonstrate this relationship.

To test whether the effect of home ownership is a wealth effect, we separated out averting behaviours that were 'movable', such as boiling water, travelling for water, and purchasing containers, from those that were not. We found that respondents who used only movable averting behaviours were less likely to own property, and more likely to have experienced more frequent service interruptions or be of Asian or Maori ethnicity.

Who treats their water? This category includes respondents who use any type of filtration, softening, ionizing or disinfection system, or boil their water. The results from the 'treat' model are similar to those from the overall averting behaviour model, due to the high proportion of averting respondents who treat their water compared to other averting behaviours and purchases. Respondents that have 'terrible'-tasting water or own their property are more likely to have treatment equipment or behaviours.

Dupont et al. (2009) find that respondents' health perceptions about their tap water are significant factors in their decision to avoid tap water. The present research lends some support to these findings, as respondents reporting feeling less than 'completely protected' are more likely to undertake some sort of treatment. Similarly, Um et al. (2002) found that perceptions of raw water quality were important determinants of averting behaviours. However, the present research did not ask respondents about their perceptions of raw water quality.

As reported by Levallois et al. (1999), Janmaat (2007), and Dupont et al. (2009), having unpleasant taste, smell or appearance experiences or being dissatisfied with the taste, smell or appearance is strongly significant. Dupont et al. (2009) and Janmaat (2007) find that people with higher incomes are more likely to treat their water. The

Table 6. Models predicting averting behaviours (standard errors in brackets).

Variable		Regression coefficient					
		Averting behaviour (any)	treat	source	boil	filter	'movable only' solutions
Taste, smell and appearance	0. Terrible	2.31***	2.03***	1.64***	0.73	1.51***	0.80
		(0.47)	(0.43)	(0.47)	(0.49)	(0.52)	(0.54)
	1. Bad	1.02***	1.06***	0.97**	0.34	0.94***	0.36
		(0.29)	(0.29)	(0.35)	(0.44)	(0.34)	(0.51)
	2. Could be better	0.68***	0.87***	0.15	0.47*	0.85***	0.41
		(0.14)	(0.14)	(0.22)	(0.23)	(0.18)	(0.27)
Perceived protection from waterborne microorganisms	0. Not at all	−0.31	−0.13	−0.67	−0.24	0.49	0.49
		(0.40)	(0.39)	(0.71)	(0.68)	(0.47)	(0.67)
	1. Somewhat	0.23	0.12	0.17	0.81***	0.13	0.80
		(0.23)	(0.24)	(0.38)	(0.36)	(0.31)	(0.47)
	2. Mostly	0.37*	0.34*	0.39	0.47*	0.54*	0.64
		(0.16)	(0.17)	(0.28)	(0.30)	(0.21)	(0.40)
Water main pressure	0. Too low	0.78	−0.08	0.99*	−1.32	0.31	0.25
		(0.43)	(0.41)	(0.43)	(1.03)	(0.45)	(0.66)
	2. Too high	−0.30	−0.43*	0.24	−0.11	−0.578**	−0.11
		(0.17)	(0.18)	(0.22)	(0.25)	(0.24)	(0.29)
Restriction duration		−0.01	−0.01	−0.00	0.01	0.01	−0.026
		(0.01)	(0.01)	(0.20)	(0.02)	(0.02)	(0.02)
Restriction frequency		0.01*	0.02*	−0.00	0.01	0.02	0.019
		(0.01)	(0.01)	(0.01)	(0.01)	(0.01)	(0.01)
No. of interruptions		0.02	0.00	0.05*	0.03	−0.04	0.06***
		(0.02)	(0.02)	(0.02)	(0.02)	(0.05)	(0.02)
Male		−0.08	−0.21	0.08	−0.29	−0.15	−0.46
		(0.14)	(0.14)	(0.20)	(0.23)	(0.17)	(0.27)
Male × interruptions		−0.08	−0.10	0.05	0.04	−0.22*	0.017
		(0.05)	(0.06)	(0.05)	(0.05)	(0.09)	(0.07)
Income		0.03	0.04	−0.02	−0.09*	0.07*	−0.06
		(0.02)	(0.02)	(0.04)	(0.04)	(0.03)	(0.05)
Fluoride_dissatisfied		0.12	0.15	−0.26	−0.08	0.08	−0.02
		(0.17)	(0.17)	(0.26)	(0.26)	(0.20)	(0.30)
Own_property		0.67***	0.58***	0.37	−0.21	0.85***	−0.60***
		(0.14)	(0.14)	(0.21)	(0.21)	(0.19)	(0.24)
Ethnicity	2. Maori	0.61*	0.54*	0.52	−0.54	−0.15	0.77***
		(0.25)	(0.26)	(0.34)	(0.30)	(0.34)	(0.35)
	3. Pacific Island	−0.50	−1.05*	−0.84	−0.33	−1.58*	0.25
		(0.38)	(0.46)	(0.79)	(0.55)	(0.74)	(0.67)
	4. Asian	0.16	0.02	−0.92	0.93	−0.76	1.43***
		(0.43)	(0.45)	(1.02)	(0.60)	(0.67)	(0.58)
	5. African	0.52	1.12	-	-	2.06	-
		(1.09)	(1.28)	-	-	(1.31)	-
	6. Other	0.36	0.25	−0.34	0.18	−0.60	0.42
		(0.24)	(0.24)	(0.32)	(0.37)	(0.33)	(0.42)
D_kids		−0.08	−.20	0.23	0.42	−0.49*	−0.05
		(0.18)	(0.18)	(0.28)	(0.24)	(0.23)	(0.30)
Num_hh		−0.04	0.02	−0.13	0.05	0.05	−0.04
		(0.07)	(0.07)	(0.11)	(0.09)	(0.08)	(0.11)
Constant		−1.21***	−1.71***	−2.50***	−2.83***	−2.95***	−2.70***
		(0.27)	(0.29)	(0.42)	(0.47)	(0.35)	(0.57)
No. observations		1230	1230	1228	1228	1230	1228
Chi-squared		101	101	56	63	112	58
Model sig.		0.0000	0.0000	0.0001	0.0000	0.0000	0.0000
Link test							
_hat		(0.000)	(0.000)	(0.004)	(0.095)	(0.000)	(0.165)
_hatsq		(0.783)	(0.607)	(0.602)	(0.862)	(0.594)	(0.704)

$^{*}p < 0.05$, $^{**}p < 0.01$, $^{***}p < 0.001$.

present research, however, finds that it is the combined impact of income and perception that influences averting behaviour.

Who sources additional water?[8] Like those who treat water, respondents who source alternate water tend to rate the taste, smell and appearance of their water very low. However, they are otherwise very different from those who treat water. They are more likely to perceive that their pressure is too low or that they have a higher number of service interruptions lasting more than four hours. Um et al. (2002) found that households with greater numbers of people were more likely to source alternate water, but the present study does not support that; perceptions were found to be more important than socio-demographic variables.

Who boils their water? Respondents who boil their water are more likely to feel less than completely protected from microorganisms, experience more service interruptions, and have lower incomes. That the other variables are significant is also intuitive: boiling water kills and inactivates most types of waterborne microorganisms. Service interruptions are also linked to more waterborne illness (Hunter, Chalmers, Hughes, & Syed, 2005; Nygård et al., 2007); boiling water offers protection against such illness. For people with lower incomes, boiling water appears to be a lower-cost alternative to filtration or other treatment devices.

Who filters their water? Dupont et al. (2009) and Um et al. (2002) find that people with higher incomes and with tap water–related health concerns are more likely to filter their water. This research finds the same. As reported by Dupont et al. (2009) and Levallois et al. (1999), residents perceiving the taste, smell and appearance of their tap water as poor are significantly more likely to filter their water.

Furthermore, Dupont et al. (2009) find that being a male with children in one's household makes people more likely to treat/filter their water. In the present study 'male' is not a significant predictor, which is not surprising given that treatment/filtration is a household decision. However, for all models, we interacted 'male' with perceived service quality attributes, to test whether men's perception had a different effect towards averting behaviours than women's. We found this interaction significant only for the number of service interruptions, and here the results suggested that in a household in which a male perceives more service interruptions it is less likely that a filter would be purchased. Having children also makes it less likely that a filter would be purchased.

Um et al. (2002) find that higher expectations for future water quality and few 'odd' experiences with tap water[9] make people less likely to filter their water. This research did not ask about households' expectations for future water quality.

This research also finds that property owners are more likely to filter their water. As discussed for respondents who undertake averting behaviours and purchases generally, this is presumably because filtration can be costly (which also explains why respondents with higher incomes are more likely to use filtration equipment) and can be difficult to move from one home to the next.

Overall, the research found that respondents with 'terrible' water aesthetics are twice as likely to undertake averting behaviours as respondents who report 'bad' water aesthetics, and four times as likely as respondents who report 'could be better' aesthetics. Having 'terrible' water aesthetics also made respondents twice as likely to undertake averting behaviours as owning the property.

The above models identify demographic and perception factors correlated with the purchase of equipment to improve household water.

Discussion

The averting behaviour and purchase approach with follow-up questioning allowed respondents to describe what they value about PWSS performance changes. Our descriptive analysis showed that respondents valued a reduction or increase in pressure, for example, because of its impacts both on shower enjoyment and on longevity and operation of fittings and appliances. This approach also showed that even though 29% of respondents reported that they have excellent-tasting water, they still had some form of treatment. Hence, excellent water aesthetics does not imply that households will not spend money on treatment. If water suppliers and policy makers wish to provide high water quality, they should recognize the many reasons why households have treatment devices. These reasons include not only the taste, smell and appearance of their water, but also perceived protection from microbes, desire to remove 'chemicals' from their water, and other perceived health qualities.

The research also identified five performance attributes that are not typically seen in the literature: control over water rates; ability to irrigate gardens and secure water for stock; protection of appliances and fittings; protection from chemicals; and protection of babies' health.

Our predictive logit models showed that the consumers that are most likely to engage in averting behaviours and purchases include property owners, those who rate the taste, smell, and appearance of their tap water as less than excellent, and those who experience more frequent water restrictions.

Most surprising, however, is that lower-income households spend more time and money to improve the taste, smell and appearance of their water. If an economically efficient water supply solution includes a standard level of quality plus household-level customization, then an equitable water supply solution might be one that subsidizes household-level alternatives that have higher upfront costs but lower long-term costs.

Applying the research findings to the six research questions set out in the introduction, the answers are as follows.

If a household is less satisfied with their quality of service, do they spend more customizing it? No. There was no statistical difference between the estimated cost of averting behaviours and purchases in a less satisfied household and that of a more satisfied household. This finding reflects a flaw highlighted by Birol et al. (2008) – that averting expenditures are valued in discrete step changes, and lower-level solutions are frequently not available.

If a household is less satisfied with their quality of service, are they more likely to undertake an averting behaviour or purchase? Yes. However, the results were only statistically significant for improvements in taste, smell and appearance.

Based on averting expenditures, would there be significant and non-negligible marginal benefits from improving the quality of service? Yes, for taste, smell and appearance, for perceived protection from chemicals, and for hardness. The marginal benefits of improving taste, smell and appearance to a level perceived as excellent are approximately \$20–60 per household. For protection from chemicals, marginal benefits are

$10–44 per household, and for improvement in hardness, an average of $8 per household.

Do lower-income households spend less than higher-income households customizing their quality of service through averting behaviours and purchases? No. Lower-income households spend more, in absolute as well as relative terms. This is probably because they cannot afford lower-cost long-term solutions that have higher investment costs. However, this finding also depends on how the policy maker values people's time; a significant proportion of the estimated costs are based on the time it takes to boil water.

Are lower-income households less likely to undertake averting behaviours and purchases? No. Lower-income households are more likely to improve the taste, smell and appearance of their drinking water. This finding may be due to low-income households having a more 'genuine' assessment of what is considered 'terrible', while higher-income households might perceive it as terrible, but not so terrible that it warrants doing something about it; but this suggestion should be considered tentative only.

Are there differences in economic burden between lower-income and higher-income households? Yes. A lower-income household is more likely to undertake averting behaviours and purchases when they rate the taste, smell and appearance of their water as poor. An economically efficient water supply system would consider these additional costs.

Conclusion

Water suppliers are facing pressure to maintain low water service prices while maintaining an acceptable quality of service. This article shows that even at a relatively high perceived quality of service, households undertake averting behaviours and purchases to 'customize' their service quality. This customization 'tops up' the quality of the centrally supplied service. Provided that the household-level customization costs less than the centrally provided equivalent, averting behaviours and purchases may serve as a form of alternative, or complementary, service delivery model. However, when water utilities deliver lower quality than households want, poor households incur more costs than well-off households. Lower-income households may suffer more when service quality is reduced, due to their inability to fund solutions that are more cost-effective in the long run. If lower-income households are indeed experiencing lower quality of service, then some may not be able to afford any averting behaviour or purchase. These households will be forced to live with lower-quality water, potentially exposing them to greater health and water security risks.

This research shows that nearly half of households invest large amounts in customizing their level of service. It also shows that averting behaviours and purchases, in developed nations such as New Zealand at least, are most predictable with respect to improvements in aesthetics, including hardness, and perceived protection from chemicals. The results suggest that despite pressures from regulators and consumers to provide a higher quality of service, households invest in household level improvements *even when service quality attributes are acceptable*. Relying on household-level customization may be particularly economically superior if these households are likely to invest in such equipment and behaviours over a wide range of service qualities (e.g., they may continue to purchase equipment until all service qualities are *consistently* excellent).

In New Zealand, households are significantly more likely to invest in household equipment or behaviours when they own their residence, or dislike the taste, smell, or appearance of their water. These findings may suggest that people in developed nations feel that the fundamental attributes, such as health or safety, are reasonably well provided for by their service providers, but that there are large marginal benefits to be achieved by investing in aesthetic attributes where consumers do not also have other concerns, such as water safety.

These findings may provide some insight for developing nations, too; even when water is supplied, and is safe and reliable, consumers may continue to undertake behaviours and purchases to improve taste, smell and appearance – or if they do not invest in such improvements due to cost or time constraints, then they may well experience residual dissatisfaction.

Since consumers are already investing privately in improvements in the quality of their water supply services, then a possible strategy for maintaining affordable services is to provide a lower quality of service, and leave it to end users to customize those aspects of water service. From a social equity perspective, however, service affordability should be evaluated by including private household investments.

Notes

1. With the exception of the Russian study (Larson and Gnedenko, 1999), in which water quality was not considered safe.
2. Distribution costs were neglected because inadequate information existed on which to base such estimates. Public source points were also neglected because they are not relevant to the New Zealand case.
3. For boiling water only. It was not possible to estimate power consumption for other equipment due to the variety of brands and models, and lack of information about volumes of water consumed.
4. For boiling water only. We did not include people's time in maintaining equipment, as this time was negligible (< 2 hours per year).
5. In this article, all dollar values are New Zealand dollars.
6. Water restrictions are considered to occur in the dry months, when there is not enough water to supply to consumers at typical demand levels.
7. Sample calculation: Column A (terrible) × Column B (terrible) – Column A (excellent) × Column B (excellent) = $59.
8. Sourcing includes tanker, container or bottled water; groundwater/bore; rainwater from roof; river, stream or spring; travelling for better or additional water; recycled water systems; and reticulation.
9. Defined as experiences that affect their enjoyment, such as rust, sediments, or taste and smell.

Acknowledgments

The authors thank Magnus Söderberg and Penny Burns for their contribution to this article.

Disclosure statement

No potential conflict of interest was reported by the authors.

Funding

This work was supported by the University of South Australia, Opus International Consultants, and TD Meloche Monnex/Engineers Canada.

ORCID

Anna Robak http://orcid.org/0000-0002-9813-3703
Henning Bjornlund http://orcid.org/0000-0003-3341-5635

References

Abdalla, C. W., Roach, B. A., & Epp, D. J. (1992). Valuing environmental quality changes using averting expenditures: An application to groundwater contamination. *Land Economics*, *68*, 163–169. doi:10.2307/3146771

Abrahams, N. A., Hubbell, B. J., & Jordan, J. L. (2000). Joint production and averting expenditure measures of willingness to pay: Do water expenditures really measure avoidance costs? *American Journal of Agricultural Economics*, *82*, 427–437. doi:10.1111/ajae.2000.82.issue-2

Adamowicz, W. L., Hanemann, W. M., Swait, J. D., Johnson, R., Layton, D., Regenwetter, M., … Sorkin, R. (2005). Decision strategy and structure in households: A "groups" perspective. *Marketing Letters*, *16*(3), 387–399. doi:10.1007/s11002-005-5900-6

Ajzen, I., Brown, T. C., & Carvajal, F. (2004). Explaining the discrepancy between intentions and actions: The case of hypothetical bias in contingent valuation. *Personality and Social Psychology Bulletin*, *30*(9), 1108–1121. doi:10.1177/0146167204264079

Alegre, H., Baptista, J. M., Cabrera, E., Jr, Cubillo, F., Duarte, P., Hirner, W., … Parena, R. (2006). *Performance indicators for water supply services*. London, UK: IWA publishing.

Altaf, M. A. (1994). The economics of household response to inadequate water supplies. *Third World Planning Review*, *16*(1), 41. doi:10.3828/twpr.16.1.m1wk8611v47009u3

Auslander, B. A., & Langlois, P. H. (1993). Toronto tap water: Perception of its quality and use of alternatives. *Canadian Journal of Public Health. Revue Canadienne De Santé Publique*, *84*(2), 99.

Aziz, S. (2007). *Valuation of avoiding arsenic in drinking water in rural Bangladesh: An averting behavior analysis* (Unpublished Doctoral). The University of Maine, Orono, ME.

Beaumais, O., Briand, A., Millock, K., & Nauges, C. (2010). *What are households willing to pay for better tap water quality? A cross-country valuation study*. Paris: Documents de travail du Centre d'Economie de la Sorbonne.

Bertrand, M., & Mullainathan, S. (2001). Do people mean what they say? Implications for subjective survey data. *American Economic Review*, *91*(2), 67–72. doi:10.1257/aer.91.2.67

Beshears, J., Choi, J. J., Laibson, D., & Madrian, B. C. (2008). How are preferences revealed? *Journal of Public Economics*, *92*(8), 1787–1794. doi:10.1016/j.jpubeco.2008.04.010

Birol, E., Koundouri, P., & Kountouris, Y. (2008). Using economic valuation techniques to inform water resources management in the southern european, mediterranean and developing countries: A survey and critical appraisal of available techniques. In P. Koundouri (Ed.), *Coping with water deficiency: From research to policymaking: With examples from Southern Europe, the Mediterranean and developing countries*. Netherlands: Springer.

Bontemps, C., & Nauges, C. (2006). To drink or not to drink (tap water)? The impact of environmental quality on consumer's choices. *Cahier de Recherche*, *1*, 1–38.

Carpentier, A., & Vermersch, D. (1997). *Measuring willingness to pay for drinking water quality using the econometrics of equivalence scales* (*Fondazione Eni Enrico Mattei Working Paper, no 92/97*). Milan: Fondazione Eni Enrico Mattei.

de Franca Doria, M. (2010). Factors influencing public perception of drinking water quality. *Water Policy*, *12*(1), 1–19. doi:10.2166/wp.2009.051

Delpla, I., Benmarhnia, T., Lebel, A., Levallois, P., & Rodriguez, M. J. (2015). Investigating social inequalities in exposure to drinking water contaminants in rural areas. *Environmental Pollution, 207*, 88–96. doi:10.1016/j.envpol.2015.08.046

Dupont, D. P. (2005). Tapping into consumers' perceptions of drinking water quality in Canada: Capturing customer demand to assist in better management of water resources. *Canadian Water Resources Journal, 30*(1), 11–20. doi:10.4296/cwrj300111

Dupont, D. P., Adamowicz, W. L., & Krupnick, A. J. (2009). Differences in water consumption choices in Canada: The role of socio-demographics, experiences, and perceptions of health risks. *Journal of Water and Health, 8*(4), 671–686. doi:10.2166/wh.2010.143

European Commission. (2008). *Guide to cost-benefit analysis of investment projects*. Brussels: European Union.

Fleming, C. M., & Bowden, M. (2009). Web-based surveys as an alternative to traditional mail methods. *Journal of Environmental Management, 90*(1), 284–292. doi:10.1016/j.jenvman.2007.09.011

Genius, M., Hatzaki, E., Kouromichelaki, E. M., Kouvakis, G., Nikiforaki, S., & Tsagarakis, K. P. (2008). Evaluating consumers' willingness to pay for improved potable water quality and quantity. *Water Resources Management, 22*(12), 1825–1834. doi:10.1007/s11269-008-9255-7

Hahn, M. B., Riederer, A. M., & Foster, S. O. (2009). The Livelihood Vulnerability Index: A pragmatic approach to assessing risks from climate variability and change–A case study in Mozambique. *Global Environmental Change, 19*(1), 74–88. doi:10.1016/j.gloenvcha.2008.11.002

Hajkowicz, S., Young, M. D., Wheeler, S., Hatton MacDonald, D., & Young, D. (2000). *Supporting Decisions Understanding Natural Resources Management Development Techniques: A Report to the Land and Water Resources Research and Development Corporation*. Adelaide: Commonwealth Scientific and Industrial Research Organisation.

Haq, M., Usman Mustafa, U., & Ahmad, I. (2008). Household's willingness to pay for safe drinking water: A case study of Abbottabad district. *The Pakistan Development Review, 46*(2), 1137–1153.

Harrington, W., Krupnick, A. J., & Spofford, W. O. (1989). The economic losses of a waterborne disease outbreak. *Journal of Urban Economics, 25*(1), 116–137. doi:10.1016/0094-1190(89)90048-X

Hatton MacDonald, D., Morrison, M., & Barnes, M. (2010). Willingness to pay and willingness to accept compensation for changes in urban water customer service standards. *Water Resources Management, 24*(12), 1–14.

Hunter, P. R., Chalmers, R. M., Hughes, S., & Syed, Q. (2005). Self-reported diarrhea in a control group: A strong association with reporting of lowpressure events in tap water. *Clinical Infectious Diseases, 40*(4), e32–e34. doi:10.1086/427750

InterNACHI. (n.d.). *InterNACHI's standard estimated life expectancy chart for homes*. Retrieved from https://www.nachi.org/life-expectancy.htm

Jagals, P., & Rietveld, L. (2011). Estimating the costs of small-scale water-supply interventions. In J. Cameron, P. Hunter, P. Jagals, & K. Pond (Eds.), *Valuing Water, Valuing Livelihoods: Guidance on social cost-benefit analysis of drinking-water interventions, with special reference to small community water supplies*. London: IWA Publishing.

Janmaat, J. (2007). A little knowledge...: Household water quality investment in the annapolis valley. *Canadian Journal of Agricultural Economics/Revue Canadienne D'agroeconomie, 55*(2), 233–253. doi:10.1111/j.1744-7976.2007.00090.x

Jardine, C. G., Gibson, N., & Hrudey, S. E. (1999). Detection of odour and health risk perception of drinking water. *Water Science & Technology, 40*(6), 91–98.

Larson, B. A., & Gnedenko, E. D. (1999). Avoiding health risks from drinking water in Moscow: An empirical analysis. *Environment and Development Economics, 4*(4), 565–581. doi:10.1017/S1355770X99000339

Larson, B. A., Minten, B., & Razafindralambo, R. (2006). Unravelling the linkages between the millennium development goals for poverty, education, access to water and household water use in developing countries: Evidence from Madagascar. *The Journal of Development Studies, 42*(1), 22–40. doi:10.1080/00220380500356258

Laughland, A. S., Musser, L. M., Musser, W. N., & Shortle, J. S. (1993). The opportunity cost of time and averting expenditures for safe drinking water. *Journal of the American Water Resources Association, 29*(2), 291–299.

Laughland, A. S., Musser, W. N., Shortle, J. S., & Musser, L. M. (1996). Construct validity of averting cost measures of environmental benefits. *Land Economics, 72*, 100–112. doi:10.2307/3147160

Levallois, P., Grondin, J., & Gingras, S. (1999). Evaluation of consumer attitudes on taste and tap water alternatives in Quebec. *Water Science and Technology, 40*(6), 135–139.

Majuru, B., Suhrcke, M., & Hunter, P. R. (2016). How do households respond to unreliable water supplies? A systematic review. *International Journal of Environmental Research and Public Health, 13*(12), 1222. doi:10.3390/ijerph13121222

McConnell, K. E., & Rosado, M. A. (2000). Valuing discrete improvements in drinking water quality through revealed preferences. *Water Resources Research, 36*(6), 1575–1582. doi:10.1029/2000WR900043

New Zealand Treasury. (2008). *Public sector discount rates for cost benefit analysis.* Wellington: New Zealand Government.

New Zealand Treasury. (2009). *Infrastructure: Facts and issues.* Wellington: New Zealand Government.

Nygård, K., Wahl, E., Krogh, T., Atle Tveit, O., Bohleng, E., Tverdal, A., et al. (2007). Breaks and maintenance work in the water distribution systems and gastrointestinal illness: A cohort study. *International Journal of Epidemiology, 36*, 873–880. doi:10.1093/ije/dym029

OECD. (2010). *Pricing water resources and water and sanitation services.* Paris: Author.

Olsen, S. B. (2009). Choosing between internet and mail survey modes for choice experiment surveys considering non-market goods. *Environmental and Resource Economics, 44*(4), 591–610. doi:10.1007/s10640-009-9303-7

Pattanayak, S. K., Yang, J. C., Whittington, D., & Bal Kumar, K. C. (2005). Coping with unreliable public water supplies: Averting expenditures by households in Kathmandu, Nepal. *Water Resources Research, 41*(2), 1–11. doi:10.1029/2003WR002443

Renzetti, S. (1999). Municipal water supply and sewage treatment: Costs, prices, and distortions. *The Canadian Journal of Economics/Revue Canadienne d'Economique, 32*(3), 688–704. doi:10.2307/136444

Robak, A. (2013). *Identifying and monetising the impacts of potable water supply infrastructure investments in New Zealand using stated and revealed preferences* (Doctoral dissertation). University of South Australia.

Statistics New Zealand. (2013a). Household composition, for households in occupied private dwellings, 2001, 2006, and 2013 Censuses. Retrieved from http://nzdotstat.stats.govt.nz/wbos/Index.aspx?DataSetCode=TABLECODE8165

Statistics New Zealand. (2013b). Household Use of Information and Communication Technology: 2012. Retrieved from http://www.stats.govt.nz/browse_for_stats/industry_sectors/information_technology_and_communications/HouseholdUseofICT_HOTP2012/Commentary.aspx

Turgeon, S., Rodriguez, M. J., Theriault, M., & Levallois, P. (2004). Perception of drinking water in the Quebec City region (Canada): The influence of water quality and consumer location in the distribution system. *Journal of Environmental Management, 70*(4), 363–373. doi:10.1016/j.jenvman.2003.12.014

Um, M. J., Kwak, S. J., & Kim, T. Y. (2002). Estimating willingness to pay for improved drinking water quality using averting behavior method with perception measure. *Environmental and Resource Economics, 21*(3), 285–300. doi:10.1023/A:1014537330423

USEPA. (2010). Sustainable Infrastructure for Water & Wastewater Retrieved 28/03/10. Retrieved from http://www.epa.gov/waterinfrastructure/

Whitehead, J. C., & Van Houtven, G. (1997). *Methods for valuing the benefits of the safe drinking water act: Review and assessment* (Working Papers 9705). Department of Economics, East Carolina University, Greenville, NC.

Willis, K. G., Scarpa, R., & Acutt, M. (2005). Assessing water company customer preferences and willingness to pay for service improvements: A stated choice analysis. *Water Resources Research*, *41*(2). doi:10.1029/2004WR003277

Wutich, A., & Ragsdale, K. (2008). Water insecurity and emotional distress: Coping with supply, access, and seasonal variability of water in a Bolivian squatter settlement. *Social Science & Medicine*, *67*(12), 2116–2125. doi:10.1016/j.socscimed.2008.09.042

Quality matters: incorporating water quality into water access monitoring in rural Malawi

Sarah L. Smiley

ABSTRACT

The Sustainable Development Goals offer an ambitious plan to achieve universal access to water that is safe and affordable. This article uses data from 139 household surveys from 13 villages in rural Malawi and Petrifilm quality testing of 27 drinking water sources to highlight areas where the goals' proposed monitoring framework might not fully capture water quality. Households make complex water decisions and sometimes choose to use unsafe sources. Households generally do not use water treatment but universally practise water storage. This article argues for improved monitoring to more effectively and accurately measure sustainable water access.

Introduction

In 2015, the United Nations' Sustainable Development Goals (SDGs) introduced an ambitious development policy to end poverty. One of these goals is to 'achieve universal and equitable access to safe and affordable drinking water for all' by the year 2030 (UN, n.d.). Specifically, this goal seeks to increase access to safely managed water, which is water located on premises, available when needed, and free from faecal, fluoride and arsenic contamination (WHO, 2017). This emphasis on safely managed water recognizes the limitations of prior development policy, specifically the UN's Millennium Development Goals (MDGs). The MDGs sought to reduce the number of people without access to safe water, free from faecal contamination. They measured access by proximity to a source of safe water: if a person lived within a reasonable distance of the source, they had access. The MDGs received criticism for not considering the cost, quality, or availability of that water or the burden of fetching it (Dar & Khan, 2011; Martínez-Santos, 2017; Smiley, 2016).

The SDGs acknowledge the limitations of the past focus on water source and instead offer a broad policy that considers the accessibility, availability and quality of drinking water (WHO, 2017). This more comprehensive approach to water access is notable, but it also presents challenges for monitoring. What indicators will effectively and accurately track progress towards sustainable development? There are many possible indicators for monitoring water access; the challenge is to determine which ones will provide a comprehensive framework (Kayser, Moriarty, Fonseca, & Bartram, 2013).

This article uses data from rural Malawi to illustrate the complex challenges of monitoring water access. It highlights sources, storage and treatment, and forms of contamination as three areas of water quality that are not fully considered in the proposed monitoring framework. It argues that additional indicators should be incorporated into quality monitoring to better measure real water access.

The World Health Organization (WHO) and the United Nations International Children's Emergency Fund's (UNICEF) Joint Monitoring Program (JMP) for Water Supply and Sanitation tracked water access progress for the MDGs and will continue to do so for the SDGs. The JMP's report on safely managed water indicates how accessibility, availability and quality will be defined and what types of data will be used for monitoring. This article highlights three aspects of this monitoring that do not fully measure quality. First, the JMP assumes that certain types of sources provide safe water. It classifies sources as 'improved' or 'unimproved' based on whether they have the potential to deliver uncontaminated water; improved sources include piped water, boreholes and protected wells. These sources are not tested but just assumed to provide safe water based on their construction style. But quality testing of improved sources finds that a significant number do not provide water that is free from faecal contamination (Bain et al., 2012; Shaheed, Orgill, Montgomery, Jeuland, & Brown, 2014). Also, monitoring focuses only the main source of drinking water, though most people use multiple sources of water for a variety of reasons, including preference, convenience, and changes in availability (WHO, 2017). Second, the JMP focuses on water quality at the point of source, though quality may deteriorate by the time the water is consumed due to inadequate treatment or storage (Gundry et al., 2006). Third, the JMP considers faecal contamination its highest monitoring priority. It therefore measures quality based on the absence of *E. coli* even though this absence does not rule out contamination; there are other types of pathogens that might also contaminate water (WHO, 2017). The SDGs offer an ambitious goal for universal access to safely managed water, but achieving this goal will require effective monitoring. Through a case study of rural Malawi, this article identifies areas where the proposed monitoring framework may not adequately capture the complexity of water access. Monitoring must be comprehensive and use the correct indicators to accurately measure sustainable water access.

Water in Malawi

Malawi's National Water Policy articulates a vision of 'water and sanitation for all, always' (Government of Malawi, 2005, p. 2), and it has made progress towards increasing water access. According to the JMP, the percentage of Malawians using an improved source of water increased from 42% in 1990 to 90% in 2015 (UNICEF & WHO, 2015). In the country's rural areas, the percentage increased from 36% to 89% (UNICEF & WHO, 2015). Data on the number of Malawians with access to safely managed water are not yet available, but the JMP notes that in much of the world this figure will be significantly lower than the number with access to an improved source of water (WHO, 2017). Water quality helps explain this discrepancy. Various studies in the country's urban and rural areas have tested shallow well water at the point of source and found it to be contaminated with faecal coliforms and bacteria (Chidya, Mulwafu, & Banda, 2016; Holm, Wandschneider, Felsot, & Msilimba, 2016; Mkandawire, 2008;

Msilimba & Wanda, 2013). The wells tested in Mzuzu were deemed to contain water that was unsuitable for consumption (Msilimba & Wanda, 2013). Pritchard, Mkandawire, and O'Neill (2008, p. 822) concluded that the well water they tested in Southern Malawi was unfit for human consumption: 'At best only around 50% of the wells met the lowest drinking water standard … [and] at worst nearly 90% of the well samples failed.' One explanation for this poor water quality was the proximity of these shallow wells to potential sources of contamination, especially latrines (Holm et al., 2016; Mkwate, Chidya, & Wanda, 2016; Pritchard et al., 2008). Msilimba and Wanda (2013) also highlight the problem of well proximity to trash pits, standing water and animal pens. Yet although much of the research on water quality in Malawi focuses on shallow wells, even improved sources can become contaminated; in Lilongwe, half of boreholes tested contained faecal bacteria (Chidya et al., 2016).

In addition to point-of-source contamination, water quality can deteriorate by the point of use, especially during transportation and household storage. In a refugee camp in southern Malawi, some of this contamination came from the hands of water fetchers (Roberts et al., 2001). In rural Malawi, Lindskog and Lindskog (1988) highlighted increases in contamination during household storage. Knowing that the country's water is potentially contaminated, whether at point of source or point of use, it is especially important to understand attitudes to water treatment. Studies suggest that approximately one-third of Malawian households treat their water (Holm, Singini, & Gwayi, 2016; Mkwate et al., 2016). Msilimba and Wanda (2013) found that people perceived contaminated well water to be of drinkable quality and did not do any water treatment. These low levels of treatment occur even though the use of chlorine can reduce the risk of diarrhoea by 25% or more (Wood, Foster, & Kols, 2012). This article builds on this research on water in Malawi to highlight aspects of quality that should be incorporated into monitoring frameworks to ensure that people are actually consuming safe water.

Research methods

This study used two methods to understand water quality in rural Malawi: household surveys and Petrifilm quality testing. This project received ethical approval from the Institutional Review Board at the author's university and Malawi's National Commission for Science and Technology. Thirteen villages were selected in an area of Central Lilongwe District in the Kalolo Traditional Authority. All of these villages sit in the service catchment area of a hospital run by a Malawian NGO. This organization works in these villages to provide healthcare and public health education and arranges for repairs to some boreholes each year. The NGO provided logistical support for this study in the form of transportation to villages and arranged for the health surveillance associates (HSAs) based at the hospital to assist with this project. The HSAs facilitated introductions to village leaders and helped obtain participant consent.

The 13 survey villages were selected purposefully from the larger hospital service catchment area (68 total villages) to obtain a range of population and water characteristics. The selected villages ranged in size from 8 households to 325 households. Their water characteristics included villages with a functioning borehole, with a non-functioning borehole, and lacking a borehole. Within each village, households were purposefully selected to obtain a spatially representative sample of households from across the entire village. The sole exception to this sampling method was the village with just eight households, in which all were surveyed.

For the first method, 139 qualitative household surveys collected information on water source, treatment and storage. For all surveys, the researcher was accompanied by a research assistant from the study area and an HSA. Together the HSA and research assistant explained the study, asked the survey questions, and translated the responses. This was a verbal survey conducted in Chichewa, the predominant language spoken in the study area. All participants provided verbal consent and received a small monetary incentive of 500 kwacha, which was approximately USD 0.70 at the time of the research. The survey was administered to the household member responsible for water collection.

For the second method, samples from 27 drinking water sources were collected and tested for total coliforms using Petrifilm plates. Petrifilms are a commercial product from the 3M Company originally designed to test food and beverage quality but recently being used to analyze water quality (Pearson et al., 2008). These are disposable plates that offer an alternative to traditional Petri dish testing; since they are 'sample ready' (3M, n.d.) rather than requiring any agar preparation, they are easier to use in a low-resource setting such as Malawi. They are also relatively inexpensive, and easy to store, and are considered effective and accurate for water contamination screening (Pearson et al., 2008; Vail et al., 2003). When compared to other forms of water quality testing, Petrifilms have been found to provide accurate and consistent results (Schraft & Watterworth, 2005; Vail et al., 2003; Wohlsen, Bates, Vesey, Robinson, & Katouli, 2006).

For this quality testing, 50 mL water samples were collected from sources identified as drinking water sources by participants. Samples were collected in sterile containers that were first rinsed with source water to minimize cross-contamination. Although the project aimed to sample all the sources identified as drinking water, that goal was not achievable; for example, one well was in a neighbour's fenced yard, and that person was not home to grant access at the time of the survey. While collecting water samples, observations were also made on the proximity of potential contamination sources such as latrines, trash, and animal pens as well as the general condition and cleanliness of the source. After collection, samples were kept refrigerated until the Petrifilm plates were prepared. The length of refrigeration depended on the availability of the testing laboratory; some samples were plated within 24 hours, while others were refrigerated for several days. Preparation requires the addition of 1 mL of water to the film. A trained laboratory technician prepared three films for each water sample. After preparation, films were incubated for 24 hours (±2 hours) at 35 °C (±1 °C) in an incubator in the hospital's laboratory. After incubation, at least one laboratory technician read each film, and photographs of all films were taken. The design of the plate traps gas produced by coliforms under the film, and the plate's gel changes colour when coliforms grow; these gas bubbles, and the darker colour, are then counted. This study tested for total coliforms as one possible indicator of faecal contamination in drinking water sources.

It is important to note some limitations of this testing method, all of which could be overcome in future studies. First, the length of time between sample collection and plating varied. Second, testing was conducted only for total coliforms and not followed up with testing for *E. coli* to confirm faecal contamination. Although *E. coli* is a subset of total coliforms, this broad category contains both faecal and environmental species (WHO, 2008). Thus, total coliforms include organisms that might normally be present in water, so they are not a perfect indicator of faecal contamination; rather, they indicate *possible* contamination (Pearson et al., 2008). Total coliform testing should therefore be used as a preliminary screening tool to be followed up with additional tests (Vail et al., 2003). Third, only point-of-source water was tested, which does not guarantee that the water consumed was of the same quality.

Results

Households in this study area of Malawi rely on four types of sources for drinking water: boreholes, piped water, shallow wells and surface water. Photograph examples of each type are included as Figures 1–4. Boreholes are covered and protected sources that use a hand-pump to bring groundwater to the surface. These are mechanically dug, and pipes protect the water as it is brought to the surface. Unlike in many urban areas, piped water is not distributed directly to the home, nor is it necessarily treated. In this area, the piped water is mechanically pumped groundwater used for irrigation on commercial tobacco farms. There are public standpipes near the farm offices that are used by farmworkers and nearby households. Shallow wells are hand-dug and can be either protected or unprotected; in this study, all the shallow wells encountered were unprotected, meaning they had no permanent and impregnable type of cover. Most of these wells were inadequately covered with sticks or metal sheets. Some were either partially or fully lined with bricks or concrete, but generally they were mud-walled. Water is brought to the surface using a bucket or pail that is lowered by rope into the well. The surface water in this area comes from wetlands, called *dambo* in Chichewa. During the rainy season these areas contain standing water, and during the dry season people dig holes in these areas that then fill with groundwater. This research was conducted during the dry season, so this surface water was not flowing but stagnant.

Figure 1. Borehole.

Figure 2. Piped water source.

Figure 3. Unprotected shallow well.

Figure 4. Surface water source (*dambo*).

The two most common sources of drinking water were boreholes and unprotected shallow wells, accounting for 48% and 33% of household use, respectively. Surface water and piped water accounted for 13% and 6%, respectively. Thus, only 54% of surveyed households used a source of water considered improved (boreholes and piped water), although this designation does not mean they definitely provide safe water.

The majority of households, 68%, did not treat their drinking water; of those that did treat, less than half treated their water all of the time. These treatment measures included chlorine (including the commercial chlorine product Water Guard) and boiling; no households reported the use of ceramic filters or ultraviolet purification. All households surveyed store drinking water, and 72% indicated that they always covered this stored water. The degree of covering varied, however. Some households used plastic buckets with secure lids, while others used buckets covered by just a basket. Of the remaining households, 5% indicated that they only occasionally covered their stored water, while the rest never did.

Twenty-seven drinking water sources were tested for total coliforms using Petrifilms. These sources included 15 shallow wells, 8 boreholes, 2 surface water sources and 2 piped water sources. Differences were found between borehole and shallow well quality. None of the boreholes contained total coliforms, while all of the unprotected shallow wells did. One piped water source and one surface water source each contained total coliforms, but in much lower amounts than found in the shallow wells. Colony forming units (CFUs) per 1 mL of water in unprotected shallow wells ranged from a low of 1 to a high of 400, with an average of 67 CFUs. The average CFU count per 1 mL of water for surface water was 2.8, and for piped water 0.8.

Discussion

Water source

In one village all the surveyed households used surface water, and that village lacked a borehole. It would seem that simply installing a borehole there would allow all of its households to use an improved source of water. Yet the reality is that households do not always use a borehole instead of another water source, nor do households always use the closest source of water. Rather, there is a complex and heterogeneous set of water behaviours that guide decision making (Shaheed, Orgill, Montgomery, et al., 2014). For example, 59% of surveyed households had a borehole in their village, 48% of households used a borehole, and 40% of households used a borehole in their own village.

This discrepancy has two explanations. First, not all boreholes function. In two villages the borehole worked but households complained that water flowed very slowly due to a declining water table; they believed that if the borehole was drilled deeper it would function properly. In a third village the borehole was completely broken. Problems with functionality are certainly not limited to this study area in Malawi. Mkwate et al. (2016) found that 76% of households in their entire rural study area used boreholes, but in one particular area 72% of the households used unprotected shallow wells. They attributed this discrepancy in part to high levels of non-functional water sources. Issues such as these illustrate inequity within populations, something the SDGs seek to eliminate through universal and equitable water access.

Second, some villages have a properly functioning borehole that households purposefully choose not to use. For some households the borehole was further away than a well, and for other households the borehole was more expensive than surface water since it had required a financial contribution for its initial construction. Interestingly, some households purposefully choose a source that is not the closest. A common explanation was that a neighbouring village's borehole had shorter queues than their own village borehole. Thus, some households choose to increase their walking time to minimize their waiting time. The decision to use other villages' water sources was not limited to boreholes; in total 14% of the surveyed households used a water source in a different village, including boreholes, shallow wells, piped supplies, and surface water. Quality monitoring must take into account this complex decision-making process. It cannot be assumed that households always use an available borehole, nor can it be assumed that all sources function properly.

Water treatment and storage

The majority of households do not regularly treat their drinking water, and this lack of treatment is not unique to this study area; just 20% of Malawi's rural households use some form of water treatment (NSO, 2015). Surveyed households offered four explanations for their non-treatment. The most commonly offered reason was a lack of supplies and resources for water treatment, including chlorine and fuel. In the recent past, the NGO's hospital distributed free chlorine to these villages through a national government programme. When the programme was scaled back, the hospital did not continue to fund this distribution, and households did not locate a new chlorine supply. Although Water Guard could be an alternative to the hospital's chlorine, it was not used as a regular option. During this study, Water Guard was out of stock in most stores, and households commented that they could not afford it. Thus, even if households wanted to treat water, many were unable to do so. A previous study on Water Guard in Malawi also found that the major reason offered for non-treatment of water was cost (Wood et al., 2012). In spite of these findings, Water Guard is generally considered affordable, with a bottle capable of treating 1000 L of water costing approximately the same as a 500 mL bottle of drinking water (Holm et al., 2016). Income disparities in rural areas affect access to health-related products like Water Guard since the poorest households might not have access to stores, besides lacking the income necessary to purchase the product (Rheingans, Dreibelbis, & Freeman, 2006). Still there are clear reasons to incentivize the use of Water Guard. In a Malawi antenatal programme, women used it more consistently when it was provided for free for a longer period of time, since the longer free use allowed women to overcome the cost barrier, see the benefits, and adjust to the altered taste of treated water (Wood et al., 2012). Likewise, some households cited a lack of charcoal or wood to boil their drinking water. This study area contains few forests, so there was little available wood to forage,[1] and charcoal is imported from elsewhere in the country and so it is prohibitively expensive. Thus households lacked the logistical and financial ability to obtain fuel. This inability to pay is not exclusive to Malawi. Gilman and Skillicorn (1985) found that the cost of fuel for households in Bangladesh would harm their ability to pay for food; thus, for those households without a free source of fuel, boiling was cost-prohibitive.

Second, some households did not see a need to treat borehole or piped water since they thought it was already clean and safe. Yet, quality testing of improved sources finds that a significant number do not provide water free from faecal contamination (Bain et al., 2012; Shaheed, Orgill, Montgomery, et al., 2014). Third, some households suggested that they were unaware that boiling was a treatment option. When that answer was given by survey respondents, the HSA explained how long to boil water for it to be effectively treated. Fourth, some households expressed concern at the time and effort it took to treat water, especially by boiling. One household could not boil water because it only had one pot and therefore could not both treat water and cook food.

It is understandable that all surveyed households store water, since they all fetch water from external sources, but this storage has important consequences for water quality. Water quality deteriorates during storage; water that is safe and clean at the point of source is not necessarily clean at the point of use (Gundry et al., 2006). The source type is important, with lower point-of-use contamination found in piped supplies, but still these supposedly clean sources do become contaminated (Lindskog & Lindskog, 1988; Shaheed, Orgill, Ratana, et al., 2014). Even water that has been boiled and then stored has been found to contain bacteria (Pickering et al., 2010; Shaheed, Orgill, Ratana, et al., 2014). This contamination of stored water can be attributed to three factors. First, contaminants can be introduced by air into uncovered sources or when a cover is removed. Roberts et al. (2001) in particular noted that wind could introduce animal waste; since many surveyed households have animals, this type of contamination is certainly possible. Second, contaminants could be introduced as water is removed from the storage container. Generally small plastic cups are used to retrieve water from the larger storage containers. Household members may drink directly from these cups and then dip them back into the container without sanitizing them. Stored water can also become contaminated by hands; as people dip the cup into the container, they may introduce bacteria or dirt into the water (Roberts et al., 2001; Shaheed, Orgill, Ratana, et al., 2014). Third, and related to the first two factors, the container itself could be dirty from repeated use without cleaning. In areas with inadequate water quantity, containers are often not regularly rinsed out. Together, the data on treatment and storage demonstrate a need for additional monitoring at the point of water consumption.

Quality testing

This study included preliminary testing for total coliforms. Although this type of test does not confirm faecal contamination, it indicates potential contamination and demonstrates a need for more quality testing. The only type of source that did not contain any total coliforms was the borehole. The coliforms found in piped water are especially noteworthy given that it is considered an improved source that should provide water free from contamination. The piped water source was connected to a hose that was tied off when not in use (Figure 2). It is likely that the low level of coliform growth is from the hose as it is touched by water fetchers and placed directly into water containers.

Unprotected shallow wells contained the highest CFU counts, and three specific wells stood out. For these three wells, the number of total coliforms was so great that the count had to be estimated. For counts of over 150 CFUs, 3M (n.d.) recommends counting the colonies in a representative square and multiplying by 20; the Petrifilm plates are designed with an underlying grid to facilitate counting. The well with the most total coliform growth (400 CFUs per mL of water) was in a village without a functioning borehole; thus, village residents relied entirely either on shallow wells or a borehole in a neighbouring village. This well was completely uncovered and near a dirt road. All the wells in this study area were unprotected, but most had at least some form of inadequate cover, such as some sticks or a piece of metal. Just as wind can carry contaminants into uncovered storage containers, it could do that with this well. The shallow well with the second-highest levels of contamination (340 CFUs per mL of water) was in a village with no borehole, although some households used a private borehole at a mosque in a neighbouring village. This well yielded brown water, and households complained that it produced scant supplies (Figure 3). It was cleaned out during the survey time period, and workers excavated a large pile of mud and debris, including a discarded shoe. The shallow well with the third-highest levels of contamination (220 CFUs per mL of water) was in a village with two boreholes. At the time of these surveys, only one of these boreholes was functioning. Households that originally used the non-functioning borehole now used this shallow well; they were reluctant to use the working borehole since they had not contributed to its construction fund. These three shallow well examples demonstrate the complexity of understanding water access. In a village with a borehole, some households still obtain drinking water from wells, even though several acknowledged its poor quality.

These water sources were also observed during the sample collection process. These observations were a variation of the sanitary inspections advocated by Pearson et al. (2008) to complement their Petrifilm testing. Specifically, the proximity of animals, trash, standing water, and latrines were noted during sample collection, and it was noted whether these hazards were within 20 m of the water source. Only one of these water sources was a good distance from such hazards, and this shallow well still contained total coliforms in spite of its isolated location. Sixty-three percent of water sources were either located near animal pens or had free-roaming animals near them.

These animals included chickens, cows, goats and pigs. The presence of animals can especially impact uncovered shallow wells or surface water, since animal faeces can easily get into them. Two-thirds of the sources were within approximately 20 m of a latrine. Given that the Malawi government states that latrines should be at least 100 m from a water source (Malawi Standards Board, 2005), this proximity is concerning. All households surveyed in this study used pit latrines, which could be another source of water contamination. Also, many water sources were near other issues of concern, including standing water, trash, cars, roads, brick-making facilities, and bars. These preliminary Petrifilm tests show the potential for faecal contamination in all types of sources, including those considered improved. Together with the study area's low levels of treatment and universal storage, they demonstrate the challenges with monitoring water quality. A comprehensive framework should incorporate monitoring beyond the point of source and should not assume that improved sources are safe or that they are the preferred water source.

Conclusion

The SDGs are an ambitious attempt to achieve universal and equitable access to safely managed drinking water and consider issues of accessibility, affordability, availability and quality. These goals should be commended for recognizing the need for a more comprehensive approach to water access that moves beyond source location. To truly achieve this universal and equitable access, however, the SDGs must use effective and appropriate monitoring. Using a case study of rural Malawi, this article highlights three areas of water quality that are not fully incorporated into the proposed monitoring framework. Indicators related to sources, storage and treatment, and forms of contamination should be considered as part of water quality monitoring to better track real water access.

An improved comprehensive framework should incorporate indicators related to the following four aspects of water quality. First, monitoring must take into account the complex decision-making process related to water sources. Households do not always use an improved source, or the closest water source, nor do they always use the same one. The broader concerns of the SDGs – accessibility, affordability and availability – all help explain these decisions. For example, the ultimate goal of safely managed water requires water service on premises, but basic water access requires total collection time (including walking and waiting) from an improved source of water to be less than 30 minutes. Households make water decisions to minimize this collection time, either to use a closer but unimproved source to shorten the walk or to use a farther source to shorten the wait. Likewise, a non-functioning source may cause a household to use a lower-quality or more distant source. To help explain water choices, a proposed framework should include indicators that measure distance and time differences between the closest source of improved water and the source a household actually uses. It should track the functionality of sources, including changes in availability based on seasonal precipitation. It should also collect information on all sources a household uses rather than just the main source of drinking water.

Second, monitoring must take a broader approach to water quality by looking at all possible sources of water as well as the point of consumption. Until universal access to safe water is achieved, monitoring must acknowledge the use of unimproved sources, especially given that some households prefer them over improved sources. Certainly this acknowledgement does not suggest that more boreholes should not be constructed or that broken boreholes should not be repaired. This is a core mission of development agencies, including the local NGO working in this study area. But a comprehensive monitoring programme must look at the entire water landscape, including the point of consumption. Indicators should compare point-of-consumption quality to source quality to understand the degree of post-collection contamination. As part of this expanded framework, it is important that monitoring occur at various times during the year. This study area has distinct wet and dry seasons, and water quality varies from the dry to the rainy season (Lindskog & Lindskog, 1988). A limitation of this study is that it provides a snapshot of quality only during the dry season. Regular monitoring conducted across seasons would provide a more accurate measure of quality.

Third, the monitoring framework should expand its definition of water affordability. The JMP focuses on regular water expenses such as the per-unit purchase price and admits that it ignores other expenses, such as construction and maintenance (WHO, 2017). Monitoring must also include expenses related to treatment and storage. Consistent treatment is necessary to guarantee the safety of water, including water from improved sources, yet many households are unable to afford treatment supplies. Also, secure and proper storage is necessary to protect water quality (Roberts et al., 2001), but improved storage containers might also be cost-prohibitive. For monitoring frameworks to measure whether households are consuming safe water, indicators must include all costs related to quality.

Fourth, a comprehensive monitoring framework should consider types of contamination beyond *E. coli*. The JMP recognizes that other types of pathogens can indicate faecal contamination and also requires that safely managed water is free from certain high-priority types of chemical contamination (WHO, 2017). Thus monitoring must include indicators of a broader range of potential contaminants. There is clear evidence of the benefits of safe water. Water-related diseases account for a majority of illness and death, especially of children, in developing-world countries such as Malawi (Mkwate et al., 2016; Pritchard et al., 2008). Diarrhoea is one of these illnesses linked to the consumption of contaminated water (Wolf et al., 2014). According to the Malawi government, in 2014 24% of children under age 5 had had an episode of diarrhoea within the past two weeks (NSO, 2015). Conversations with survey participants found that diarrhoea was common in this study area and affected children more frequently than adults. Although diarrhoea can have many causes, its prevalence seems at least in part attributable to poor water quality, so effective monitoring using comprehensive indicators is important.

The SDGs are ambitious, but improved access to safe water is possible. Yet the ultimate goal of universal and equitable access to safely managed water cannot be achieved without effective and appropriate monitoring to ensure that people are actually consuming safe water rather than just using an improved source. This article highlights sources, storage and treatment, and forms of contamination as three areas that need more consideration in the monitoring framework. It argues that more indicators should be incorporated into quality monitoring to better measure real water access.

Note

1. There is a large stack of wood visible in the background of Figure 2. This piped water source is at the office of a commercial tobacco farm, and this wood is used in the tobacco curing process. This wood is not available to nearby residents, nor was any firewood observed in the study area away from this farm.

Acknowledgments

Thank you to Alison Norris for introducing me to the study area.

Disclosure statement

No potential conflict of interest was reported by the author.

Funding

The Kent State University Research Council and the Kent State Department of Geography funded this research. The GIS Health and Hazards Lab at Kent State University purchased the Petrifilm plates.

ORCID

Sarah L. Smiley http://orcid.org/0000-0003-4181-517X

References

3M. (n.d.). 3M™ Petrifilm™ Coliform count plates. Retrieved from http://www.3m.com/3M/en_US/company-us/all-3m-products/~/3M-Petrifilm-Coliform-Count-Plates?N=5002385+3293785161&rt=rud

Bain, R. E., Gundry, S. W., Wright, J. A., Yang, H., Pedley, S., & Bartram, J. (2012). Accounting for water quality in monitoring access to safe drinking-water as part of the Millennium Development Goals: Lessons from five countries. *Bulletin of the World Health Organization*, *90*, 228–235A. doi:10.2471/BLT.00.000000

Chidya, R. C., Mulwafu, W. O., & Banda, S. C. (2016). Water supply dynamics and quality of alternative water sources in low-income areas of Lilongwe City, Malawi. *Physics and Chemistry of the Earth*, *93*, 63–75. doi:10.1016/j.pce.2016.03.003

Dar, O., & Khan, M. (2011). Millennium development goals and the water target: Details, definitions and debate. *Tropical Medicine & International Health*, *16*, 540–544. doi:10.1111/j.1365-3156.2011.02736.x

Gilman, R., & Skillicorn, P. (1985). Boiling of drinking-water: Can a fuel-scarce community afford it? *Bulletin of the World Health Organization*. *63*, 157–163. Retrieved from http://www.who.int/bulletin/en/

Government of Malawi. (2005). *National water policy*. Lilongwe: Ministry of Irrigation and Water Development.

Gundry, S. W., Wright, J. A., Conroy, R., Du Preez, M., Genthe, B., Moyo, S., ... Potgieter, N. (2006). Contamination of drinking water between source and point-of-use in rural households of South Africa and Zimbabwe: Implications for monitoring the Millennium Development Goal for water. *Water Practice & Technology*, *1*(2), 1–9. doi:10.2166/wpt.2006.032

Holm, R., Singini, W., & Gwayi, S. (2016). Comparative evaluation of the cost of water in northern Malawi: From rural water wells to science education. *Applied Economics, 48*(47), 4573–4583. doi:10.1080/00036846.2016.1161719

Holm, R., Wandschneider, P., Felsot, A., & Msilimba, G. (2016). Achieving the sustainable development goals: A case study of the complexity of water quality health risks in Malawi. *Journal of Health, Population and Nutrition, 35*(1). doi:10.1186/s41043-016-0057-x

Kayser, G. L., Moriarty, P., Fonseca, C., & Bartram, J. (2013). Domestic water service delivery indicators and frameworks for monitoring, evaluation, policy and planning: A review. *International Journal of Environmental Research and Public Health, 10*(10), 4812–4835. doi:10.3390/ijerph10104812

Lindskog, R. U., & Lindskog, P. A. (1988). Bacteriological contamination of water in rural areas: An intervention study from Malawi. *Tropical Medicine and Hygiene, 9*(1), 1–7. Retrieved from https://www.ncbi.nlm.nih.gov/pubmed/3346930

Malawi Standards Board. (2005). *Borehole and shallow well water quality – Specification MS 733:2005*. Blantyre: Malawi Bureau of Standards.

Martínez-Santos, P. (2017). Does 91% of the world's population really have "sustainable access to safe drinking water"? *International Journal of Water Resources Development, 33*, 514–533. doi:10.1080/07900627.2017.1298517

Mkandawire, T. (2008). Quality of groundwater from shallow wells of selected villages in Blantyre District, Malawi. *Physics and Chemistry of the Earth, 33*, 807–811. doi:10.1016/j.pce.2008.06.023

Mkwate, R. C., Chidya, R. C., & Wanda, E. M. (2016). Assessment of drinking water quality and rural household water treatment in Balaka District, Malawi. *Physics and Chemistry of the Earth*. doi:10.1016/j.pce.2016.10.006

Msilimba, G., & Wanda, E. M. (2013). Microbial and geochemical quality of shallow well water in high-density areas in Mzuzu City in Malawi. *Physics and Chemistry of the Earth, 66*, 173–180. doi:10.1016/j.pce.2013.07.002

NSO. (2015). *Malawi MDG endline survey 2014*. Zomba, Malawi: Author.

Pearson, A. L., Roberts, M. C., Soge, O. O., Ivanova, I., Mayer, J. D., & Meschke, J. S. (2008). Utility of EC 3M™ Petrifilm™ and sanitary surveys for source water assessment in Nyabushozi County, south-western Uganda. *Water SA, 34*, 279–284. Retrieved from http://www.wrc.org.za

Pickering, A. J., Davis, J., Walters, S. P., Horak, H. M., Keymer, D. P., Mushi, D., ... Boehm, A. B. (2010). Hands, water, and health: Fecal contamination in Tanzanian communities with improved, non-networked water supplies. *Environmental Science & Technology, 44*, 3267–3272. doi:10.1021/es903524m

Pritchard, M., Mkandawire, T., & O'Neill, J. G. (2008). Assessment of groundwater quality in shallow wells within the southern districts of Malawi. *Physics and Chemistry of the Earth, 33*, 812–823. doi:10.1016/j.pce.2008.06.036

Rheingans, R., Dreibelbis, R., & Freeman, M. C. (2006). Beyond the Millennium Development Goals: Public health challenges in water and sanitation. *Global Public Health, 1*(1), 31–48. doi:10.1080/17441690500443139

Roberts, L., Chartier, Y., Chartier, O., Malenga, G., Toole, M., & Rodka, H. (2001). Keeping clean water clean in a Malawi refugee camp: A randomized intervention trial. *Bulletin of the World Health Organization, 79*, 280–287. Retrieved from http://www.who.int/bulletin/en/

Schraft, H., & Watterworth, L. A. (2005). Enumeration of heterotrophs, fecal coliforms and Escherichia coli in water: Comparison of 3M Petrifilm plates with standard plating procedures. *Journal of Microbiological Methods, 60*, 335–342. doi:10.1016/j.mimet.2004.10.008

Shaheed, A., Orgill, J., Montgomery, M. A., Jeuland, M. A., & Brown, J. (2014). Why "improved" water sources are not always safe. *Bulletin of the World Health Organization, 92*, 283–289. doi:10.2471/BLT.13.119594

Shaheed, A., Orgill, J., Ratana, C., Montgomery, M. A., Jeuland, M. A., & Brown, J. (2014). Water quality risks of 'improved' water sources: Evidence from Cambodia. *Tropical Medicine & International Health, 19*(2), 186–194. doi:10.1111/tmi.12229

Smiley, S. (2016). Water availability and reliability in Dar es Salaam, Tanzania. *The Journal of Development Studies*, *52*(9), 1320–1334. doi:10.1080/00220388.2016.1146699

UN. (n.d.). Sustainable development goals: Goal 6, ensure access to water and sanitation for all. Retrieved from http://www.un.org/sustainabledevelopment/water-and-sanitation

UNICEF, & WHO. (2015). *Progress on sanitation and drinking water: 2015 update and MDG assessment*. Geneva: WHO.

Vail, J. H., Morgan, R., Merino, C. R., Gonzales, F., Miller, R., & Ram, J. L. (2003). Enumeration of waterborne Escherichia coli with Petrifilm plates: Comparison to standard methods. *Journal of Environmental Quality*, *32*, 368–373. doi:10.2134/jeq2003.3680

WHO. (2008). *Guidelines for drinking-water quality: Incorporating the first and second Addenda, Volume 1* (3rd ed.). Geneva: WHO.

WHO. (2017). *Safely managed drinking water: Thematic report on drinking water 2017*. Geneva: Author.

Wohlsen, T., Bates, J., Vesey, G., Robinson, W. A., & Katouli, M. (2006). Evaluation of the methods for enumerating coliform bacteria from water samples using precise reference standards. *Letters in Applied Microbiology*, *42*, 350–356. doi:10.1111/j.1472-765X.2006.01854.x

Wolf, J., Prüss-Ustün, A., Cumming, O., Bartram, J., Bonjour, S., Cairncross, S., . . . Higgins, J. P. T. (2014). Systematic review: Assessing the impact of drinking water and sanitation on diarrhoeal disease in low- and middle-income settings: Systematic review and meta-regression. *Tropical Medicine & International Health*, *19*, 928–942. doi:10.1111/tmi.12331

Wood, S., Foster, J., & Kols, A. (2012). Understanding why women adopt and sustain home water treatment: Insights from the Malawi antenatal care program. *Social Science & Medicine*, *75*, 634–642. doi:10.1016/j.socscimed.2011.09.018

Radiation knowledge and willingness to buy bottled water from regions near the Fukushima Daiichi Nuclear Power Plant

Kentaka Aruga

ABSTRACT

The study determined whether consumers informed about radiation have a higher willingness to buy (WTB) bottled water from regions near the Fukushima Daiichi Nuclear Power Plant (Fukushima Daiichi). The study also investigated what types of radiation knowledge are effective in mitigating negative images about bottled water from these regions. We found that consumers with radiation knowledge have a higher WTB than those uninformed about radiation, and some types of radiation knowledge have a positive effect on WTB. Our study indicates the importance of enhancing radiation knowledge to improve WTB bottled water from regions near the Fukushima Daiichi.

Introduction

Following the Great East Japan Earthquake of 2011, the Fukushima Daiichi Nuclear Power Plant (hereafter, Fukushima Daiichi) was hit by a tsunami that forced the plant into a station blackout, a situation where the plant loses its entire electricity supply, including the emergency diesel generators (Aruga, 2017a). This total station blackout was the main cause of the partial meltdown of reactors at the Fukushima Daiichi, and substantial amounts of radioactive materials were released and dispersed into the sea and air in the surrounding areas. Various food products harvested near the Fukushima Daiichi were found to be radioactively contaminated, and drinking water was no exception. In April 2011, radioactive materials that exceeded the Japanese safety standards for radioactive caesium were detected, even in tap water from the Kanto district (Ohta et al., 2012), a regional area on average 200 km away from the Fukushima Daiichi.

To survive this situation, the Japanese government instituted, under the Food Sanitation Act, one of the strictest regulations in the world on the radioactivity level of food and water in April 2012 (Consumer Affairs Agency [CAA], 2013). However, concerns persisted regarding the risk of radiation contamination of products from regions near the Fukushima Daiichi, and some people refused to buy products from these regions even when they were declared free of radiation contamination. These consumer reactions are driving down the prices of products from regions near the Fukushima Daiichi. To cope with this issue, local governments in these regions are trying to restore consumer confidence and improve the negative images of their products to compete against products from other regions.

Recently, many researchers have conducted consumer surveys to investigate consumer reactions towards products from regions near the Fukushima Daiichi (Aruga and Wakamatsu, 2018; Ito & Kuriyama, 2017; Ujiie, 2012). These previous studies investigating factors affecting consumer perceptions of these products focused only on agricultural and seafood products. Up until this study, no studies have investigated consumer perceptions of drinking water, although some studies have addressed consumer awareness of water safety issues. For example, Chatterjee et al. (2017) and Logar and Brouwer (2017) identify people's willingness to pay (WTP) to improve the quality of water, and Larue et al. (2017) reveal the WTP to reduce water pollution. This study is different from these previous studies because it not only determines consumer reaction to water with risk of radiation contamination but also identifies how types of radiation knowledge affect consumer decision to buy water from regions near the Fukushima Daiichi.

Recently, studies have shown that consumers with radiation knowledge react more positively towards buying food from regions near the Fukushima Daiichi when compared to those uninformed about radiation knowledge (Aruga, 2017b; Lee et al., 2017). Furthermore, Shimizu et al. (2017) presented evidence that providing an education programme for young students to enhance radiation knowledge can reduce their fear of radiation. In 2014, the Japanese government distributed new reading materials about radiation to elementary, junior, and high schools to foster knowledge about this subject. To create and implement education programmes on radiation, which will help reduce negative ideas about products from regions near the Fukushima Daiichi, it is imperative to understand what types of radiation knowledge are most effective at helping these products to be judged fairly by consumers. However, until now, few studies have investigated what types of radiation knowledge are important for improving consumer willingness to buy (WTB) products from regions near the Fukushima Daiichi.

To address this issue, this study has two objectives. The first objective is to identify factors affecting consumer WTB drinking water from regions near the Fukushima Daiichi. We investigate this question by using a consumer survey to identify consumer perceptions of bottled water produced in these regions. The WTB used in our study is the same concept as the conventional willingness to accept (WTA), but we did not ask the actual price that the consumer was willing to pay and instead asked about the discount rate at which the consumer would buy. We chose this approach because previous studies have shown that consumers tend to avoid buying food from regions near the Fukushima Daiichi even if it is offered at less than half the price of food from other regions (Aruga and Wakamatsu, 2018; Aruga, 2017b).[1] The second objective of this study is to examine how differences in the types of radiation knowledge affect consumer WTB bottled water from regions near the Fukushima Daiichi. For this purpose, we compare the gap in WTB among consumers with different types of radiation knowledge. We investigate how five different types of radiation knowledge influence consumer WTB bottled water from these regions. The types of knowledge investigated in this study are radiation rays, radioactive decay, the difference between Sievert (Sv) and Becquerel (Bq), natural radiation, and the effects of radiation on the human body. These types of radiation knowledge are chosen based on a booklet distributed by the Japanese government (CAA, 2013), which provides information about the risk of food and beverage contaminated with radioactive materials.

Most previous studies identified negative impacts on consumer WTB products from regions near the Fukushima Daiichi (Aruga and Wakamatsu, 2018; Ito & Kuriyama, 2017; Ujiie, 2012), but Frank and Schvaneveldt (2016) find that some consumers behave positively towards such products because they feel empathy and identify with disaster victims. As the fear of water contamination from regions near the Fukushima Daiichi persists among Japanese consumers, it is important to examine what factors influence consumer WTB bottled water from these regions. The results will provide useful information to policymakers and people working in the bottled water business, and they will assist regions near the Fukushima Daiichi in restoring consumer confidence in their bottled water products.

This study also reveals whether radiation knowledge can help reduce the negative image that prevails around bottled water products from regions near the Fukushima Daiichi, and it identifies what type of radiation knowledge is effective at mitigating negative reactions. Therefore, this study provides important information for implementing education programs to enhance consumer knowledge of radiation to improve consumer WTB bottled water from regions near the Fukushima Daiichi. As this is the first study to investigate the relationship between radiation knowledge and consumer perceptions of bottled water at risk of radiation contamination, the study is valuable to any region in the world located near a nuclear power plant, as these regions might face a similar problem as the case of regions near the Fukushima Daiichi in the future.

Methods

Our choice model is drawn from Lancaster (1966), showing that consumer utility is derived more by the characteristics or attributes that the goods possess than the goods themselves. Since McFadden (1974), application of such choice models with a random utility model (RUM) became very popular because while it is unrealistic for an empirical researcher to create a complete consumer utility function by knowing consumer's complete attributes, RUM can capture the unobserved attributes in its nondeterministic component (Manski, 1977). Hence, the consumer choice behaviour in this study is also analysed with a random utility model. The model is estimated with a probit model. Defining U_{ik} as the utility of individual i when choosing choice k from K alternatives, a random utility is generally written as:

$$U_{ik} = \beta' X_{ik} + \varepsilon_{ik} \tag{1}$$

where X_{ik} is the vector of attributes affecting the respondent's deterministic utility, β is the vector of parameters for the attributes, and ε_{ik} represents an unobserved random factor following a normal distribution with zero mean.

We performed a survey to identify factors and types of radiation knowledge affecting consumer WTB bottled water from regions near the Fukushima Daiichi. To identify the WTB water, we defined a region 100 km away from the Fukushima Daiichi as a region near the Fukushima Daiichi and a region 300 km away as one that is distant from the Fukushima Daiichi; we then had the respondents compare the bottled water from these regions. We defined the 100 km region as a region near the Fukushima Daiichi because although the capital city of the Fukushima prefecture is located between 50–100 km from the Fukushima Daiichi and was never designated as an evacuation area, it is considered to be near the Fukushima Daiichi by many Japanese who live far away from that area. We defined 300 km as far because if a region is 300 km away from the Fukushima Daiichi, it is more than 300 km from the Tokyo metropolitan area, which is home to more than one quarter of the entire Japanese population.

After showing the survey respondents a Google Map with 100 km, 200 km, and 300 km concentric circles around the Fukushima Daiichi, we asked them to choose between bottled water from a region 100 km away from the Fukushima Daiichi and that produced in a region 300 km away from the Fukushima Daiichi. We first asked the respondents to consider being at a grocery store and seeing two bottled waters, one from a region 100 km away from the Fukushima Daiichi and the other from a region 300 km away from the Fukushima Daiichi. Then, we told the respondents that the two bottles were sold next to each other and were offered at the same price. We also explained that both products had been tested for radiation and were proven risk-free based on Japanese standards on radioactive substances in drinking water.

In this hypothetical setup, our WTB variable takes one if the respondents agreed to buy bottled water from a region 100 km away from the Fukushima Daiichi at the same price as the bottled water produced 300 km away from the Fukushima Daiichi and zero otherwise. Of the total 1709 survey respondents obtained from all parts of Japan in this study, 47% of the respondents agreed to buy bottled water from a region near the Fukushima Daiichi and the remaining 53% declined to buy the water without any discounts.

Based on the above assumption, our econometric model is analysed using equation (2).

$$WTB^*_{ik} = X^{'}_{ik}\beta + \varepsilon_{ik} \tag{2}$$

where $WTB_{ik} = \begin{cases} 1 & if\ WTB^*_{ik} > 0 \\ 0 & otherwise \end{cases}$, WTB^*_{ik} denotes an indicator variable for the unobservable latent WTB, and X_{ik} is a vector of attributes that have a potential impact on our WTB water from regions near the Fukushima Daiichi. The suffix k in equation (2) represents one of the two choices presented to our survey respondents: whether to buy or not buy bottled water from regions 100 km away from the Fukushima Daiichi at the same price as water from a region 300 km away.

The variables investigated in our study that identify factors potentially affecting the WTB bottled water from regions near the Fukushima Daiichi and the variables related to radiation knowledge are incorporated as independent variables in equation (2). To consider multicollinearity, we performed a cross-correlation analysis among all these independent variables and ensured that all the cross-correlation values were below 0.4 prior to their inclusion in the model. These relatively strict criteria for correlation are based on Suzuku et al. (2008). The results of the cross-correlation values can be seen in Table A1 of the Appendix. To meet this condition, we used four econometric models where independent variables other than the radiation knowledge variables were the same, but the radiation knowledge variables were different.

First, because the cross-correlation values among the radiation knowledge variable *cancer* and other independent variables were all below 0.4, *cancer* was included in all four models. This model with *cancer* and the independent variables presented in Table 1 became our base model. Second, to identify whether having any of the types of radiation knowledge investigated in our study affects consumer WTB, we created a model with only the variable *Ignorant* added to this base model. This is how Model 1 is built. Third, Model 2 was created based on the cross-correlation values shown in Table A1 to test the effect of the *Radiation Type* variable. Fourth, Model 3 is used to investigate the influence of the variables *Decay* and *Sv & Bq* on the WTB. Finally, because the variable *Natural* was correlated with most of the radiation knowledge variables, we created Model 4, which includes *Natural* and *Decay* other than the independent variables in the base model. All four models were estimated with a probit regression.

Table 1 shows the list of variables examined in our study other than the radiation knowledge variables. *Trust* captures the level of confidence the respondents put in the Japanese safety standard for radioactive material concentrations in food after 2012. *Safety-Indifferent* is a variable that represents respondents' awareness of safety issues when buying food products. This variable reveals the difference between respondents who are concerned about food safety and those who are indifferent. *Support* reflects respondents' awareness of the need to support the disaster region. This variable represents the respondents' levels of willingness to help the disaster regions recover from the damages suffered from the Great East Japan Earthquake. The higher the value is, the higher the respondent's willingness to support the disaster region. *Children* is a dummy variable that equals one if the respondent lives with a child who is fifteen years old or younger and zero if they do not. *Age, education* and *gender* are demographic variables as defined in the table. Finally, *distance* considers the geographical distance of the respondents' residence from the Fukushima Daiichi. Aruga (2017b) indicated that consumers living away from Fukushima Daiichi have a lower WTB agricultural products from regions near the Fukushima Daiichi compared to those residing closer to the Fukushima Daiichi, so we aim to determine if similar attitudes persist for bottled water.

Table 1. Characteristic and demographic variables.

Trust	Trust the Japanese safety standard for radioactive material concentrations in food, based on a scale of 1 to 10: 1 = do not trust the standard at all to 10 = highly trust the current standard.
Safety-Indifferent	Indifferent about safety when buying food products: 1 = yes, 0 = no.
Support	Willing to help the disaster regions recover from damages after the Great East Japan Earthquake, based on a scale from 1 to 10: 1 = not willing to help at all to 10 = very much willing to help.
Children	1 = Living with at least one child who is fifteen years old or younger, 0 = Do not live with children who are fifteen years old or younger.
Age	1 = 20s, 2 = 30s, 3 = 40s, 4 = 50s, 5 = 60 or older.
Education	1 = earned a high school degree or less, 2 = earned a bachelor's degree, 3 = earned a master's or doctoral degree.
Gender	1 = male, 0 = female.
Distance	Distance of the residence from the Fukushima nuclear plant: 1 = less than 100 km, 2 = 100 km to 200 km, 3 = 200 km to 300 km, 4 = 300 km to 400 km, 5 = greater or equal to 400 km

Table 2. Radiation knowledge variables.

Ignorant	Does not possess any of the radiation knowledge asked in the survey: 1 = yes, 0 = no.
Type	Knows about the three types of radiation rays, alpha, beta, and gamma rays, and their differences: 1 = yes, 0 = no.
Decay	Knows that radioactive isotopes continue to decay until they become a stable isotope: 1 = yes, 0 = no.
Sv & Bq	Knows that Sievert (Sv) represents the degree of influence on the human body, while Becquerel (Bq) is used to measure the strength of radioactivity: 1 = yes, 0 = no.
Natural	Knows that apart from human activity, radiation exists naturally, as in cosmic rays and earth rays: 1 = yes, 0 = no.
Cancer	Knows that when an additional amount of radiation received exceeds 100 mSv, the probability of developing cancer in a lifetime increases by approximately 0.5%: 1 = yes, 0 = no.

Table 2 illustrates the radiation knowledge variables used in our study. The variables in the table are chosen based on basic knowledge about radiation enumerated in a booklet created by the Japanese government after the Fukushima disaster (CAA, 2013). This booklet was intended to provide Japanese residents with information about the risks of radioactively contaminated food and beverage. The knowledge related to *type, decay, sv & bq, natural, cancer* variables in the table was displayed as a list in a questionnaire about which knowledge the respondents possessed. If the respondents knew about any of the knowledge, they were asked to check the box displayed next to the knowledge. If the respondents did not know any of the knowledge items listed, they checked the box saying I do not know any of the items.

In Table 2, *Ignorant* is a binary variable that is one when the respondents had none of the five types of radiation knowledge presented in the table and zero if they possessed at least one. This variable captures whether respondents with radiation knowledge have different WTB compared with those without any radiation knowledge. *Type* is a variable that reveals the difference in the WTB between respondents who have and do not have knowledge regarding the three types of radiation rays. *Decay* identifies the gap in WTB between knowledgeable respondents and those who are uninformed about radioactive decay. *Sv & Bq* distinguish respondents with and without knowledge of Sievert and Becquerel. The variable *Natural* determines the difference in the WTB bottled water between respondents who know and do not know about naturally occurring radiation. Finally, *Cancer* is a variable that considers the difference between respondents who possess and do not possess knowledge about the probability of developing cancer when exposed to radioactive material.

Our data sample was gathered through an online survey performed during Jan. 30–4 February 2014, with the help of Nikkei Research Inc. We received a total of 1,709 responses from people living in all parts of Japan. The survey was distributed randomly based on the distribution of the Japanese population by prefecture in 2013.

Table 3 depicts the sample description of variables other than the radiation knowledge variables investigated in this study. The demographic variables, including age, education,

Table 3. Sample descriptions of characteristic and demographic variables.

Variables		Percentage and average	No. of respondents
WTB with no discounts	Yes	47%	797
	No	53%	912
Trust	Average of the index	5.4	na
Safety-Indifferent	Yes	18%	300
	No	82%	1409
Support	Average of the index	6.2	na
Children	Has at least one child	27%	455
	Has no children	73%	1254
Age	20s	15%	264
	30s	23%	387
	40s	16%	268
	50s	18%	312
	60 or older	28%	478
Education	Earned a high school degree or less	45%	761
	Earned a bachelor's degree	48%	822
	Earned a master's or doctoral degree	7%	126
Gender	Male	52%	897
	Female	48%	812
Distance	less than 100 km	3%	49
	100 km to 200 km	6%	103
	200 km to 300 km	29%	493
	300 km to 400 km	5%	85
	Greater or equal to 400 km	57%	979

gender, and distance, are similar to the Japanese population census in 2013 because the survey respondents were randomly drawn based on this census. As seen in the *distance* variable, the reason for having more respondents who live between 200 km and 300 km and greater than or equal to 400 km away from the Fukushima Daiichi is because most of the Japanese population is concentrated in three metropolitan areas: Tokyo, Nagoya, and Osaka.[2]

Table 4 shows the distribution of our sample data for radiation knowledge variables. First, it is discernible that more than 80% of the survey respondents had some type of knowledge about radiation. Second, the majority of the survey respondents were uninformed about the three types of radiation rays and the difference between Sievert and Becquerel. Finally, more than 60% of the respondents were knowledgeable about radiation decay, the existence of naturally occurring radiation, and the risk of developing cancer from radiation exposure.

Table 4. Sample descriptions of radiation knowledge variables.

Variables		Percentage	No. of respondents
Iqnorant	Yes	17.79%	304
	No	82.21%	1405
Type	Yes	46.58%	796
	No	53.42%	913
Decay	Yes	68.11%	545
	No	31.89%	1164
Sv & Bq	Yes	48.10%	887
	No	51.90%	822
Natural	Yes	68.58%	1172
	No	31.42%	537
Cancer	Yes	77.30%	388
	No	22.70%	1321

Results and discussion

As explained in the methods section, four econometric models were used for the probit estimation. From Table 5, it is discernible that all characteristics and demographic variables other than *Education* had some impact on the WTB bottled water from regions near the Fukushima Daiichi. The results of all four models show that *Trust, Safety-Indifferent, Support, Age*, and *Gender* have a positive impact on WTB. The result of the *Trust* variable indicates that compared with consumers who are distrustful of the Japanese safety standard for radioactive material concentrations in food, those with a higher level of confidence in the safety standard have a higher WTB water from regions near the Fukushima Daiichi. The finding that *Safety-Indifferent* positively affects WTB suggests that if people do not care about food safety in general, they tend to answer affirmatively when asked about buying bottled water from regions near the Fukushima Daiichi. The result for *Support* tells us that consumers who are willing to help the disaster regions have a higher WTB than those indifferent about the disaster regions. This finding might suggest that consumers with empathy towards the disaster regions are more likely to buy bottled water from these regions, thinking that their consumer behaviour will support the impacted areas. This type of altruistic consumption was also observed in previous studies investing other agricultural products (Lee et al., 2017). Finally, the results of *Age* and *Gender* imply that older, male consumers tend to have a higher WTB water from regions near the Fukushima Daiichi compared with young, female consumers. These results are also consistent with previous studies, and this is perhaps related to the risk of developing cancer via radiation exposure from contaminated food being lower for elderly males.

On the other hand, Table 5 also indicates that *Children* and *Distance* have a negative effect on WTB. Respondents who have a child in their household show a comparatively lower WTB bottled water from regions near the Fukushima Daiichi, which is an expected result because the risk of developing cancer from consuming radioactively contaminated products is higher for children. This result implies that adults with children want to

Table 5. Estimation results.

	Model 1			Model 2			Model 3			Model 4		
	Coeff.		Z-stat	Coeff.		Z-stat	Coeff.		Z-stat	Coeff.		Z-stat
Constant	−1.72	***	−7.64	−1.80	***	−8.07	−1.77	***	−7.88	−1.86	***	−8.21
Trust	0.20	***	11.38	0.20	***	11.46	0.20	***	11.30	0.20	***	11.36
Safety-Indifferent	0.31	***	3.47	0.29	***	3.31	0.29	***	3.22	0.33	***	3.64
Support	0.09	***	4.80	0.09	***	4.76	0.09	***	4.92	0.09	***	4.66
Children	−0.19	***	−2.59	−0.19	***	−2.58	−0.20	***	−2.69	−0.19	**	−2.46
Age	0.05	**	1.99	0.05	**	1.97	0.05	**	2.15	0.05	*	1.96
Education	0.03		0.58	0.03		0.47	0.01		0.25	0.01		0.14
Gender	0.31	***	4.46	0.29	***	4.18	0.28	***	3.93	0.28	***	4.04
Distance	−0.07	**	−2.57	−0.07	**	−2.57	−0.07	**	−2.54	−0.07	**	−2.47
Ignorant	−0.18	**	−2.01		na			na			na	
Type		na		0.14	**	2.11		na			na	
Decay		na			na		0.21	***	2.59	0.14	*	1.83
Sv & Bq		na			na		−0.05		−0.66		na	
Natural		na			na			na		0.17	**	2.24
Cancer	−0.11		−1.36	−0.11		−1.39	−0.11		−1.33	−0.13		−1.64
McFadden R^2	0.1134			0.1135			0.1145			0.1165		
Number of Obs.		1709			1709			1709			1709	

***, **, and * denote significance at the 1%, 5%, and 10% levels.

protect their children from the risk of bottled water from regions near the Fukushima Daiichi. The finding that *Distance* has a negative influence on the WTB indicates that consumers living farther away from the Fukushima Daiichi have lower WTB bottled water from these regions. This is perhaps because consumers living away from the Fukushima Daiichi tend to have fewer chances to buy products from regions near the Fukushima Daiichi (Aruga, 2017b), and hence, are uninformed about these products. As the Ministry of Agriculture, Forestry and Fisheries (MAFF) of Japan promotes local production for local consumption, the Japanese government is currently subsidizing local governments to increase consumption of products from their own regions (Ministry of Agriculture, Forestry and Fisheries [MAFF], 2019), and recently, the number of local grocery stores specializing in products from their regions has been increasing throughout Japan. For people living away from the Fukushima Daiichi, ignorance about products from regions near the Fukushima Daiichi likely led to a negative reaction towards buying bottled water from regions near the Fukushima Daiichi.

Next, we discuss the results for the radiation knowledge variables. Model 1 in Table 5 presents the results of whether having radiation knowledge has an influence on WTB bottled water from regions near the Fukushima Daiichi. As the variable *Ignorant* is negatively significant at 5% in the table, consumers uninformed about radiation knowledge have a lower WTB compared with those with at least one of the types of radiation knowledge examined in this study. This finding shows that dispersing radiation knowledge can help improve consumer WTB bottled water from regions near the Fukushima Daiichi.

Finally, the results of Models 2, 3, and 4 in the table demonstrate the different impacts on WTB of the types of radiation knowledge investigated in this study. The results suggest that all radiation knowledge variables with statistically significant effects on the WTB had positive effects. This means that programs to teach people about the types of radiation rays, radioactive decay, and naturally occurring radiation can improve consumer WTB bottled water from regions near the Fukushima Daiichi. This finding implies that these types of radiation knowledge can have positive effects on WTB and can reduce people's fear about the persistent risks of water products from regions near the Fukushima Daiichi. On the other hand, radiation knowledge of Sievert and Becquerel and knowledge related to the probability of developing cancer from radiation exposure did not have a significant effect on the WTB. This tells us that these types of radiation knowledge will not be effective at improving WTB bottled water from regions near the Fukushima Daiichi.

Conclusions

This study investigated factors affecting consumer WTB bottled water from regions near the Fukushima Daiichi and revealed types of radiation knowledge that are effective at increasing consumer WTB bottled water from these regions.

Our tests identifying factors affecting consumer WTB bottled water from regions near the Fukushima Daiichi revealed that increasing consumer confidence in the Japanese safety standard for radioactive material concentrations in food can improve consumer WTB. We also found that consumers eager to help the disaster regions recover from damage after the Great East Japan Earthquake react positively towards buying bottled water from regions near the Fukushima Daiichi. We expect such consumers are motivated by social salience or altruism and are thinking that they can help affected regions through consumption. On the other hand, people living with a child and those living away from the Fukushima Daiichi had lower WTB bottled water from regions near the Fukushima Daiichi. We believe this was due to the tendency for parents with a child to be particularly concerned about their child consuming water at risk of radiation contamination and that people who live away from the Fukushima Daiichi are uninformed about bottled water from regions near the Fukushima Daiichi.

Finally, the most intriguing findings of our study are, first, that respondents with radiation knowledge have a higher WTB compared with those uninformed about radiation. Second, all types of radiation knowledge with statistically significant effects on the WTB, such as radiation rays, radioactive decay, and natural radiation, showed a positive influence on the WTB. These results imply the importance of implementing education programs to enhance people's radiation knowledge so that they can perceive and understand the risk of consuming water products from regions near the Fukushima Daiichi in a more unbiased manner.

We hope that these radiation education programs will deepen people's radiation knowledge and help water products from regions near the Fukushima Daiichi to be fairly considered and selected at stores. These changes will reduce the effects of false rumours causing reputational damage to these regions, and hence, our study indicates the importance of enhancing radiation knowledge to change the negative perceptions people have towards products from regions near the Fukushima Daiichi.

In future research, we need to determine why radiation knowledge of radiation rays, radioactive decay, and natural radiation had a positive impact on the WTB, while knowledge related to Sievert and Becquerel and the probability of developing cancer from radiation exposure did not have statistically significant effects. Identifying the reasons behind these differences among the types of radiation knowledge regarding their effects on the WTB would be highly valuable when creating effective education programs to enhance radiation knowledge.

Notes

1. We would also like to note that the main purpose of this study is not to estimate how consumers react to changes in price but to understand what factors influence consumer behaviour when choosing between water produced near the Fukushima Daiichi and water produced away from the Fukushima Daiichi. This is one of the reasons we use the term WTB and not the conventional WTA or willingness to pay (WTP).
2. The Tokyo metropolitan area is 200–300 km away from the Fukushima Daiichi, and the Nagoya and Osaka areas are located more than 400 km away from the Fukushima Daiichi. Therefore, our sample contains few respondents living 100–200 km and 300–400 km away from the Fukushima Daiichi.

Disclosure statement

No potential conflict of interest was reported by the author.

Funding

This work was supported by JSPS KAKENHI Grant Numbers JP25712025 and JP18K12763.

ORCID

Kentaka Aruga http://orcid.org/0000-0003-3033-272X

References

Aruga, K. (2017a). *Consumer reaction, food production and the fukushima disaster: Assessing reputation damage due to potential radiation contamination*. Springer International Publishing AG.

Aruga, K. (2017b). Consumer responses to food produced near the Fukushima nuclear plant. *Environmental Economics and Policy Studies*, *19*(4), 677–690. https://doi.org/10.1007/s10018-016-0169-y

Aruga, K., & Wakamatsu, H. (2018). Consumer perceptions toward seafood produced near the Fukushima nuclear plant. *Marine Resource Economics*, *33*(4), 373–386. https://doi.org/10.1086/698998

Chatterjee, C., Triplett, R., Johnson, C. K., & Ahmed, P. (2017). Willingness to pay for safe drinking water: A contingent valuation study in Jacksonville, FL. *Journal of Environmental Management*, *203*, 413–421. https://doi.org/10.1016/j.jenvman.2017.08.008

Consumer Affairs Agency (CAA). (2013). *Food and radiation Q & A*. Retrieved February 4, 2017, from https://www.caa.go.jp/disaster/earthquake/understanding_food_and_radiation/material/pdf/130902_food_qa_en.pdf

Frank, B., & Schvaneveldt, S. J. (2016). Understanding consumer reactions to product contamination risks after national disasters: The roles of knowledge, experience, and information sources. *Journal of Retailing and Consumer Services*, *28*, 199–208. https://doi.org/10.1016/j.jretconser.2015.08.005

Ito, N., & Kuriyama, K. (2017). Averting behaviors of very small radiation exposure via food consumption after the Fukushima nuclear power station accident. *American Journal of Agricultural Economics*, *99*(1), 55–72. https://doi.org/10.1093/ajae/aaw078

Lancaster, K. T. (1966). A new approach to consumer theory. *The Journal of Political Economy*, *74*(2), 132–157. https://doi.org/10.1086/259131

Larue, B., West, G. E., Singbo, A., & Tamini, L. D. (2017). Risk aversion and willingness to pay for water quality: The case of non-farm rural residents. *Journal of Environmental Management*, *197*, 296–304. https://doi.org/10.1016/j.jenvman.2017.03.050

Lee, D., Seo, S., Song, M. K., Lee, H. K., Park, S., Jin, Y. W., & Mousseau, T. A. (2017). Factors associated with the risk perception and purchase decisions of Fukushima-related food in South Korea. *PloS one*, *12*(11), e0187655. https://doi.org/10.1371/journal.pone.0187655

Logar, I., & Brouwer, R. (2017). The effect of risk communication on choice behavior, welfare estimates and choice certainty. *Water Resource Economics*, *18*, 34–50. https://doi.org/10.1016/j.wre.2016.11.004

Manski, C. (1977). The structure of random utility models. *Theory and Decision*, *8*(3), 229–254. https://doi.org/10.1007/BF00133443

McFadden, D. (1974). Conditional logit analysis of qualitative choice behavior. In P. Zarembka (Ed.), *Frontiers of econometrics* (pp. 105–142). Academic Press.

Ministry of Agriculture, Forestry and Fisheries (MAFF). (2019). *Chisan chishou no suishin ni tsuite (About local production for local consumption)*. Retrieved June 14, 2019, fromhttp://www.maff.go.jp/j/shokusan/gizyutu/tisan_tisyo/attach/pdf/index-58.pdf

Ohta, T., Mahara, Y., Kubota, T., Fukutani, S., Fujiwara, K., Takamiya, K., Yoshinaga, H., Mizuouchi, H., & Igarashi, T. (2012). Prediction of groundwater contamination with ^{137}Cs and ^{131}I from the Fukushima nuclear accident in the Kanto district. *Journal of Environmental Radioactivity*, *111*, 38–41. https://doi.org/10.1016/j.jenvrad.2011.11.017

Shimizu, Y., Iida, H., Nenoi, M., & Akashi, M. (2017). Importance of supporting school education on radiation after the Fukushima Daiichi Nuclear Power Plant accident. *Journal of Health Education Research & Development*, *5*(1), 1–4. https://doi.org/10.4172/2380-5439.1000214

Suzuku, N., Olson, D. H., & Reilly, E. C. (2008). Developing landscape habitat models for rare amphibians with small geographic ranges: A case study of Siskiyou mountains salamanders in the western USA. *Biodiversity and Conservation*, *17*(9), 2197–2218. https://doi.org/10.1007/s10531-007-9281-4

Ujiie, K. (2012). Consumer's evaluation on radioactive contamination of agricultural products in Japan: Decomposition of WTA into a part due to radioactive contamination and a part due to area of origin. *Journal of Food System Research*, *19*(2), 142–155. https://doi.org/10.5874/jfsr.19.142

Appendix

Table A1. Cross-correlation table for radiation knowledge variables.

	Ignorance	Radiation type	Decay	Sv & Bq	Natural	Cancer
Ignorance	1.00					
Radiation Type	−0.43	1.00				
Decay	−0.32	0.50	1.00			
Sv & Bq	−0.48	0.46	0.39	1.00		
Natural	−0.69	0.40	0.32	0.42	1.00	
Cancer	−0.25	0.24	0.21	0.26	0.18	1.00

Because the cross-correlation values among the radiation knowledge variables and other independent variables were all less than 0.4, the table only illustrates the results of the radiation knowledge variables.

Farmer perceptions regarding irrigation with treated wastewater in the West Bank, Tunisia, and Qatar

Anne Dare and Rabi H. Mohtar

ABSTRACT

Heads of households, farmers, and experts in wastewater and agriculture in the West Bank, Tunisia and Qatar were engaged in consultations to understand their farming/gardening practices, perceived benefits and risks of using treated wastewater, willingness to pay for various qualities of water, safe handling practices, and knowledge of local organizations and regulatory agencies. Most view wastewater as unsafe for reuse in agriculture, and users feel local monitoring and oversight are insufficient. Improving transparency of wastewater management and providing extension about the true risks and benefits of this practice will improve the success and safety of future reuse projects.

Introduction

No matter where one goes in the world, the reuse of treated wastewater carries a stigma. Across cultures, people generally consider it dirty or unsafe, making it no surprise that there is resistance among farmers and the public to connecting this 'unsatisfactory' source of water to the food supply through irrigation. However, the degree to which people are willing to accept irrigation with treated wastewater varies from one location to another. For example, in a survey of farmers in Jordan and Tunisia, over 50% of respondents were accepting of using treated (reclaimed) wastewater for both restricted and unrestricted irrigation; but the remaining farmers cited numerous obstacles, including distrust of water quality, marketing concerns, and perceived health risks (Abu Madi, Braadbaart, Al-Sa'ed, & Alaerts, 2003). In a study in Qatar, only 8% of respondents approved of treated wastewater reuse in farming (Ahmad, 1991). This range of approval exemplifies the complexity of defining farmer's perception of the practice.

In an analysis of seven surveys conducted on the acceptability of reclaimed water use in the United States, Bruvold (1988) concluded that two main factors inform opinion regarding the use of reclaimed water: the degree of human contact with the water and a combination of 'health, environmental, treatment, distribution, and conservation' concerns (p. 48). When we look across cultures and geographic regions, many factors influence the perception of treated wastewater, including availability of water, culture, climate, societal valuation of water, and religious beliefs. Perception also depends on one's understanding of an issue. Farmers may be misinformed or lack knowledge about the use of treated wastewater in agriculture, and many misconceptions have been perpetuated about the actual risks and benefits associated with the practice. In order to better understand the education and extension needed to promote treated wastewater irrigation, there must be an understanding of the factors hindering acceptance.

This study links various aspects of farmer perception of treated wastewater reuse in communities in the West Bank (Palestinian Territories), Tunisia and Qatar. These locations were selected as part of a larger study on the challenges and limitations of wastewater reuse in agriculture, and represent a spectrum of economic and political conditions across the Middle East and North Africa (Dare et al., 2017). A local partner in each location (a non-governmental organization working in the water sector in the West Bank, a governmental organization in Tunisia, and a research institute in Qatar) provided research guidance and field support to this work. This article explores the key questions on (1) the factors limiting or driving interest in treated wastewater reuse among agricultural producers, (2) the perceptions around safe handling practices, (3) farmers' willingness to pay for treated wastewater compared with other sources of irrigation water, and (4) actual versus perceived roles played by local organizations in providing oversight to treated wastewater reuse programmes.

Methodology

Through review of published literature on perception of reclaimed water, and following the advice of the local partners, a set of questions were developed to guide consultations in each study location. Due to the nuances of how water is managed, the specific technologies employed in each location, and the farmers' experiences with water reuse, guiding questions were modified and each consultation took its own direction. In compliance with an institutional review board–approved protocol for research with human subjects, over several weeks in 2013, 20 heads of household in rural communities in the West Bank and 13 farmers in Tunisia participated in consultations (semi-structured interviews). Participants were known entities to the local partners: they were either beneficiaries/end users of water reuse projects or neighbours of users. In Qatar, differences in how agriculture is practised (farms are typically owned by wealthy landowners and operated by hired labour) led the researcher to conduct consultations with seven 'experts', including researchers, community-organization leaders, and ministry staff in the fields of water, agriculture and food security, instead of interviewing landowners or farm labour. In each location, questions focused on current irrigation practices, sources of irrigation water, crops grown, the value placed on various sources of irrigation water, satisfaction with services provided by organizations working in the water and agriculture sectors, and the ways the household receives information about public services, including information about agricultural extension. The questions are listed in the Appendix.

Research settings

West Bank, Palestinian Territories

Over the last 40 years, ways of life and agrarian practices have undergone major changes as Palestinians have acclimatized to the political constraints on land, water, technology and other resources. In that time, a notable shift has taken place in the West Bank from an agrarian economy to a service-based economy. Despite this shift, 6.2% of the West Bank's gross domestic product was accounted for by agricultural activities in 2011, and 11.9% of the population was employed in the agricultural sector (Central Intelligence Agency, 2013; Palestinian Central Bureau of Statistics, 2013). Food insecurity is a concern in the West Bank, where 40% of the population is classified as food-insecure or vulnerable to food insecurity (Palestinian Central Bureau of Statistics, 2016). That figure is expected to rise as food prices continue an upward trend, wages fail to increase at a similar rate, and the population continues to increase. The West Bank relies on limited quantities of groundwater from springs and wells, and an allotment of water from the Israeli water company, Mekorot, to meet the domestic, agricultural and industrial demands.

The communities targeted by this study were Bil'in, Kharbatha al-Misbah, Deir Qaddis and Bani Zeid, in the Ramallah/Al-Bireh Governorate. These communities were selected because of their proximity to the main office of the local partner in this study, the Palestinian Hydrology Group, and because residents in these communities had been beneficiaries of projects supported by the partner, which made it easier to approach these individuals for the consultations. Of the 20 households participating in consultations, eight had household greywater treatment units and were actively irrigating their home gardens with treated greywater, six were in communities where greywater treatment units had been installed but did not have their own units, and six were in a community (Bani Zeid) with a centralized municipal wastewater treatment system that was being considered for reuse applications. Many participants had personal experience with treated and untreated greywater reuse, but as there are few cases of treated wastewater reuse in the West Bank, none had personal experience with the practice. Participants were selected who would have relatively informed opinions about the topic, so we consider this a purposive sample of the informed population.

Gabès Governorate, Tunisia

The World Food Programme (2011) estimated that around 7% of Tunisia's population was food-insecure as of April 2011. Regional civil unrest continues following the 2010–2011 political revolutions in Tunisia and Libya, where a depressed economy, largely dependent on tourism, continues to influence the cost of living. As of 2012, 8.9% of Tunisia's gross domestic product was accounted for by agricultural activities, and 18.3% of the population was employed in the agricultural sector (Central Intelligence Agency, 2013). Access to improved drinking water sources and sanitation services is nearly universal in Tunisia.

Participants in this study were farmers in the Dissa and El Hamma agricultural areas in Gabès Governorate. These communities were selected because of their proximity to the Gabès office of the Institute of Arid Lands (the local partner in this study), and the fact that these locations were actively irrigating agricultural areas with treated wastewater from Gabès City and El Hamma wastewater treatment plants. Again, this was a purposive sample, as only farmers actively engaged in irrigation with treated wastewater were interviewed.

Qatar

Qatar's energy-based economy is largely dependent upon the 90% of their population who are expatriates: guest workers, or foreign nationals working for organizations in Qatar. Qatar is obliged to support the lifestyle expectations of the expatriate population. Qatar is an arid country, receiving only 74 mm of rainfall annually. Access to improved water and sanitation services is universal in Qatar. Ninety-nine percent of the nation's potable water comes from desalinated water, of which there is only a 48-hour reserve, but there are plans to increase the reserve to 7–14 days (Darwish & Mohtar, 2013). Qatar imports 95% of its food, while the modest agricultural sector makes a negligible contribution to the nation's gross domestic product. The Qatar National Food Security Programme's master plan calls for increasing domestic food production to 40% by 2024 (Law, 2014).

Results

A breakdown of the demographics of the interview participants is provided in Table 1.

In the West Bank (n = 20) and Tunisia (n = 13), all participants were asked the same set of four questions and were asked to rate their opinion of each question on a scale of 1 (very unfavourable) to 5 (very favourable). Treated wastewater was defined for the participants as municipal wastewater, including that from homes, businesses and industry, treated by mechanical and/or chemical processes. The results are outlined in Table 2 and illustrate the differences in opinion on some key issues related to wastewater reuse between study locations. Additional results from the entirety of the consultations, including from Qatar, are presented thematically in the following sections.

Limiting and driving factors

Participants were asked about their primary concerns about using treated wastewater, and reasons they might be motivated to use treated wastewater for irrigation on their home gardens or farms. These results are presented in Figure 1. The Palestinians selected for consultations seem to have more concerns and less motivation for treated wastewater irrigation than the Tunisian farmers, who identified a number of specific motivations, including increased agricultural productivity, source of income, and reduced fertilizer inputs.

Table 1. Participant demographics.

	West Bank, Palestinian Territories	Tunisia	Qatar
Number of participants	20	13	7
Gender	Male 13, female 7	Male 13, female 0	Male 6, female 1
Age	Range 25–85; avg. 42.5	Range 25–64; avg. 44	n/a
Education	Primary (reads) 4, secondary 9, post-secondary 7	Primary (reads) 3, secondary 9, post-secondary 1	Post-secondary 7
Primary occupation	Homemaker 8, farmer 3, teacher 3, government 2, technician/engineer 2, unemployed/retired 2	Farmer 12, mechanic 1	Government 4, consultant 1, university 1, non-profit 1
Number of members in household	Range 1–17, avg. 7	Range 1–7, avg. 4	n/a

Table 2. Average responses from explicit questions asked of consultation participants.[a]

	Palestine (n = 20)	Tunisia (n = 13)
Assuming they are safe, what is your opinion of crops irrigated with treated wastewater being sold in the same market as crops grown with freshwater?	1.9	3.5[b]
What is your opinion of the water sector (quality and quantity available) in the West Bank/Gabès Governorate?	1.1	2.3
Do you trust (have confidence in) the local authorities to be responsible for water and sanitation infrastructure?	3.3	1.9
What do you think the opinion of the general population of the West Bank/Tunisia is regarding irrigation of agricultural products with treated wastewater?	1.8	1.8

[a] Results presented are on a scale of 1 (very unfavourable) to 5 (very favourable).
[b] For this location, this question was asked in such a way that it only included crops that are currently approved for irrigation with treated wastewater according to Tunisian national law (NT 106.02) – fodder, industrial, and tree crops.

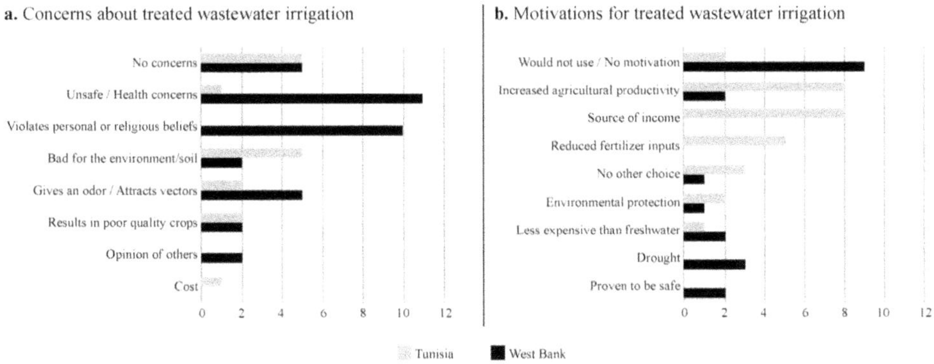

Figure 1. Results from consultations with households/farmers in the West Bank (n = 20) and Tunisia (n = 13).

In Qatar, treated wastewater is used to irrigate fodder crops at two industrial farms, and landscapes at Qatar University, along highways, and in green spaces in Doha. From consultations with experts in Qatar, two main themes emerged with regard to whether treated wastewater should be considered for irrigation. The first theme was that treated wastewater reuse could have negative health impacts, so this water should be used to fulfil the water demands of other industries (e.g., cement, industrial cooling, and landscape irrigation) before resorting to using it in agriculture. The second theme was that transparency with regard to how water and agriculture decisions are made in Qatar needs to improve. With increased transparency, treated wastewater could be made available to farmers through a carefully monitored programme to prove its potential and acceptability.

Adoption of safe handling practices

Participants currently using treated greywater or wastewater (those in the West Bank and Tunisia) were asked to describe the safety measures they use when handling these waters. We assume in this case that the safety practices used here are analogous to those which may be employed should treated wastewater be made available. Of the 10 participants in the West Bank currently using raw or treated greywater on their home gardens, 2 individuals did not handle the water differently than they would handle freshwater, while the remaining participants used drip irrigation (4), and/or avoided contact with skin by using gloves (5).

In Tunisia, 6 of the 13 participants had been trained in safe handling of treated wastewater. All six of those participants employed some kind of safe handling practice on their farms. Among all the respondents from Tunisia, 3 farmers did not use any kind of safe handling practices, while the others wore boots (9), used localized irrigation (1), or washed with soap after irrigating (1).

In Qatar, due to the limited applications of treated wastewater irrigation and the nature of the consultations, this issue was not explored.

Cost and willingness to pay

Participants in each location were asked whether there were any conditions under which they would consider using treated wastewater in their home gardens or farms. If they answered in the affirmative, they were asked how much they would be willing to pay for treated wastewater relative to freshwater. In the West Bank, 11 of the 20 participants indicated that there was no condition under which they would accept the use of treated wastewater. Of the remaining nine participants, only three provided a direct answer, and they indicated that they would be willing to pay approximately the same amount that they pay for freshwater (around NIS 2, or USD 0.58. per cubic metre).

In Tunisia, national law dictates a set price for treated wastewater of 0.02 dinar (USD 0.01) per cubic metre, so farmers were asked about their satisfaction with that price. Of the 13 farmers engaged in consultations in Tunisia, 11 were not satisfied with the current rate and cited reasons such as poor quality and management of water. Four interviewees believe they should receive the water free of charge, and three advocated a specific rate of 0.5 dinar (USD 0.3) per hour of irrigation.

In Qatar, the two industrial farms using treated wastewater currently receive this water free of charge. Since all the utilities in Qatar (e.g., water and electricity) are subsidized and treated wastewater is not a valuable commodity, the experts indicated that it was quite unlikely that anything could be charged for this water.

Role of local organizations

Participants in the West Bank and Tunisia were asked to list the organization(s) responsible for treated wastewater reuse policies and oversight, and to identify those who *should be* responsible for policies and oversight. Table 3 includes the actual organizations involved, and the frequency of responses.

Discussion

Limiting and driving factors

In the West Bank, the most commonly cited concerns with using treated wastewater are that it is perceived as generally 'unsafe' and that it violates personal or religious beliefs

Table 3. Organizations responsible for treated wastewater reuse policy and oversight.

		Participant responses	
	Actual organizations	*Who is responsible for treated wastewater reuse policy and oversight?*	*Who should be responsible for treated wastewater reuse policy and oversight?*
West Bank	Ministry of Agriculture Palestinian Water Authority Palestinian Standards Institute	Don't know 5 Palestinian Water Authority 5 Municipalities / local government 4 Ministry of Environment 3 Ministry of Agriculture 3 Ministry of Health 3 Palestinian Authority 1 Palestinian Hydrology Group 1	Don't know 5 Ministry of Environment 5 Ministry of Agriculture 5 Ministry of Health 4 Palestinian Water Authority 3 Municipalities / local government 2 Palestinian Authority 1 Other 2
Tunisia	Ministry of Agriculture, through the Regional Department for Agricultural Development (CRDA) Ministry of Environment and Land Use Planning, through the National Office of Sanitation (ONAS) National Agency for the Protection of the Environment (ANPE)	CRDA 10 ONAS 3 No one 1	CRDA 9 ONAS 5 New organization linking CRDA and ONAS 3 Local government 1 Farmers' union 1
Qatar	Ministry of the Environment Public Works Authority (Ashghal)	*Not applicable*	*Not applicable*

about appropriate sources of water for irrigation (about half of the individuals engaged in consultation). It should be noted, though, that some of these individuals had experience with using treated greywater, but none had experience with using treated wastewater. These issues were not raised in Tunisia, among participants with years of experience of working with treated wastewater. Instead, Tunisian farmers referenced things they had actually experienced, such as degradation of soil quality (specifically, issues with soil salinization), issues with management, and challenges with growing particular kinds of crops. Farmers in Tunisia who had actual experience with using treated wastewater on their farms were generally positive about the opportunities that treated wastewater presented, including higher agricultural productivity and less need for fertilizer inputs. In contrast, some of the more commonly cited motivations for using treated wastewater in the West Bank stemmed from necessity as opposed to opportunity: drought, cost and availability.

Lending creditability to this perception, Dare et al. (2017) evaluated treated wastewater quality from the greywater and wastewater treatment systems used by the consultation participants. Results from that study indicated that faecal coliform counts (an indicator of biological contamination and public health risk) and electrical conductivity (a measure of salinity, which can have negative impacts on soil health and crop production potential) were elevated in the treated greywater and treated wastewater effluents sampled in the West Bank and in treated wastewater effluents sampled in Tunisia.

In the Arab world, religion is a commonly cited reason for opposing the use of treated wastewater, though it has been demonstrated to adhere to religious policy. Scholars of Islam have debated the use of treated wastewater and concluded that it is suitable for use in irrigation; *fatwa*s (rulings on a point of Islamic law by a recognized authority) have been issued accordingly (Farooq & Ansari, 1983). The Quran makes reference to water quality on multiple occasions and indicates that impurities in water can be diluted to be made more pure (Wilson & Pfaff, 2008). Furthermore, Islam acknowledges that things (e.g., water) can also change form, from impure to pure, through metamorphosis.

Another notable difference between the responses of Palestinian and Tunisian farmers is their openness to crops grown with treated wastewater being sold in the same markets as those grown with freshwater. In the West Bank, this question received an average response of 1.9 out of 5, while in Tunisia the average response was 3.7 out of 5. Again, we are considering a different population in Tunisia: one with experience growing and selling crops irrigated with treated wastewater, while in the West Bank, markets are stocked with vegetables grown in Israel using treated wastewater. When asked further about this question, Palestinian interviewees acknowledged that they were already buying and consuming treated-wastewater-irrigated products, but they were still uncomfortable with the concept and strove for 'better' conditions for themselves.

In contrast, there is very little motivation for treated wastewater reuse in Qatar. According to consultations, the public is reluctant to adopt treated wastewater as an alternative to freshwater despite the high quality of treated effluent available and the huge public expenditures dedicated to water production and wastewater management in the country. Being a wealthy country with high expectations for standard of living, there is a societal preference for fresh or desalinated water over reusing treated wastewater in agriculture.

Adoption of safe handling practices

Encouraging users of treated wastewater and greywater to embrace safe handling practices is a challenge for the organizations providing oversight of treated wastewater reuse programmes. From the results, respondents in both the West Bank and Tunisia admitted to not using any kind of protective barrier between their skin and the irrigation water. Irrigation water in these locations does not undergo disinfection, so this careless attitude represents a health risk for farmers. To ensure farmworker safety, extension to farmers and monitoring programmes must keep pace with organizational efforts to implement more water reuse projects.

Cost and willingness to pay

The power of choice can play a major role in the adoption of treated wastewater as a resource for irrigation. Despite the environmental and economic benefits of treated wastewater reuse, farmers are generally unwilling or not required to pay for treated wastewater in the Middle East and North Africa. Public perception can impact the economics of a product: when not considered desirable, the value of the product to the consumer is low. Studies of farmers across the West Bank have indicated that they would be willing to pay nothing, or perhaps a small fee, for treated wastewater (Abu-Madi, 2009; Al-Sa'ed & Mubarak, 2006; Shaheen, 2003). However, farmers have been willing to make the switch to wastewater whenever pressure on conventional water sources increased, rendering those sources inaccessible or more costly (Jimenez &

Asano, 2008). In Israel and Jordan, the two successful cases of treated wastewater reuse in the Middle East and North Africa, the power of choice was removed, and treated wastewater was integrated into the national water strategy. Jordan went so far as to enact legislation requiring the reuse of treated wastewater.

Role of local organizations

Across all the study locations there is a concern that treated wastewater is unsafe and/or that the authorities in charge of water and sanitation services cannot be trusted to deliver treated wastewater that is safe for use in agriculture. This issue of trust seems to be more prevalent where there is more uncertainty about the future of the government, such as in the West Bank, where the government lacks complete authority, and in Tunisia, which at the time of this study was led by a transitional government.

The public's trust in local authorities (ministries, agencies, various levels of government, the private sector, etc.) to adequately monitor and maintain wastewater treatment facilities is a major hurdle in gaining acceptance for treated wastewater reuse projects. In societies with new or transitional governments, or under military occupation, there is uncertainty about the future and a significant disruption from the norm, which makes it difficult for citizens to be optimistic about delivery of services, or the commitment of the government to act in favour of the people. Uslaner (2002) describes a close connection between optimism in public projects and trust in governments.

In more established governments, trust also remains an issue. In this case, the involvement of stakeholder groups such as non-governmental organizations and conservation groups can be an effective approach to building trust and credibility around treated wastewater reuse projects. Oftentimes the only message people hear about water reuse projects are examples of failure (Jeffrey & Temple, 1999; Wegner-Gwidt, 1991). To change the mood around treated wastewater reuse, efforts can also be made to provide potential users with clear, easy-to-understand information about successful reuse projects through popular media and extension services.

Conclusions

Water scientists and engineers have been effective at developing technologies to treat wastewater to a point where it is suitable for reuse, but social scientists have a way to go in lessening the stigma and convincing farmers and the public of the merits of treated wastewater irrigation. As environmental and social pressures increasingly stress conventional sources of irrigation water, the need to find safe ways of using lesser-quality waters for irrigation will grow. Furthermore, stricter policies for the discharge of treated effluent into ambient waters are forcing communities to either invest in advanced technologies or consider land application of this water. Adoption of this practice has been and will continue to be slow, but it is prudent to provide farmers with the resources and knowledge to safely incorporate treated wastewater into their irrigation schemes. Extension initiatives and dissemination of low-cost, basic diagnostics for water quality assessment could improve farmworker safety and increase confidence in this practice.

Acknowledgments

We appreciate the kind contributions of laboratory and field assistance by Eng. Abdalrazzaq Abu Rahma, Dr Ayman Rabi, Dr Amjad Assi, Dr Rachid Boukchina, Faiza Khalfali, Dr Basem Shomar, Nora Kuiper and Candace Rowell. A. Dare is thankful to the Purdue University Center for Global Food Security's Borlaug Fellows in Global Food Security grant programme, which made the fieldwork for this study possible, and to the Purdue University Global Engineering Program and the Department of Agricultural and Biological Engineering for their support. We acknowledge the Palestinian Hydrology Group, Institute of Arid Lands (Tunisia), and Qatar Environment and Energy Research Institute for their cooperation in this study.

Funding

This work was supported by a Borlaug Fellows in Global Food Security grant [205786] from the Purdue University Center for Global Food Security.

References

Abu Madi, M., Braadbaart, O., Al-Sa'ed, R., & Alaerts, G. (2003). Willingness of farmers to pay for reclaimed wastewater in Jordan and Tunisia. *Water Science and Technology: Water Supply*, *3*(4), 115–122. Retrieved from http://ws.iwaponline.com/content/3/4/115.article-info

Abu-Madi, M. (2009). Farm-level perspectives regarding irrigation water prices in the Tulkarm district, Palestine. *Agricultural Water Management*, *96*(9), 1344–1350.

Ahmad, S. (1991). Public attitude towards water and water reuse. *Water Science and Technology*, *23*, 2165–2170. Retrieved from http://wst.iwaponline.com/content/23/10-12/2165

Al-Sa'ed, R., & Mubarak, S. (2006). Sustainability assessment of onsite sanitation facilities in Ramallah-Albireh district with emphasis on technical, socio-cultural and financial aspects. *Management of Environmental Quality: An International Journal*, *17*(2), 140–156.

Bruvold, W. H. (1988). Public opinion on water reuse options. *Journal of the Water Pollution Control Federation*, *60*(1), 45–49. Retrieved from http://www.jstor.org/stable/25043456

Central Intelligence Agency. (2013). World factbook. Retrieved from https://www.cia.gov/library/publications/the-world-factbook/

Dare, A. E., Mohtar, R. H., Jafvert, C. T., Shomar, B., Engel, B., Boukchina, R., & Rabi, A. (2017). Opportunities and challenges for treated wastewater reuse in the West Bank, Tunisia, and Qatar. *Transactions of the ASABE*, *60*(5), 1563–1574.

Darwish, M. A., & Mohtar, R. H. (2013). Qatar water challenges. *Desalination and Water Treatment*, *51*(1–3), 75–86.

Farooq, S., & Ansari, Z. I. (1983). Wastewater reuse in Muslim countries: An Islamic perspective. *Environmental Management*, *7*(2), 119–123.

Jeffrey, P., & Temple, C. (1999). Sustainable water management: Some technological and social dimensions of water recycling. *Sustainable Development International*, *1*, 63–66. Retrieved from http://infohouse.p2ric.org/ref/22/21992.pdf

Jimenez, B., & Asano, T. (2008). *Water reclamation and reuse around the world. Water reuse: An international survey of current practice, issues and needs*. London: IWA.

Law, B. (2014). Qatar scales back ambitions amid financial constraints. *BBC News*. Retrieved from http://www.bbc.com/news/world-middle-east-26140986

Palestinian Central Bureau of Statistics. (2013). Population statistics. Retrieved from http://www.pcbs.gov.ps/

Palestinian Central Bureau of Statistics. (2016). Socio-economic & food security survey 2014. Retrieved from http://fscluster.org/sites/default/files/documents/sefsec2014_report_all_web.pdf

Shaheen, H. Q. (2003). Wastewater reuse as means to optimize the use of water resources in the West Bank. *Water International*, *28*(2), 201–208.

Uslaner, E. M. (2002). *The moral foundations of trust*. Cambridge: Cambridge University Press.
Wegner-Gwidt, J. (1991). Winning support for reclamation projects through pro-active communication programs. *Water Science and Technology*, *24*(9), 313–322. Retrieved from http://wst.iwaponline.com/content/24/9/313
Wilson, Z., & Pfaff, B. (2008). Religious, philosophical and environmentalist perspectives on potable wastewater reuse in Durban, South Africa. *Desalination*, *228*, 1–9.
World Food Programme. (2011). Secondary data analysis of the food security situation in Tunisia. Retrieved from http://documents.wfp.org/stellent/groups/public/documents/ena/wfp236106.pdf

Appendix

Guiding questions for consultations with study participants:

1. What is your relationship to the farm (own, rent, part-time labour)?
2. What is the area of the farm?
3. What is the area of the farm planted in an annually harvested crop?
4. What is the area of the farm irrigated?
5. What is the source of irrigation water?
6. How much irrigation water used per month/year?
7. If the farm had access to additional water, would you (or the decision maker of the farm) increase the amount of irrigated land?
8. Which crops are grown (including specialty or tree crops)?
9. Are any livestock raised on the farm (types and number)?
10. Are any livestock products produced or sold?
11. Does the farm use synthetic fertilizers?
12. Does the farm apply animal manure or animal wastewaters as fertilizer?

To understand the primary factors limiting or driving interest in treated wastewater reuse among agricultural producers (Objective 1), participants were asked to rate their opinion of the following questions on a scale of 1 (very unfavourable) to 5 (very favourable):

13. Assuming they are safe, what is your opinion of crops irrigated with treated wastewater being sold in the same markets with crops grown with freshwater?
14. What is your opinion of the quality and quantity of water available in your location?
15. What is your opinion of the food security situation in your location?
16. Do you trust (have confidence in) the local authorities to be responsible for water and sanitation infrastructure?
17. What do you think the opinion of the general population of your location is regarding irrigation of agricultural products with treated wastewater?
18. Describe any conditions under which you would be amenable to irrigating agricultural products with treated wastewater.
19. How much should be charged for treated wastewater (e.g., relative to potable water)?
20. What do you believe are the benefits, if any, of using treated wastewater for irrigation?
21. What do you believe are the negative aspects, if any, of using treated wastewater for irrigation?

To identify what safety measures are commonly practised by farmers using treated wastewater (Objective 2), the following questions were asked:

22. Are there any crops you refuse to irrigate with treated wastewater?
23. If you use treated wastewater, what kind of irrigation method do you use with it (e.g., drip irrigation, flood)?

24. If you use treated wastewater, what safety measures do you follow when handling the water (e.g., wear gloves, boots)?

To determine the participants' willingness to pay for treated wastewater, compared with other sources of irrigation water (Objective 3), the following questions were asked:

25. If you currently use treated wastewater, how much do you pay?
26. What is the most you would pay for treated wastewater, as compared to freshwater?

To identify the roles local organizations, play in providing oversight of treated wastewater reuse programmes and opportunities for better extension and outreach to farmers (Objective 4), the following questions were asked:

27. What organizations do you believe *are* responsible for the policies and oversight of treated wastewater reuse?
28. What organizations do you believe *should be* responsible for the policies and oversight of treated wastewater reuse?
29. What organizations working in the water, agriculture and sanitation sectors do you know about?
30. Are you interested in learning more about treated wastewater reuse?
31. What media type would you most use to learn more about using treated wastewater?

Issues and challenges of reclaimed water usage: a case study of the dragon-shaped river in the Beijing Olympic Park

Xiaoqin Zhou, Zifu Li, Chad Staddon, Xuejun Wu and Han Song

ABSTRACT

Increased use of reclaimed water could be one of the solutions to Beijing's growing water shortage, particularly for non-potable (e.g. landscaping) purposes. The dragon-shaped river, a large artificial waterscape built on the site of the 2008 Beijing Olympic games, offers a useful case study of the issues and challenges attendant on wastewater reclamation and reuse. Of particular interest is the use of phytoremediation techniques for bioremediation of nutrient loads. This article presents the results of ongoing monitoring which indicate that phytoremediation is succeeding in preventing eutrophication, though some challenges, notably management of aquatic biomass, remain.

Introduction

With increasing population, urbanization and climate change, water shortages are now appearing all over the world, including in China's capital city, Beijing. In an average year, Beijing has 2680 million m^3 of available water from existing sources, or about 123.8 m^3 per capita, making Beijing a water-stressed city by international standards (National Bureau of Statistics of China, 2016; Staddon, 2010). Water experts are therefore developing multiple alternative water resources to help address this imbalance. Currently there are four main water resources for water supply into Beijing: surface water, groundwater, the South-North Water Transfer Scheme and reclaimed water, accounting for 13%, 55%, 10% and 22% of total available resources, respectively (Jensen & Yu, 2016). The 22% of water supply provided through wastewater reclamation in 2013 amounted to around 860 million m^3 and is expected to rise to 1200 million m^3 by 2020 as new schemes are completed. Wastewater reuse is clearly a key part of Beijing's future water balance.

Reclaimed water reuse is receiving more and more attention from water specialists, with well-known and successful systems in place in locations as diverse as Windhoek, Namibia, and Singapore (Bixio et al., 2006; Jensen & Yu, 2016; Staddon, 2010). These are systems for direct potable reuse of treated wastewater, however, and there are many,

many more examples of non-potable wastewater reuse. Non-potable wastewater reuses can include irrigation, industrial cooling water, car washing, landscape irrigation, and ecological/environmental supplementation during periods of low base flow. Reclaimed water quality can be an issue, however. Physical and chemical contaminants such as pH, chemical oxygen demand (COD), total phosphorus (TP) and total nitrogen (TN), as well as microbial contamination, need to be reduced to below regulatory norms before even non-potable use is possible (Bixio et al., 2006; Huertas et al., 2008; Salgot, Huertas, Weber, Dott, & Hollender, 2006). In 2000, quality standards for reclaimed water for non-potable use were published by the Chinese government, and these standards have been revised several times since (Yi, Jiao, Chen, & Chen, 2011). Current standards stipulate that BOD_5 (biological oxygen demand calculated by the five-day method) should be below 6 mg/L, TN should be below 15 mg/L, and TP should not exceed 0.5 mg/L. Even within these standards, however, there remains a risk of eutrophication in certain circumstances (Soyupak, Mukhallalati, Yemi, Bayar, & Yurteri, 1997). Moreover, the complex structure of water quality regulation (with multiple departments exercising control over different sources of wastewater and different potential end uses) makes regulatory compliance challenging (Jensen & Yu, 2016). Therefore, reclaimed water reuse faces both technical and regulatory challenges (Liu & Persson, 2013).

As host of the 29th Olympic Summer Games in summer 2008, China strongly promoted the concept of a high-tech and environmentally sustainable games, and the games venues themselves were designed to showcase Chinese developments in sustainable urbanization, including reclaimed water use and rainwater harvesting (Jensen & Yu, 2016). In particular, the main Olympic site master plan included a specially built wastewater treatment plant, linked to a series of slow-flow water bodies which together form a semi-natural ecosystem – the dragon-shaped river – which terminates in the Dragon Lake, now considered a beauty and biodiversity hotspot. In this study, we focus on the use of phytoremediation to address eutrophication pressures in the dragon-shaped river.

An overview of the water quality situation of the dragon-shaped river at the Beijing Olympic Park

The Beijing Olympic Park was the main location for the games, and many of its structures (such as the Bird's Nest and the Water Cube) have become iconic buildings. Of particular interest to water managers and scholars is the artificial river/lake complex, shaped like a dancing dragon, which was designed to provide a semi-natural structure integrating the Olympic site along a north-south axis. As shown in Figure 1, the head of the dragon forms the central water feature of the Olympic Forest Park at the northern end of the site; the body runs down the eastern side of the park; and the tail coils around the National Stadium (the Bird's Nest) at the southern extremity. The watershed in the Olympic Park is the world's largest artificial landscape using reclaimed water as the sole water source, with a total water surface area of 84.2 hm^2, and a total water storage capacity of 1.3 million m^3.

Of this total surface area, the main dragon-shaped river and lake structures occupy 48.5 hm^2 and have a total water storage capacity of 0.75 million $metres^3$. The river itself is 2.7 km in length and 20–125 m wide, and its depth varies from about 0.6–1.2 m. The

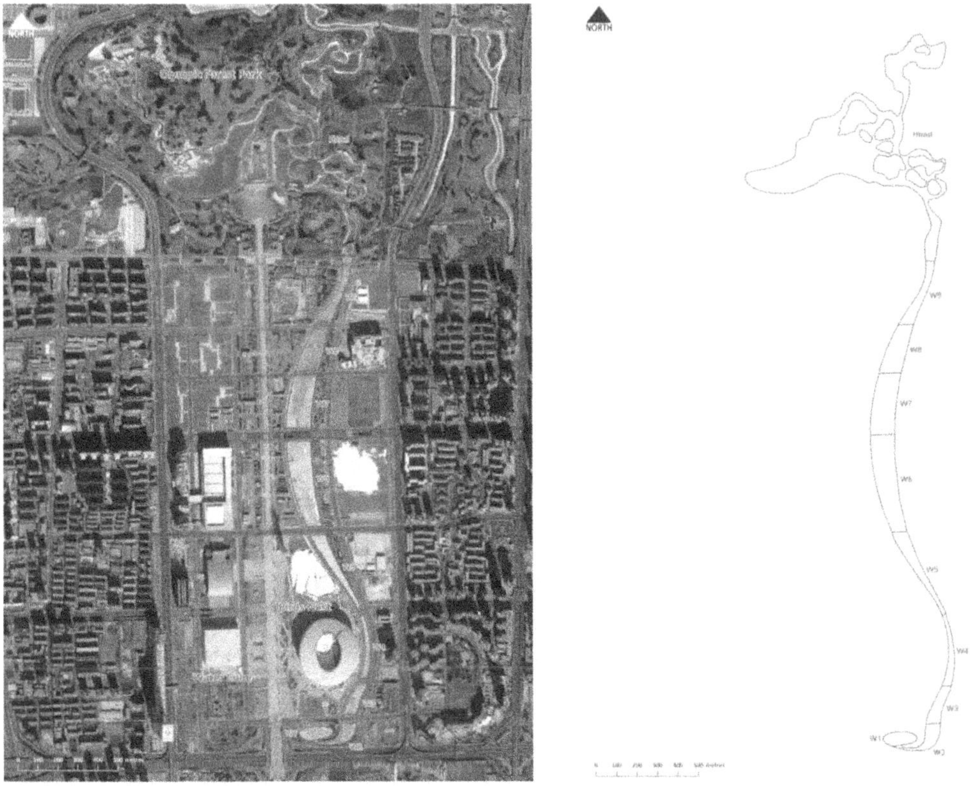

Figure 1. Main Beijing Olympics site, showing dragon-shaped river system, with Dragon Lake and Olympic Forest Park at top.
Source: Paul Satchell, UWE, Bristol.

river is divided into nine individual sections by eight road bridges, with each section connected by an underground culvert. Water flows northwards from the first section (the tail of the dragon, W1 in Figure 1) to the last section (the neck of the dragon, W9). The water gradually deepens from the edges inward to the centre, but only to a maximum depth of 1.24 m at its deepest point in section W8.

The amount and frequency of reclaimed water inflow varies considerably, for example between 2400 and 3500 m^3/d. Despite this variability, key nutrients (TN, TP, NH_3-N) in the reclaimed water usually meet the wastewater quality standards for 'scenic water' in China (Table 1). During normal operation, no untreated wastewater is discharged into the river. Non-point contamination does arrive in surface water

Table 1. Main chemical parameters of reclaimed water at the W1 inlet (see Figure 1 for location).

Sampling date	NH_3-N (mg/L)	Total nitrogen (mg/L)	Total phosphorus (mg/L)	Chemical oxygen demand (mg/L)	Turbidity (NTU)
20 September 2010	0.35	11.0	0.04	3.96 (Cr)	18.0
21 October 2010	0.21	2.28	0.04	4.3 (Cr)	4.79
21 June 2011	1.6	11.0	0.2	7.96 (Cr)	21.0

Note: Cr = chromium. COD-Cr is a standard measurement method for COD value which uses K2Cr2O7 as oxidant to determine COD value.

runoff, but this contributes only 4.04% of the TN, 15.2% of the TP, 8.17% of the COD and 14.47% of the NH_3-N loading. The issues related to non-point source contamination are not specifically discussed in this article.

By 2010, signs of eutrophication were appearing: floating (organic) material was reported in the river, and there were growing numbers of complaints about the smell from the water. The most likely explanation for this was relatively high levels of nitrogen and phosphate, combined with the shallow water level, the low velocity of water flow from W1 to W9, and sufficient sunshine reaching the bottom of the shallow sections to create perfect conditions for algal and microbial overgrowth, along with the foul smell – in other words, eutrophication was occurring.

To improve the situation, aquatic plants were placed in all nine sections to promote phytoremediation of nutrients and thereby reduce eutrophication. The process had three steps: cleaning the river channel of algal biomass; planting submerged plants according to a carefully planned scheme; and the operation, monitoring and management of the resulting enhanced water system. In the first step, the river channel was cleared of weeds and microorganisms were added to reactivate bottom sediments and form a benthic microbial degradation system. In the second step, submerged plants were selected and planted in the nine sections in carefully calculated ratios and densities. The selection of the plants was based on the measured contamination levels and a review of the secondary literature on phytoremediation. This literature (e.g. Gross, Erhard, & Iványi, 2003; Hilt & Gross, 2008) suggests that plants such as *Ceratophyllum demersum* L. (hornwort) and *Najas marina* L. (spiny water nymph) release allelochemicals that inhibit anaerobic decomposition, a primary cause of foul smells and turbidity. These plants, and also plants from the Elodea (waterweed) and Potamogetonaceae (pondweed) families, can be assembled into clumps or complexes that work together in the river, with different allelochemicals released by different plants, so that most kinds of harmful algae are inhibited (Körner & Nicklisch, 2002). Such constructed plant communities can also inhibit the production of harmful types of cyanobacteria. In the final step, a monitoring programme was put in place to assess the contribution of each basin to phytoremediation of target pollutants (TN, TP, COD).

The ecological engineering was completed by the end of 2010. The monitoring began two years later, and from April 2013 to July 2014, samples were taken twice per month in each of the nine segments (Figure 1, W1 to W9), at about 0.1 m below the surface, except during winter, when the surface freezes. The water quality parameters measured include COD, TN, ammonia (NH_3-N) and TP, and sample analyses were conducted in accordance with the methods described in the state publication, 'Water and Wastewater Monitoring and Analysis Method' (State Environmental Protection Administration of the People's Republic of China, 2002).

Measured impact of phytoremediation

Landscape effects before and after planting

The landscape's scenic value improved considerably after the introduction of the submerged plants. Before, the water had looked green, with grey suspended solids and poor transparency, as well as the much-commented-upon foul smells.

Afterwards, the water was clear enough to see directly to the bottom, and both surface scum and foul smells largely disappeared. Since the implementation of the submerged plant communities, no return of eutrophic conditions was observed, even in hot weather. The submerged plants seem to have reduced turbulence in the water bodies, supported sedimentation and prevented sediment resuspension, thus increasing transparency (Horppila & Nurminen, 2003). They also improved circulation of energy and gases (particularly oxygen) between water, sediments and plants, thus forming a healthier and more integrated system with higher phyto and bio remediative abilities.

Main contaminants in the dragon-shaped river

After 12 months of monitoring the water quality in the nine sections, the results were as follows. Both TN and TP decreased progressively from section W1 to section W9, though the results for TN were the most impressive. This gradient of TN and TP concentrations was in accordance with expectations – reclaimed water fed in at the beginning of section W1 flows along from W2 to W9, with nutrient levels decreasing in each successive section (Figure 2). For TP concentration, as it was initially low (less than 0.4 mg/L) in the reclaimed water in W1, measured reductions were smaller. The average TP concentrations in the sections from W1 to W9 were normally below 0.1 mg/L, which meets the requirement of third-class surface water quality in China (the standard for third-class water is TP ≤ 0.2 mg/L), indicating that the main contaminant in the dragon-shaped river is nitrogen.

A closer look at TN concentration change from W1 and W9 was undertaken. The TN values obtained during the 12-month monitoring period fluctuated considerably. One reason may be non-point pollution from rainfall and resultant surface water runoff, especially from nutrient application by horticulturalists and landscape

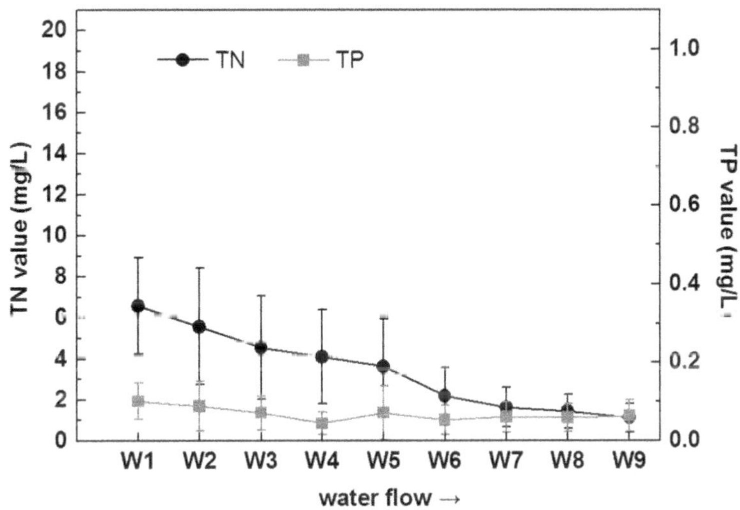

Figure 2. Total nitrogen (TN) and total phosphorus (TP) change from section W1 to section W9 (see Figure 1 for locations).

managers. Seasonal effects related to the submerged plant system could also explain some of this variation. Beijing's climate is characterized by a typical north temperate semi-humid continental monsoon climate, with hot and rainy summers and cold and dry winters. Both spring and autumn are relatively short. The TN concentration was relatively low in warmer summer temperatures. It appears that the sunshine and high temperatures of summer are good for plant growth, and therefore for nutrient absorption and transformation. TN levels were usually higher at other times of the year.

Overall, we conclude that the installation of engineered plant communities has been successful in reducing eutrophication pressures in the dragon-shaped river. There is however a consequence of considerable importance for the sustainability of the new system. Harvesting of aquatic plant biomass is very important in maintaining macrophytic plant communities' purification potency. Thus, regular management of submerged plants is essential, or decomposing submerged macrophytes will release nitrogen back into the water, reducing nitrogen removal efficiency (Feng et al., 2008; Vymazal, 2002; Vymazal & Kröpfelová, 2009) and increasing COD and BOD_5. Timely harvest in summer could also contribute to better nutrient removal by promoting additional fresh growth, and therefore bioremediation.

To determine the bioavailability of different forms of nitrogen, ammonia levels were also monitored at each sampling point. Ammonia concentrations were very low compared with TN concentrations in all sections. Ammonia concentrations in all the samples were under, and usually well under, 1 mg/L, which meets the requirement of third-class water according to the regulatory surface water standard used in China. It is already recognized that submerged macrophytes have a better nitrogen removal capacity compared to emergent macrophytes, because they can absorb nitrogen from both sediment and water through root and body (Kadlec, 2006; Taguchi & Nakata, 2009). In the dragon-shaped river, substances like nitrogen (TN and NH_3-N) and phosphorus in reclaimed water can survive as necessary nutrients for submerged plants to grow and develop, some (such as ammonia) can be absorbed by plants directly, and some can be utilized metabolically, reducing residual contaminant volumes.

The oxygen concentration decreased with depth; thus the availability of dissolved oxygen at the bottom is limited. With the submerged plants in the water system, oxygen can be transferred by plant root, forming an aerobic-anoxic-anaerobic environment in the root zones, which provides conditions for the formation of nitrite and nitrate from nitrogen. Therefore we conclude that initially TN concentrations (in W1, W2 and W3) are partly transformed by high nitrite and nitrate production, with ammonia concentrations remaining low.

Compared with nitrogen and phosphorus, COD levels in the water system were not similarly reduced from section W1 to section W9. COD concentrations from W1 to W9 were similar, and even showed some increase as surface runoff entered the system. The highest COD concentration was in W6 (106 mg/L), but observed values were mainly in a range of 20–40 mg/L. COD removal is mainly dependent on the activity of microorganisms by the processes of aerobic or anaerobic degradation (Sani, Scholz, & Bouillon, 2013; Vymazal, 2002), but there are not enough filters in the submerged plant ecosystem for the formation of the necessary biofilm, so little or no COD is removed. Based on the current ecosystem, we suggest that a constructed wetland could be built along the river basin to improve COD levels and reduce environmental risk (Vymazal & Kröpfelová, 2009; Zhang et al., 2014).

From the basic water parameter analysis in the dragon-shaped river, it can be concluded that the water quality was improved by the implementation of engineered aquatic plant communities, but it is still hard to meet the surface water standard, and there is still a risk of eutrophication related to surface runoff along the system. Nevertheless, with these aquatic plants in place the reclaimed water can safely be used to augment flows through the dragon-shaped river and into Dragon Lake. Therefore, the submerged plants played a key dual role in the dragon-shaped river: removing nutrients, as well as inhibiting the growth of harmful algae.

Issues and challenges in reclaimed water reused for landscape values

The dragon-shaped river is a good example of reclaimed water implementation in China, and its experience can be compared with similar projects elsewhere in China and around the world. However, our study shows that there are still persistent challenges, particularly relating to sustained nutrient removal across the broader spectrum (nitrates, nitrites, etc.) and the sustainable management of resulting plant biomass. As can be seen from the water quality analysis of the whole dragon-shaped river, little or no COD was removed through planting, so there is certainly a need for further re-engineering of the system with a view to reducing COD levels. Moreover, harvesting of submerged plants is an effective way to physically remove bio-accumulated nutrients (and other contaminants, e.g. metals) from the river. Indeed, regular removal of biomass is necessary, otherwise submerged macrophytes will eventually release nitrogen back into the water during decomposition. Final disposal options for this submerged plant biomass (up to 10 tonnes during the summer season) are still under development (Feng et al., 2008). The freshly harvested plants are rich in water, with moisture content of more than 90%, and could be used as feedstock for anaerobic digestion-based methane recovery. Another option might be co-composting with other wastes, not least because the harvested materials are rich in compost-friendly nutrients (N, P, K). Without careful attention these system wastes could create potentially damaging leachate runoff and a perfect breeding ground for mosquitoes and flies. Study of the optimal management of harvested biomass from the dragon-shaped river is currently being undertaken by the authors of this article.

Another problem follows from the under-recognized challenge of integrating wastewater reuse schemes into existing 'conventional' treatment systems or treated water discharge management systems. With respect to linking alternative water supply systems to conventional systems, one development path could be to disaggregate distribution of potable and non-potable supplies by creating a complete 'dual' reticulation system. This is of course enormously expensive and disruptive. Another development path could be to upgrade treatment of wastewater to the point where it meets Chinese government standards for drinking water, but this is as yet a long way off. It is also not certain that increasing production of non-potable standard treated wastewater will necessarily compensate for increased abstractions and pollutant discharges and reduced natural inflows in the natural environment.

Conclusions

From this case study of Beijing's dragon-shaped river system, it can be seen that without phytoremediation water quality struggles to meet the national environmental quality standards for surface water and actually tends towards eutrophic conditions. As more and more artificial lakes and rivers start to receive reclaimed water as a supplement to natural inflows, phytoremediation may be a useful option for avoiding eutrophication and attendant aesthetic problems. It is environmentally friendly and cost-effective to employ an engineered plant-based treatment system, and the management of the dragon-shaped river has been a success. Currently, however, there are no specific standards or regulations for such approaches to water quality management. There is also a need for design and engineering guidance appropriate to Chinese urban conditions. Thus, there is an urgent need for evidence-based guidelines for reclaimed water and sustainable landscape management, including water quality control and landscape impact evaluation, as well as life cycle–focused management of macrophytes in the ecological system. There is also an emerging debate about how such 'decentralized' solutions can integrate with 'centralized' solutions to best achieve water quality outcomes (Jensen & Yu, 2016) – the experience of the dragon-shaped river could be quite instructive.

Acknowledgement

We gratefully acknowledge support from the International Water Security Network, funded by Lloyd's Register Foundation, a charitable foundation helping protect life and property by supporting engineering-related education, public engagement, and the application of research.

Disclosure statement

No potential conflict of interest was reported by the authors.

ORCID

Chad Staddon http://orcid.org/0000-0002-2063-8525

References

Bixio, D., Thoeye, C., De Koning, J., Joksimovic, D., Savic, D., & Wintgens, T. (2006). Wastewater reuse in Europe. *Desalination*, *187*(1–3), 89–101. doi:10.1016/j.desal.2005.04.070

Feng, Z., Wen, J., Jiang, F., Deng, Z., & Lu, X. (2008). Construction and management of artificial ecosystem of the dragon-shaped olympic water system in Beijing City (in Chinese). *Wetland Science and Management*, 4 (2), 33–36.

Gross, E. M., Erhard, D., & Iványi, E. (2003). Allelopathic activity of Ceratophyllum demersum L. and Najas marina ssp. intermedia (Wolfgang) Casper. *Hydrobiologia, 506-509*(1–3), 583–589. doi:10.1023/B:HYDR.0000008539.32622.91

Hilt, S., & Gross, E. M. (2008). Can allelopathically active submerged macrophytes stabilise clear-water states in shallow lakes? *Basic and Applied Ecology*, 9(4), 422–432. doi:10.1016/j.baae.2007.04.003

Horppila, J., & Nurminen, L. (2003). Effects of submerged macrophytes on sediment resuspension and internal phosphorus loading in Lake Hiidenvesi (southern Finland). *Water Research*, *37*(18), 4468–4474. doi:10.1016/S0043-1354(03)00405-6

Huertas, E., Salgot, M., Hollender, J., Weber, S., Dott, W., & Khan, S. (2008). Key objectives for water reuse concepts. *Desalination*, *218*(1–3), 120–131. doi:10.1016/j.desal.2006.09.032

Jensen, O., & Yu, X. (2016). Wastewater reuse in Beijing: An evolving hybrid system. *International Journal of Water Resources Development*, *32*(4), 590–610. doi:10.1080/07900627.2016.1148589

Kadlec, R. H. (2006). Free surface wetlands for phosphorus removal: The position of the everglades nutrient removal project. *Ecological Engineering*, *27*(4), 361–379. doi:10.1016/j.ecoleng.2006.05.019

Körner, S., & Nicklisch, A. (2002). Allelopathic growth inhibition of selected phytoplankton species by submerged macrophytes1. *Journal of Phycology*, *38*(5), 862–871.

Liu, S., & Persson, K. M. (2013). Situations of water reuse in China. *Water Policy*, *15*, 705–727.

National Bureau of Statistics of China. (2016). *China statistical yearbook 2016*. Beijing, China: China Statistics Press.

Salgot, M., Huertas, E., Weber, S., Dott, W., & Hollender, J. (2006). Wastewater reuse and risk: Definition of key objectives. *Desalination*, *187*(1–3), 29–40. doi:10.1016/j.desal.2005.04.065

Sani, A., Scholz, M., & Bouillon, L. (2013). Seasonal assessment of experimental vertical-flow constructed wetlands treating domestic wastewater. *Bioresource Technology*, *147*, 585–596. doi:10.1016/j.biortech.2013.08.076

Soyupak, S., Mukhallalati, L., Yemi, E. D., Bayar, A., & Yurteri, C. (1997). Evaluation of eutrophication control strategies for the Keban Dam reservoir. *Ecological Modelling*, 97(1–2), 99–110. doi:10.1016/S0304-3800(96)00077-4

Staddon, C. (2010). *Managing Europe's Waters: 21st century challenges*. Aldershot: Ashgate.

State Environmental Protection Administration of the People's Republic of China. 2002. *Water and wastewater monitoring and analysis methods*. Beijing: Chinese Environmental Science Press (in Chinese).

Taguchi, K., & Nakata, K. (2009). Evaluation of biological water purification functions of inland lakes using an aquatic ecosystem model. *Ecological Modelling*, *220*(18), 2255–2271. doi:10.1016/j.ecolmodel.2009.05.007

Vymazal, J. (2002). The use of sub-surface constructed wetlands for wastewater treatment in the Czech Republic: 10 years experience. *Ecological Engineering*, *18*(5), 633–646. doi:10.1016/S0925-8574(02)00025-3

Vymazal, J., & Kröpfelová, L. (2009). Removal of organics in constructed wetlands with horizontal sub-surface flow: A review of the field experience. *Science of the Total Environment*, *407*(13), 3911–3922. doi:10.1016/j.scitotenv.2008.08.032

Yi, L., Jiao, W., Chen, X., & Chen, W. (2011). An overview of reclaimed water reuse in China. *Journal of Environmental Sciences*, *23*(10), 1585–1593. doi:10.1016/S1001-0742(10)60627-4

Zhang, D. Q., Jinadasa, K. B. S. N., Gersberg, R. M., Liu, Y., Ng, W. J., & Tan, S. K. (2014). Application of constructed wetlands for wastewater treatment in developing countries – A review of recent developments (2000–2013). *Journal of Environmental Management*, *141*, 116–131. doi:10.1016/j.jenvman.2014.03.015

The potential impact of water quality on the spread and control of COVID-19 in Syrian refugee camps in Lebanon

Issmat I. Kassem and Hadi Jaafar

The Syrian civil war has been raging since 11 March 2011, causing millions of civilians to seek refuge in safer countries, including Lebanon, a small Mediterranean country that is bordered by Syria to the north and east. A great number of Syrian refugees, including children, elderly, and immunocompromised individuals, have found themselves living in makeshift camps and tents that provide little protection from the elements as well as limited access to basic needs such as clean water and sanitation, robust medical support, and safe and nutritious foods. It is known that the spread of COVID-19 is most likely to be facilitated in crowded places where it is impossible to maintain 'social distance', and in communities that do not have access to good medical systems, sanitation, and hygienic products or infrastructure. Thus, the conditions that facilitate the spread of COVID-19 in disenfranchised populations such as those in the Syrian civil war refugee camps in Lebanon are of significant concern (Kassem, 2020).

As of 4 June 2020, there are no reports of cases or outbreaks of COVID-19 in Syrian refugee camps in Lebanon, which has recorded about 1306 cases and 28 deaths nationwide in a population of over 6 million. However, this does not mean that the camps have been spared . In reality, there has been little concentrated effort to sample the refugees in Lebanon due to the lack of resources and test kits. Also, there are anecdotal suggestions that the refugees may refrain from reporting cases due to unawareness or fear of stigma and deportation. The latter might have been spurred by a crackdown on unregistered refugees and strict curfews, which took place even before the onset of the pandemic (Kassem, 2020).

Currently, Lebanon is in a precarious situation, with many economic and environmental problems, of which deterioration of water quality and increasing water stress are among the gravest (Jaafar et al., 2016a, 2016b). Host to almost 1.5 million Syrian refugees, Lebanon is estimated to have the highest number of refugees per capita in the world (Jaafar et al., 2020). Many of the refugees have settled in Lebanon's fertile Bekaa Valley and the plain of Akkar (Figure 1), causing an additional shock to the water and wastewater infrastructure. More than 20% of the refugees live in informal settlements (shelters, tents, school buildings), which are mostly located close to river courses in the Bekaa and the north of Lebanon. By the end of 2016, there were 29,000 tents in the Bekaa, distributed over 2650 informal settlements. Depending on the location, these settlements receive their domestic water by trucks, through the existing water network, or from new boreholes drilled close by. Water is stored in tanks or underground reservoirs. Reportedly, only a third of all Syrian refugees have access to tap water for more than two hours per day. It is estimated that the water used by informal settlements varies between 50 and 90 litres per person per day (Jaafar et al., 2020). Furthermore, wastewater from the settlements is discharged untreated, either into the neighbouring watercourses or into septic tanks within the settlements. This ongoing humanitarian situation has increased water quality threats due to the temporary lack of sanitation and waste management measures. Local and international stakeholders, including NGOs, hosting these displaced people have not been able to create the systems required for the safe treatment and disposal of the additional sewage load and waste generated by the 150,000 refugees living in informal settlements in the Bekaa Valley (Jaafar et al., 2016a, 2016b).

Taken together, this indicates that the Syrian refugee camps in Lebanon have poor access to the clean water and sanitation that are integral in the fight against COVID-19. The situation is complicated further by (1) evidence that the water available to the camps can be of poor quality and affected by faecal pollution, and (2) the possibility of spreading the virus via raw sewage discharges into waterways that surround the camps. The latter may lead to self-inoculation within the camps and may extend the risk of exposure to the hosting communities (discussed below). These potential scenarios are causes of concern and require immediate remediation to avoid outbreaks and disease in the camps and hosting communities.

Surface water, such as rivers and irrigation water, in proximity to the refugee camps (Figure 1) has been shown to be contaminated with faecal indicators (Dagher, 2020; Hmede et al., 2019). For example, in a nationwide study, we showed that faecal indicator bacteria (*Escherichia coli* and faecal coliforms) were detected in water sampled from all the rivers in Lebanon (Dagher, 2020). The density of the indicators varied between rivers, ranging between 31 and 2400 CFU of *E. coli* per 100 mL and between 86 and 3440 CFU of faecal coliforms per 100 mL. The elevated levels of indicator bacteria were not unexpected, because it has been estimated that 90% of urban water in Lebanon is contaminated, and 64% of the Lebanese do not have access to safe water on a regular basis (Baylouny & Klingseis, 2018).

Furthermore, approximately 39% of Syrian households in Lebanon use unimproved drinking water sources (Machayekhi et al., 2017). In 2019–2020, before the COVID-19 outbreak, the sewer and domestic water, including drinking and well water, in two Syrian refugee camps in the Bekaa Valley were investigated (Alhaj Sulaiman & Kassem, 2020; Sulaiman & Kassem, 2019). The camps housed 158 refugees, of whom 72 were children less than 10 years old. Camp 1 had irregular access to municipal water. Well water was frequently used for domestic needs, including drinking, when municipal water was unavailable. In Camp 2, drinking water was provided daily by trucks and stored in two large containers, while well water was used for other domestic needs. The refugees in Camp 2 suspected that their well might have been contaminated by camp sewer water, which was drained via poorly covered and relatively shallow conduits in both camps. *E. coli* was detected in the drinking and well water of Camp 1 and in the well water of Camp 2 (Sulaiman & Kassem, 2019). Further analysis showed that the *E. coli* isolates were multi-drug resistant and harboured transmissible genes that encoded resistance to last-

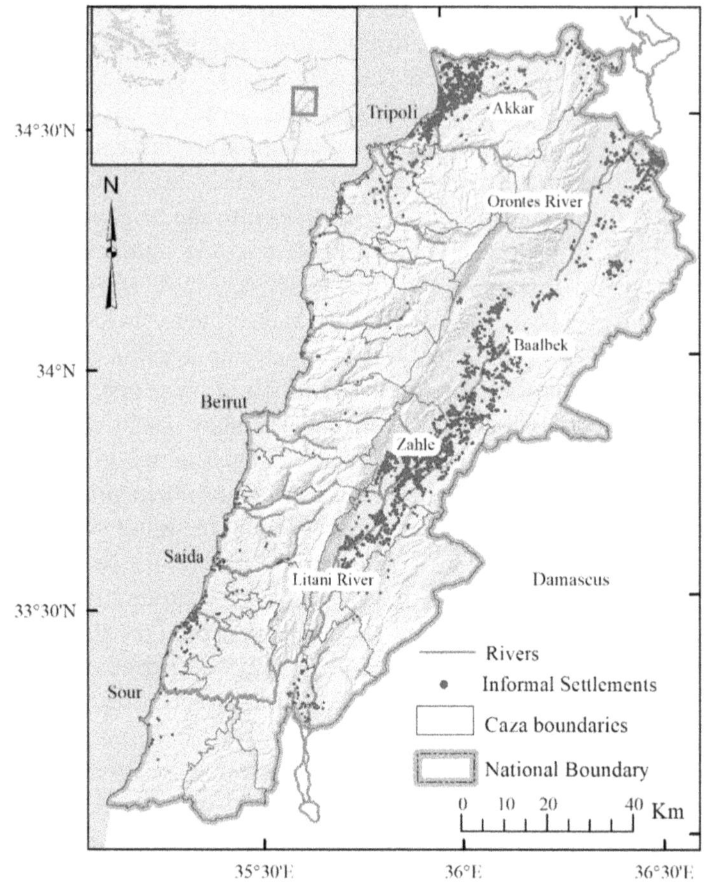

Figure 1. Distribution of Syrian war refugee camps across Lebanon and their proximity to rivers. Many of the settlements release their wastewater directly into the rivers. There is also evidence that the refugees use surface water in their camps. Note: Caza = district in Lebanon.

resort antibiotics (Sulaiman & Kassem, 2019). In addition, multi-drug-resistant *Proteus mirabilis* was detected in the well water of Camp 1 and the drinking and well water of Camp 2 (Alhaj Sulaiman & Kassem, 2020). *Proteus mirabilis* is a bacterial pathogen that can cause serious problems, mainly urinary infection and less frequently respiratory, eye and wound infections (Schaffer & Pearson, 2015), as well as bacteraemia and other life-threatening conditions (Schaffer & Pearson, 2015). Taken together, these observations indicated that the water quality was inappropriate and that the refugees were exposed to antibiotic-resistant pathogens. Consequently, water in the Syrian refugee camps can be unsuitable for drinking and hygiene, which would complicate the control of COVID-19 due to potential exposure to the virus via contaminated water (more on this below). Furthermore, the exposure of the refugees to bacterial pathogens and/or pathogens that can cause respiratory co-infections may significantly reduce their immunity and increase their vulnerability to COVID-19 and the severity of the disease (Zhu et al., 2020).

SARS-CoV-2 has been shown to replicate in the gastrointestinal tract and to be shed in faecal matter for a prolonged period (Lamers et al., 2020; Wong et al., 2020).

Interestingly, shedding of virus RNA was also documented in the stool of an asymptomatic individual (Tang et al., 2020). Another study reported the detection of elevated levels of virus RNA in hospital patients' toilet areas (Liu et al., 2020). While studies have focused on the shedding of viral RNA, few have reported the isolation of the infectious virus from the stool (Gu et al., 2020). It is important to note that the infectivity of the virus and its survival in stool and associated matrices remain under investigation. Still, a recent study clearly demonstrated that infectious SARS-CoV-2 was shed in the stool of a patient with severe COVID-19, while further analysis suggested that the infectious virus in faeces was a 'common manifestation of COVID-19' (Xiao et al., 2020). According to the authors, the detection of infectious viruses in faeces raises the 'possibility of fecal–oral transmission or fecal–respiratory transmission through aerosolized feces' (Xiao et al., 2020), which corroborated previous concerns regarding the potential faecal oral transmission of the virus (reviewed in Tian et al., 2020). Infectious SARS-CoV-2 was also detected in the urine of a patient with severe COVID-19 (Sun et al., 2020), while a recent analysis of the literature has highlighted the potential importance of the detection of SARS-CoV-2 in wastewater (Kitajima et al., 2020). In the light of these studies, there have been increasing calls for vigilance regarding the presence of the virus in faeces and wastewater and the implications for the spread of COVID-19 (Kitajima et al., 2020). Furthermore, a review of the literature showed that the waterborne 'circulation' of SARS-CoV-2 cannot be overlooked, especially in the absence of data on the persistence of the virus in water matrices (Carducci et al., 2020).

These observations and analyses are of utmost importance to the Syrian refugee camps, because many of these settlements might share common facilities, such as latrines, which may increase the risk of transmission of COVID-19 in the absence of good hygiene. Furthermore, the camps' untreated wastewater might contaminate nearby rivers. The latter may spread the virus beyond the camps or enhance the cycle of exposure within the camps that use contaminated river and other surface and ground waters. It should be equally emphasized that contaminated waters are also generated by raw sewage released from the hosting communities.

Severe economic challenges, political and civil unrest, and distancing and shutdowns due to COVID-19 in Lebanon are predicted to hurt all communities, but most of all the refugees. The prices of essential commodities have increased significantly, and much-needed assistance (medicine, food, hygienic material, clean water) is likely to be less available to the refugees. The UN's refugee agency (UNHCR) and other NGOs are trying to provide guidance (outreach) and support to the refugees. Yet the great number of refugees and their diverse needs already strain the available resources, so current efforts remain limited and fraught with difficulties. While UNHCR (2020) has called for more funding, global economic challenges might hinder a timely acquisition of funds.

Perhaps more critically, UNHCR, in collaboration with the hosting country, should acquire testing kits and recruit a qualified team of local and international experts to screen the refugees for COVID-19. As importantly, efforts should be directed to building trust relationships to allow the refugees to report potential COVID-19 cases without fear of repercussions.

Investments should be directed to provide the camps with clean and regularly monitored water by providing filters on the main water sources for the camps when possible. Further investigation is urgently required to determine the persistence and infectivity of

SARS-CoV-2 in polluted water and sewage as well as the potential of disease transmission via exposure to contaminated water matrices. This might be of critical importance in controlling COVID-19 in vulnerable communities, such as those of refugees and those in crisis zones, with poor access to water, sanitation and hygiene.

Importantly, clean water and the subsequent impact of wastewater on the health of the refugees and the hosting communities remain to be addressed and resolved sustainably. This might lack political impetus in the case of refugee camps because it might signal more permanent settlements, which might not be desirable to the hosting countries. Yet it is evident that sufficient access to clean water and sustainable treatment of wastewater are critical for the health of the refugees and the hosting communities, especially during the unfolding COVID-19 pandemic. Failure to address these issues promptly might lead to severe disease and outbreaks in these populations. We urge the global community to address the highlighted issues related to COVID-19 and water contamination and to dedicate more effort and resources to transparently and effectively test and monitor the spread of COVID-19 in refugee camps and beyond. The refugees also require more access to safe and nutritious foods and hygienic products and gear, such as disinfectants, soap, masks, and quarantine facilities. The health of the refugees and that of the hosting community are intimately linked, and it is imperative to take precautions to reduce the risk of COVID-19 and other infectious diseases in these disenfranchised populations.

Disclosure statement

The authors declare no competing interests.

ORCID

Issmat I. Kassem http://orcid.org/0000-0002-2978-9573
Hadi Jaafar http://orcid.org/0000-0003-2612-3191

References

Alhaj Sulaiman, A. A., & Kassem, I. I. (2020, January-February). First report of the plasmid-borne colistin resistance gene (mcr-1) in Proteus mirabilis isolated from domestic and sewer waters in Syrian refugee camps. *Travel Medicine and Infectious Disease*, *33*, 101482. https://doi.org/10.1016/j.tmaid.2019.101482

Baylouny, A. M., & Klingseis, S. J. (2018). *Water thieves or Political catalysts? Syrian refugees in Jordan and Lebanon*. Middle East Policy Council. Retrieved 27 May 2020 from https://mepc.org/journal/water-thieves-or-political-catalysts-syrian-refugees-jordan-and-lebanon

Carducci, A., Federigi, I., Liu, D., Thompson, J. R., & Verani, M. (2020). Making waves: Coronavirus detection, presence and persistence in the water environment: State of the art and knowledge needs for public health. *Water Research*, *179*, 115907. https://doi.org/10.1016/j.watres.2020.115907

Dagher, L. (2020). *Evaluation of the microbiological water quality in rivers across Lebanon by assessing densities of fecal indicators and antibiotic resistance profiles of Escherichia coli* [M.S. Thesis]. Department of Nutrition and Food Sciences, American University of Beirut. ST:7171.

Gu, J., Han, B., & Wang, J. (2020). COVID-19: Gastrointestinal manifestations and potential fecal-oral transmission. *Gastroenterology*, *158*(6), 1518–1519. https://doi.org/10.1053/j.gastro.2020.02.054

Hmede, Z., Sulaiman, A. A. A., Jaafar, H., & Kassem, I. I. (2019, July). Emergence of plasmid-borne colistin resistance gene mcr-1 in multidrug-resistant Escherichia coli isolated from irrigation water in Lebanon. *International Journal of Antimicrobial Agents*, *54*(1), 102–104. https://doi.org/10.1016/j.ijantimicag.2019.05.005

Jaafar, H., Ahmad, F., Holtmeier, L., & King, C. (2020). Refugees, water balance, and water stress: Lessons learned from Lebanon. *Ambio*, *49*(6), 1179–1193. https://doi.org/10.1007/s13280-019-01272-0

Jaafar, H., Archer, D., Jomaa, I., Machayekhi, D., Mansour, E., Machlab, H., & King-Okumu, C. (2016a). *Balancing water stress, human crises, and innovation under a changing dryland climate.* International Institute for Environment and Development. http://pubs.iied.org/pdfs/17413IIED.pdf

Jaafar, H., King-Okumu, C., Haj-Hassan, M., Abdalla, C., El-Korek, N., & Ahmad, F. (2016b). *Water resources within the Orontes and Litani Basins: A balance, demand, and supply analysis amid the Syrian refugee crisis.* IIED. http://pubs.iied.org/10174IIED

Kassem, I. I. (2020). Refugees besieged: The lurking threat of COVID-19 in Syrian war refugee camps. *Travel Medicine and Infectious Disease*, 101736. https://doi.org/10.1016/j.tmaid.2020.101736

Kitajima, M., Ahmed, W., Bibby, K., Carducci, A., Gerba, C. P., Hamilton, K. A., Haramoto, E., & Rose, J. B. (2020). SARS-CoV-2 in wastewater: State of the knowledge and research needs. *Science of the Total Environment*, *179*, 115907. https://doi.org/10.1016/j.scitotenv.2020.139076

Lamers, M. M., Beumer, J., van der Vaart, J., Knoops, K., Puschhof, J., Breugem, T. I., Ravelli, R. B. G., Paul van Schayck, J., Mykytyn, A. Z., Duimel, H. Q., van Donselaar, E., Riesebosch, S., Kuijpers, H. J. H., Schippers, D., van de Wetering, W. J., de Graaf, M., Koopmans, M., Cuppen, E., Peters, P. J., Haagmans, B. L., & Clevers, H. (2020, May 1). SARS-CoV-2 productively infects human gut enterocytes. *Science*, eabc1669. https://doi.org/10.1126/science.abc1669

Liu, Y., Ning, Z., Chen, Y., Guo, M., Liu, Y., Gali, N. K., Sun, L., Duan, Y., Cai, J., Westerdahl, D., Liu, X., Xu, K., Ho, K. F., Kan, H., Fu, Q., & Lan, K. (2020, April 27). Aerodynamic analysis of SARS-CoV-2 in two Wuhan hospitals. *Nature*. https://doi.org/10.1038/s41586-020-2271-3

Machayekhi, D., Pierpaoli, M., & Cancelliere, G. (2017). *Domestic water in the Bekaa Valley, Lebanon: Demand, access and institutional Aspects*. International Institute for Environment and Development. Retrieved 27 May 2020 from www.jstor.org/stable/resrep02712

Schaffer, J. N., & Pearson, M. M. (2015, October). Proteus mirabilis and urinary tract infections. *Microbiology Spectrum*, *3*(5), 5. https://doi.org/10.1128/microbiolspec.UTI-0017-2013

Sulaiman, A. A. A., & Kassem, I. I. (2019, July-August). First report on the detection of the plasmid-borne colistin resistance gene mcr-1 in multi-drug resistant E. coli isolated from domestic and sewer waters in Syrian refugee camps in Lebanon. *Travel Medicine and Infectious Disease*, *30*, 117–120. https://doi.org/10.1016/j.tmaid.2019.06.014

Sun, J., Zhu, A., Li, H., Zheng, K., Zhuang, Z., Chen, Z., Shi, Y., Zhang, Z., Chen, S.-B., Liu, X., Dai, J., Li, X., Huang, S., Huang, X., Luo, L., Wen, L., Zhuo, J., Li, Y., Wang, Y., Zhang, L., & Li, Y.-M (2020). Isolation of infectious SARS-CoV-2 from urine of a COVID-19 patient. *Emerging Microbes & Infections*, *9*(1), 991-993. https://doi.org/10.1080/22221751.2020.1760144

Tang, A., Tong, Z. D., Wang, H. L., Dai, Y. X., Li, K. F., Liu, J. N., Wu, W. J., Yuan, C., Yu, M. L., Li, P., & Yan, J. B. (2020, 6 17). Detection of novel coronavirus by RT-PCR in stool specimen from asymptomatic child, China. *Emerging Infectious Diseases*, *26*, Jun. https://doi.org/10.3201/eid2606.200301

Tian, Y., Rong, L., Nian, W., & He, Y. (2020). Gastrointestinal features in COVID-19 and the possibility of faecal transmission. *Alimentary Pharmacology & Therapeutics*, *51*(9), 843-851. https://doi.org/10.1111/apt.15731

UNHCR. *Coronavirus emergency appeal: UNHCR's preparedness and response plan.* United Nations High Commissioner for Refugees (UNHCR). Retrieved 9 April 2020 from http://reporting.unhcr.org/sites/default/files/UNHCR%20COVID-19%20Appeal%20 %20March%202020.pdf

Wong, S. H., Lui, R. N., & Sung, J. J. (2020, March 25). Covid-19 and the digestive system. *Journal of Gastroenterology and Hepatology*, *35*(5), 744–748. https://doi.org/10.1111/jgh.15047

Xiao, F., Sun, J., Xu, Y., Gersch, J., Frankel, M. B., Vallari, A., Cloherty, G., Hackett, J., Evans, J. L., Page, K., & Dawson, G. J. (2020). Infectious SARS-CoV-2 in feces of patient with severe COVID-19. *Emerging Infectious Diseases*, *26*(8), 265–272. https://doi.org/10.3201/eid2608.200681

Zhu, X., Ge, Y., Wu, T., Zhao, K., Chen, Y., Wu, B., Zhu, F., Zhu, B., & Cui, L. (2020). Co-infection with respiratory pathogens among COVID-2019 cases [published online ahead of print, 2020 May 11]. *Virus Research*, *285*, 198005. https://doi.org/10.1016/j.virusres.2020.198005

Exploring challenges in safe water availability and accessibility in preventing COVID-19 in refugee settlements

Nazifa Rafa, Sayed Mohammad Nazim Uddin and Chad Staddon

Introduction

The new coronavirus disease (COVID-19) has been declared as a pandemic, attributed to its high infectivity and transmission ability. As COVID-19 becomes widespread around the world, its impact will be experienced disproportionately, with the world's most vulnerable populations, like refugees, facing the worst of it. Currently, there are approximately 25 million refugees in the world (Volkin, 2020). Fear and caution surround refugee settlements, which tend to be severely congested, around the world as outbreaks have already broken out in many of them (Volkin, 2020).

Proper hygiene practices such as routine handwashing with soap and social distancing are the key COVID-19 prevention strategies, with the main concern to follow both measures in the context of refugee settlements. Adopting a practice of handwashing that lasts at least 20 seconds for 8–10 times per day is being recommended, with appropriate soap and drying facilities. The total amount of water required per person per day therefore can amount to 8 to 10 litres if washed under running water (Staddon et al., 2020). Thus, the control of the spread of the virus is closely related to water and hygiene, which is difficult to manage in refugee settlements, which are less than adequate and without proper access to health, water, or sanitation facilities. As it is, billions of people in the world lack access to safe and sufficient water, and thus exercise sanitation and handwashing practices that are unsatisfactory. Refugee settlements are appropriate examples that contain populations afflicted with issues of availability and accessibility to safe water. UNHCR (2000) suggests a water quantity of more than 20 litres/person/day within a distance of 200 m between refugee dwellings and water points. In addition, it also recommends that no functional hand pump or well be used by more than 250 people. For households, more than 95% of them should collect their drinking water from protected water sources (piped, protected springs, tap stands, handpumps with aprons and sanitary seals), and at least 85% of households should be using narrow-necked containers or covered containers with a tap, with a 10 litres/person potable water storage capacity. These guidelines are hardly maintained because of water-related challenges. This commentary attempts to draw attention to the water-related challenges areas may face in the prevention of COVID-19 with regards to refugee settlements.

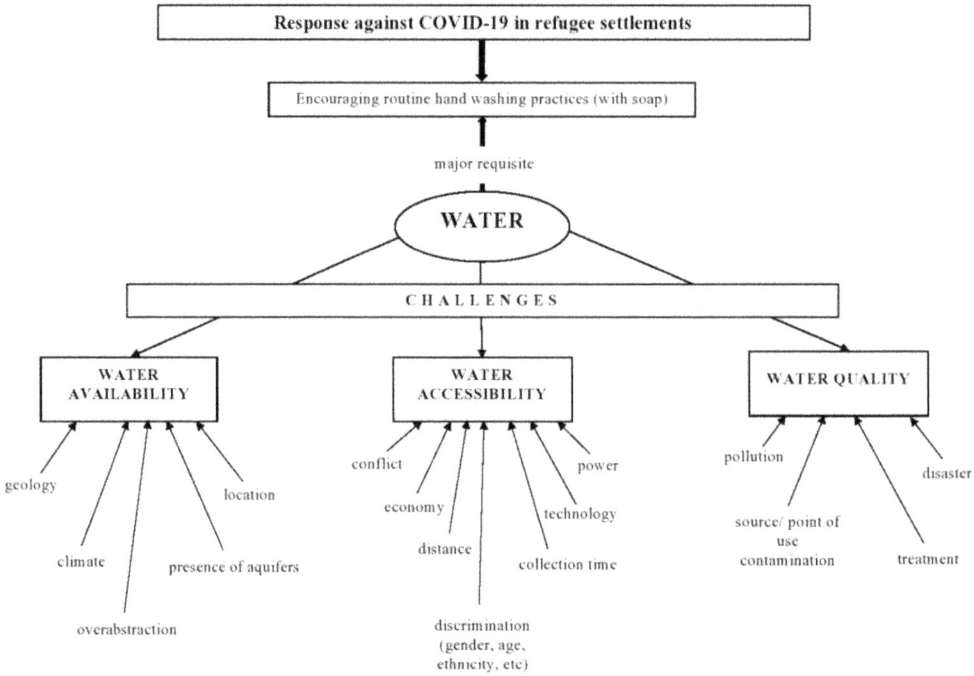

Figure 1. Challenges in safe water availability and accessibility in preventing COVID-19 in refugee settlements.

Water-related challenges in refugee settlements

Relevant articles and case studies have been reviewed to identify the major water-related challenges, and the findings have been summarized in Figure 1. As mentioned, one of the most notable measures for the prevention of COVID-19 is frequent handwashing, where one of the major requisites is water. The water-related challenges that could serve as a barrier to achieving the best forms of prevention strategies arise from water availability, accessibility, and quality.

Challenge: water availability

The rise in demand for water for food production, industries, and the growing population has led to a growing scarcity of fresh water in many parts of the world. Surface water and groundwater are being depleted at a rate faster than they can be replenished, exhausting aquifers and the base flows of rivers. Consequently, governments, corporations, and communities are occupied with ensuring the future availability and sustainability of water supplies for their citizens.

Regardless, ensuring the availability of adequate amounts of water in refugee settings are key to the protection of these marginalized peoples. While the right to drinking water applies to all humans, guaranteeing this right is especially significant in the context of refugees because of the population's vulnerability and reliance on external help. However, the availability of adequate water to sustain refugees is a challenge in these times of water stress, particularly when the host countries' nationals themselves do not receive enough water for their day-to-day activities or livelihoods.

Water availability is heavily dependent on the climate, geology, and location. Extended periods of low levels of rainfall and high temperature lead to water deficits. Consequently, most arid environments experience increased frequency of droughts compared to tropical regions. For example, in Chad and Darfur, there are huge demands for water from both refugees and local communities due to the regions being arid and having scarce water resources, giving rise to conflicts between refugees and host communities (Cronin et al., 2008). Over abstraction of water also induces water scarcity. In Lebanon, groundwater exhaustion had already been a major concern even before the migration of Syrian refugees, but the influx of refugees has led to further stress (Gassert et al., 2013).

Soil structure determines the degree of infiltration, and thus the availability of water below and above ground. Sandy soil tends to allow more infiltration of water compared to clayey soil due to its increased diameter and larger pore space. Permeable rocks can lead to less surface water but lead to the formation of aquifers, which are sources of clean water. Many regions do not possess aquifers and may thus have to turn to surface water or other alternatives, which are often unsafe to use and thus expensive to purify. For example, two of the areas Rohingya refugee camps have been built in Cox's Bazar are Teknaf and Ukhiya. While Ukhiya enjoys the benefit of several groundwater resources, Teknaf has none due to the presence of hard impermeable rocks (UNHCR, AUW & Oxfam, 2020). As a result, water scarcity is far more marked in the Rohingya refugee camps in Teknaf due to the absence of groundwater sources. The concern is whether the existing water sources in Bangladesh will be able to supply water to both its growing national and growing refugee populations to prevent COVID-19 infections.

The case of the 60,000 deaths of Rwandan refugees in the Democratic People's Republic of Congo in the summer of 1994 due to the vicious cycle of water shortage, and thus, cholera (UNHCR, 2003), already provides evidence for how water availability is closely linked with prevention of diseases. The concern of meeting the growing water needs of refugee settlements is alarming, even more so in the face of COVID-19.

Challenge: water accessibility

The availability of an adequate amount of water alone does not guarantee that the water will reach the people. Water accessibility is found to be heavily influenced by socioeconomic factors, such as gross national income (GNI), female primary completion rate, agriculture, growth of rural population, and governance indicators, such as political stability, control of corruption, or regulatory quality (Gomez et al., 2019). Often, the gross domestic product (GDP) is the major determinant of how committed countries are to develop and expand their water supply infrastructures as GDP also decides the technical capacities of the country. Water access for medical community centres for Saharawi refugees in Algerian desert face issues pertaining to the non-availability of food-grade water tanks for the institutions, the use of small containers as the main water supply, poor maintenance, and insufficient chlorine levels for microbiological decontamination (Vivar et al., 2016).

Refugee groups may also experience hindrances in their attempts to access water from host communities or from within their own communities due to social hierarchies, gender, age, etc. As the demand for water increases in times for water stress, there is usually a rise in conflicts among the groups competing for them, which can quickly become an issue about territorial rights and political power and give rise to discrimination. In addition, accessibility to water can be worsened by long distances between water points and dwellings, as well as long water collection time due to the low ratio of functional water sources available and the number of residents. In the Moria camp, the largest refugee camp in Europe, one water point is shared by 1,300 people, which can make the practice of frequent handwashing very challenging (Courrier d'Europe, 2020). NGOs are the only actors struggling to provide the camp with basic water and sanitation facilities. However, due to the hostility from some of the native inhabitants who are calling for relocating the refugees to other EU states, most of the NGOs have abandoned the camp, leaving the residents of Moria in a crisis.

Moreover, 70% to 80% of refugee groups are made of women and children who bear the burden of water collecting activities. Issues of distance and water collection time have been experienced in the Kutupalong camp for Rohingyas in Ukhia, where mostly women and young girls experienced the burden of collecting water (Akhter, 2020). As preventive measures such as handwashing for COVID-19 becomes more enforced, the water demand is expected to increase, which increases the burden of water collection on women and young girls.

Challenge: water quality

Water quality is essential to secure the good health of consumers. Inadequate water quality can have both immediate effects, usually associated with gastrointestinal diseases attributed to microbial contamination, or long term health ailments due to the continuous consumption of water containing harmful levels of contaminants like fluoride and iodine. For Saharawi refugees, even though the access and quality of water have improved significantly over the years, improved water quality controls, ensuring access to treated water for all, expansion of distribution networks, and adequate storage systems and maintenance are some of the issues that need to be addressed as many still drink and use water that has high levels of fluoride and that has been stored in tanks that have a high risk of microbial contamination (Vivar et al., 2016).

Despite the presence of plentiful water, pollution and/or contamination, such as by untreated sewage and wastewater from households, or after a natural disaster like a flood or cyclone can make the water unsafe to use. In Rohingya refugee camps, latrines facilities and waste disposals are sometimes too close to water sources (Akhter, 2020), which significantly affects the quality of water. A recent study has revealed faecal coliform and *E. coli* contamination in water samples derived from tube wells and from stored household sources in Rohingya refugee camps, suggesting during contamination having occurred during collection, transportation, and storage of water due to lack of knowledge of personal and domestic hygiene (Mahmud et al., 2019). The presence of faecal coliforms and *E. coli* denote the presence of enteric pathogens, therefore breakouts of waterborne diseases like cholera, bloody diarrhoea, typhoid, and hepatitis E have been the major concern in refugee camps worldwide.

Water treatment facilities need to be increased and improved. Improving the microbiological and chemical quality of household water by on-site or point-of-use treatment and safe storage in improved vessels can reduce waterborne diseases and long term illnesses in refugee settlements (Vivar et al., 2016).

The importance of considering the quality of water and its resulting effect on the health of refugees lies in the fact that the immunocompromised are far more susceptible to COVID-19. Waterborne diseases caused by unsafe water supplies can considerably weaken the immune system. Therefore, the guarantee of safe water both at source and at point of use is necessary to tackle COVID-19, not only to prevent infections by COVID-19 itself but also to prevent the weakening of the immune system by waterborne diseases that may make refugees susceptible to COVID-19 infection.

Conclusion

The humanitarian community and the governments of nations are working closely and proactively to address preparation and response for COVID-19 across the refugee settlements their nations are hosting. Isolation has been attempted at various refugee settlements around the world. However, measures relating to encouraging frequent handwashing practices by making safe and adequate water accessible still remain a matter of alarm and concern due to the water-related challenges discussed in this commentary, which compound the COVID-19 crisis. It is crucial to address the water-based needs of refugee populations first as placing attention towards water-based needs will be beneficial in the long-term by inhibiting the outbreaks of other waterborne diseases in refugee settlements. Overcoming the challenges of accessibility and availability of safe water is thus, paramount in the response against COVID-19 in refugee settlements.

Disclosure statement

No potential conflict of interest was reported by the authors.

ORCID

Nazifa Rafa http://orcid.org/0000-0003-0595-8253
Sayed Mohammad Nazim Uddin http://orcid.org/0000-0002-6143-5572
Chad Staddon http://orcid.org/0000-0002-2063-8525

References

Akhter, M. (2020) *Drinking water security challenges in Rohingya Refugee Camps of Cox's Bazar, Bangladesh* (Undergraduate thesis, Asian University for Women).

Courrier d'Europe. (2020, April 25) *Covid-19: What about the refugees at Europe's borders?* The New Federalist. https://www.thenewfederalist.eu/covid-19-what-about-the-refugees-at-europe-s-borders?lang=f. (accessed 16 June 2020)

Cronin, A. A., Shrestha, D., Cornier, N., Abdalla, F., Ezard, N., & Aramburu, C. (2008). A review of water and sanitation provision in refugee camps in association with selected health and nutrition indicators–the need for integrated service provision. *Journal of Water and Health*, *6* (1), 1–13. https://doi.org/10.2166/wh.2007.019

Gassert, F., Landis, M., Luck, M., Reig, P., & Shiao, T. (2013). *Aqueduct global maps 2.0, vol. 20, 2011–2012.* Water Resources Institute (WRI).

Gomez, M., Perdiguero, J., & Sanz, A. (2019). Socioeconomic factors affecting water access in rural areas of low and middle income countries. *Water, 11*(2), 202. https://doi.org/10.3390/w11020202

Mahmud, Z. H., Islam, M. S., Imran, K. M., Hakim, S. A. I., Worth, M., Johnston, D., Johnston, D., Haider, M., Islam, M. R., Hossain, F., Johnston, D., & Hossan, S. (2019). Occurrence of Escherichia coli and faecal coliforms in drinking water at source and household point-of-use in Rohingya camps, Bangladesh. *Gut Pathogens, 11*(1), 1–11. https://doi.org/10.1186/s13099-019-0333-6

Staddon, C., Everard, M., Mytton, J., Octavianti, T., Powell, W., Quinn, N., Uddin, S. M. N., Young, S. L., Miller, J. D., Budds, J., Geere, J., Meehan, K., Charles, K., Stevenson, E.G.J., Vonk, J. & Mizniak, J. (2020). Water insecurity compounds the global coronavirus crisis. *Water International,* 1–7. https://doi.org/10.1080/02508060.2020.1769345

UNHCR. (2000). *Handbook for emergencies.* United Nations High Commissioner for Refugees.

UNHCR. (2003). Three days to live. *Refugees Magazine, 3*(132), 22–23. United Nations High Commissioner for Refugees.

UNHCR, AUW, & Oxfam. (2020). HEAL in WASH: Towards a roadmap. https://drive.google.com/file/d/1kD1cj0kINk6apcv3qhOLt1E9RzADE8Wt/view

Vivar, M., Pichel, N., Fuentes, M., & Martínez, F. (2016). An insight into the drinking-water access in the health institutions at the Saharawi refugee camps in Tindouf (Algeria) after 40 years of conflict. *Science of the Total Environment, 550,* 534–546. https://doi.org/10.1016/j.scitotenv.2016.01.113

Volkin, S. (2020, 21 April) *How are refugees affected by COVID-19?* Hub, John Hopkins University. https://hub.jhu.edu/2020/04/20/covid-19-refugees-asylum-seekers/

Index

Note: Figures are indicated by *italics*. Tables are indicated by **bold**.

For Product Safety Concerns and Information please contact our EU representative GPSR@taylorandfrancis.com
Taylor & Francis Verlag GmbH, Kaufingerstraße 24, 80331 München, Germany

www.ingramcontent.com/pod-product-compliance
Ingram Content Group UK Ltd.
Pitfield, Milton Keynes, MK11 3LW, UK
UKHW052213190926
475702UK00006B/993
* 9 7 8 1 0 3 2 3 6 3 3 4 9 *